WITHDRAWN
UTSA LIBRARIES

*Dispersion Polymerization
in
Organic Media*

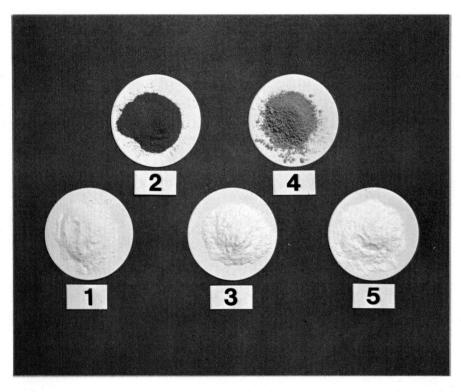

Examples of polymer powders produced by dispersion polymerization in organic media: 1, Nylon-11; 2, Terylene (with dyestuff); 3, Terylene; 4, Nylon-11 (with dyestuff); 5, Poly(methyl methacrylate)

# Dispersion Polymerization in Organic Media

*Edited by*

K. E. J. BARRETT

*Imperial Chemical Industries Limited,
Paints Division, Slough*

with a Foreword by
Professor C. H. Bamford, F.R.S.

*A Wiley–Interscience Publication*

JOHN WILEY & SONS

London · New York · Sydney · Toronto

*Library of Congress Cataloging in Publication Data:*

Barrett, Keith E.J.
Dispersion polymerization in organic media.

"A Wiley–Interscience publication."
"Based upon contributions by members of the Research and Development Department of the Paints Division of Imperial Chemical Industries Limited at Slough."
1. Polymers and polymerization. 2. Suspensions (Chemistry)
I. Imperial Chemical Industries, Ltd. Paints Division.
II. Title.
TP156.P6B293      660.2'9'93      74–5491

ISBN 0 471 05418 6

Printed in Great Britain by J. W. Arrowsmith Ltd.
Winterstoke Road, Bristol.

Copyright © 1975, by Imperial Chemical Industries Limited.

All rights reserved.

No part of this book may be reproduced by any means, nor transmitted, nor translated into a machine language without the written permission of the Copyright owner and the publisher.

*The aim is to live lucidly in a world where dispersion is the rule*

Albert Camus

# Foreword

Professor C. H. Bamford

The impact of synthetic high polymers on classical colloid science has long been familiar, so that it is of particular interest that the systems discussed in this book, which are the offspring of the fusion of polymer science and colloid science, are of relatively recent origin. Dispersions of polymers in organic media were first developed for the surface coatings industry, in which they have a unique rôle to play, but, as is so often the case, technological advances have posed theoretical problems of great interest. Chief among these is the mechanism by which essentially lyophobic colloids are stabilized in media of low polarity. Electrostatic stabilization, operative in aqueous colloids, cannot be invoked and a newer concept—'steric stabilization'—has been elaborated. As might be anticipated, studies of the preparation of these dispersions by direct polymerization yield a rich harvest for the kineticist.

These topics, together with many others, form the subject of this book. The authors and their colleagues have been responsible for many of the practical and theoretical developments and this is reflected in the authority of the treatment and the abundance of illuminating detail. I believe the book will be read with interest not only by those directly concerned with surface coatings but by polymer scientists in general, including those in universities who wish to enlarge their appreciation of the contribution that research in industry can make to advances in science.

C. H. Bamford

*Department of Inorganic, Physical*
*and Industrial Chemistry,*
*Donnan Laboratories,*
*The University of Liverpool*

# Preface

Dispersions of polymers of colloidal dimensions, prepared in water by emulsion polymerization and related techniques, have been established products in the rubber, paint, adhesives and other industries for several decades. Indirect methods have been available for converting polymers into particulate dispersions in organic liquids but the products obtained have only been of limited application. The technique of dispersion polymerization in organic media is a more recent development and now allows the direct preparation of a wide range of stable polymer dispersions of controlled particle size in the medium in which they are to be used.

Much of the research work in the development and application of dispersion polymers was carried out in industrial laboratories in Europe, North America and Australia. Consequently, the bulk of the published information is in the patent literature and no detailed treatment of the subject has been available in book form until now. The aim of the present work has been to give a systematic account of the underlying theoretical principles of the process as well as a description of procedures and applications. The development of dispersion polymers has involved a detailed study of topics of interest to both polymer and colloid chemists and although so far the main applications have been in the field of surface coatings, further developments could well extend the area of interest.

The chapters of this book are based upon contributions by members of the Research and Development Department of the Paints Division of Imperial Chemical Industries Limited at Slough who have been associated with the development of a particular aspect of the subject. Although therefore the book has been compiled from different contributions, a continuous interchange was possible during its writing so that hopefully a reasonably unified and balanced account of the subject would emerge.

I would like to acknowledge here the support and encouragement for this work given by Mr C. I. Snow and Dr F. J. Long, who were respectively Research and Development Director and Research Manager at Paints Division during the major period of development of dispersion polymers. Thanks are also due to Professor C. H. Bamford, F.R.S., of the University of Liverpool, to Professor R. H. Ottewill of the University of Bristol and to colleagues in the Research and Development Department who have read and commented upon various parts of the manuscript during the preparation of the book. Finally, special mention must be made of Mrs D. N. Hughes for her patience and precision in typing the drafts and the final manuscript.

K. E. J. BARRETT

*Slough, Bucks*
*January, 1974*

# Contents

Glossary of Symbols . . . . . . . . . xiii

1. Introduction to Dispersion Polymer Technology
   K. E. J. Barrett . . . . . . . . . 1

2. The Theoretical Basis for the Steric Stabilization of Polymer Dispersions Prepared in Organic Media
   D. W. J. Osmond and F. A. Waite . . . . . . 9

3. The Design and Synthesis of Dispersants for Dispersion Polymerization in Organic Media
   D. J. Walbridge . . . . . . . . . 45

4. Kinetics and Mechanism of Dispersion Polymerization
   K. E. J. Barrett and H. R. Thomas . . . . . . 115

5. The Preparation of Polymer Dispersions in Organic Liquids
   K. E. J. Barrett and M. W. Thompson . . . . . 201

6. The Properties of Polymer Dispersions Prepared in Organic Liquids
   D. W. J. Osmond and I. Wagstaff . . . . . . 243

7. Applications of Polymer Dispersions Prepared in Organic Media
   M. W. Thompson . . . . . . . . . 273

   Appendix: Patents on Polymers Dispersed in Organic Media
   K. E. J. Barrett . . . . . . . . . 291

   Author and Patentee Index . . . . . . . 309

   Subject Index . . . . . . . . . 315

# Glossary of Symbols

| | |
|---|---|
| $a$ | radius of a sphere |
| $A$ | Hamaker constant |
| $b$ | length of rigid chain |
| $B$ | second virial coefficient |
| $c$ | concentration |
| $c_m$ | chain transfer constant |
| $c_s$ | equilibrium solubility |
| $C$ | electrical capacity; concentration; surface area |
| $d$ | dispersant content |
| $D$ | diffusion coefficient; diameter of a sphere |
| $e$ | elementary charge |
| $E$ | energy; cohesive energy density |
| $E_0$ | zero-point energy |
| $f$ | initiator efficiency |
| $G$ | Gibbs free energy |
| $\bar{G}$ | partial molal free energy ($= \mu$) |
| $G_m$ | free energy of mixing |
| $G_v$ | free energy of volume restriction |
| $h$ | distance between surfaces of spheres or planes; Planck's constant |
| $H$ | geometric (Hamaker) function; enthalpy |
| $H_m$ | heat of mixing |
| $H_v$ | heat of vaporization |
| $J$ | Joule |
| $k$ | Boltzmann's constant; rate constant |
| $k_d$ | decomposition rate constant |
| $k_i$ | initiation rate constant |
| $k_p$ | propagation rate constant |
| $k_t$ | termination rate constant |
| $k_{tr}$ | transfer rate constant |
| $K_p$ | adsorption coefficient |
| $L$ | average distance of diffusion |
| $L_m$ | length of polymer chain |
| $M$ | mass; molecular weight |
| $\bar{M}_n$ | molecular weight, number average |
| $\bar{M}_w$ | molecular weight, weight average |
| $n$ | number of moles; number of particles; micellization number |
| nm | nanometre ($= 10^{-9}$ metre $= 10$ Å) |
| $N$ | number of chains per unit area of surface; number of particles per unit volume |

| | |
|---|---|
| $N_A$ | Avogadro's number |
| $P$ | degree of polymerization |
| $\bar{P}_n$ | degree of polymerization, number average |
| $q$ | number of elements per unit volume; number of branch points in a copolymer; number of atoms in a polymer chain |
| $Q$ | electrical charge; average number of oligomers/molecules forming a nucleus |
| $r$ | radius of a sphere, micelle core |
| $r_1, r_2$ | monomer reactivity ratios |
| $R$ | distance between centres of spheres; gas constant; mean particle diameter; reaction rate |
| $R_i$ | initiation rate |
| $R_p$ | polymerization rate |
| $S$ | entropy; dispersant concentration; solubility |
| $S_m$ | entropy of mixing |
| $S_r$ | relative supersaturation |
| $t$ | time |
| $T$ | absolute temperature (Kelvin) |
| $T_g$ | glass transition temperature |
| $T_m$ | melting temperature |
| $v$ | velocity; volume fraction of disperse phase |
| $V$ | volume; volume fraction |
| $V_m$ | molar volume |
| $V_A$ | attractive potential energy |
| $V_K$ | kinetic energy |
| $V_R$ | repulsive potential energy |
| $x$ | fractional conversion; number of polymer chains attached to backbone |
| $Z$ | number of electrons in outer shell of atom |
| $\alpha$ | Flory solvency coefficient; partition coefficient |
| $\alpha_0$ | static polarizability; Einstein coefficient |
| $\gamma$ | interfacial tension |
| $\delta$ | thickness of adsorbed polymer layer; solubility parameter |
| $\varepsilon$ | dielectric constant |
| $\zeta$ | electrokinetic potential |
| $\eta$ | viscosity |
| $\theta$ | theta temperature |
| $\kappa$ | enthalpy parameter |
| $\lambda$ | London interaction constant |
| $\mu m$ | micrometre or micron ($= 10^{-6}$ metre $= 10^3$ nanometre) |
| $\mu$ | chemical potential |
| $\mu_s$ | entropic component |
| $\nu$ | frequency; reaction chain length |
| $\Pi$ | osmotic pressure |

| | |
|---|---|
| $\rho$ | density |
| $\sigma$ | standard deviation |
| $\tau$ | mean residence time |
| $\phi$ | volume fraction; solid angle |
| $\chi$ | polymer-solvent interaction parameter |
| $\psi$ | entropy parameter |

# CHAPTER 1
# Introduction to Dispersion Polymer Technology

K. E. J. BARRETT

| | |
|---|---|
| 1.1. THE NEED FOR POLYMER DISPERSIONS IN ORGANIC MEDIA | 1 |
| 1.2. HETEROGENEOUS POLYMERIZATION PROCESSES | 3 |
| 1.3. THE TERMINOLOGY OF DISPERSION POLYMERIZATION | 4 |
| 1.4. REFERENCES | 6 |

## 1.1. THE NEED FOR POLYMER DISPERSIONS IN ORGANIC MEDIA

The development of techniques for the preparation of dispersions of polymers of controlled particle size in organic liquids has been largely motivated by the requirements of the surface coatings industry.

In general, protective pigmented compositions require a polymer of high molecular weight as the organic binder and film-former in order to achieve the most durable paint films. Since paints have to be applied in the liquid state to a range of different substrates, solutions of polymers have for a long time been the basis of thermoplastic or lacquer formulations. However, since the viscosity of such solutions rises sharply with the concentration and molecular weight of the polymer used, e.g. nitrocellulose, the need to be able to apply the solutions by brush or by spray has limited their use to relatively low concentrations and consequently a number of separate layers of polymer have to be built up in order to provide a coating of adequate thickness for surface protection.

An alternative procedure, which allows the application of higher polymer concentrations, involves the use of solutions of oligomers or polymers of low molecular weight and which contain reactive groups, which after application to the substrate increase their molecular weight by cross-linking reactions to form thermosetting or enamel-type formulations. Typical examples of this type of polymer are the alkyd resins (that is, polyesters based on alcohols and acidic materials, such as glycerol and phthalic anhydride) which contain linoleic acid and other drying-oil chains capable of cross-linking reactions induced by metal-catalysed autoxidation. A large number of compositions based on variants of this and other types of cross-linking reactions have now been utilized for film-formation[1].

At the same time, an extensive range of polymers of high molecular weight, such as poly(methyl methacrylate) and poly(vinyl chloride), has been developed by the plastics industry. However, only a limited application of these materials in paint technology has been achieved since the very properties which make them potentially useful as surface coatings also make them difficult to apply in solution by conventional techniques.

A direct method of overcoming the disadvantages of solution application is offered by the use of finely-divided polymer dispersions in water. The technique of emulsion polymerization, originally developed in the synthetic rubber industry[2], now provides aqueous dispersions of polymers, such as poly(vinyl acetate) and poly(methyl methacrylate), which serve as the basis of an extensive range of aqueous emulsion paints and similar products[3]. Emulsion polymerization allows the preparation of polymers of high molecular weight at high rates and the dispersions which result can be applied without the restrictions due to the viscosity of solution systems.

Water has a number of practical advantages as a paint medium. Its use involves a low fire-risk and it is, in addition, free of odour and toxic effects. It is also, of course, cheap to use and the manufacturing and application equipment in which dispersions are used can readily be cleaned. At the same time, the presence of water also has considerable disadvantages in paint formulations which can detract from its use in practical systems. It has a large latent heat of evaporation (580 cal/g) compared with most organic liquids ($<100$ cal/g) and a correspondingly larger input of heat is required for its evaporation during film formation. Since its evaporation rate under ambient conditions is affected by such factors as relative humidity, the application properties due to variable evaporation rates are frequently difficult to control. In practice, this defect is frequently counteracted by the addition of various water-miscible organic diluents to modify the evaporation rate of the aqueous component but often at the expense of introducing toxicity. The single boiling point of water greatly limits the scope of the usual paint formulating procedures which are normally able to make use of organic liquids of a wide range of boiling points (50–300 °C) in order to obtain the required rate of evaporation during and after application of the film former. In addition, as compared with most organic liquids, the freezing point of water is high and although this can be depressed to some extent by the use of suitable additives, it can still complicate the storage and transport of emulsion polymers in cold conditions. Other factors militate against the use of water—the exclusion of the use of conventional plasticizing aids, which are normally water-insoluble and are used to reduce the glass transition temperature of the polymer during application and assist the process of film formation, limits formulation procedures. In addition, the dispersants and other additives used in water-paint formulations tend to confer a residual sensitivity to water on the final paint film and generally degrade its properties.

It follows, therefore, that special efforts have been made in the surface coatings industry to develop methods for preparing polymer dispersions in non-aqueous media which make use of the advantages of dispersions as such but without the concomitant disadvantages of water as the continuous phase.

## 1.2. HETEROGENEOUS POLYMERIZATION PROCESSES

Since polymer dispersions in organic liquids have a potential advantage over the more conventional aqueous dispersions or organic solution systems for surface coatings, a number of indirect methods have been developed for their preparation. The methods used have all involved the subsequent conversion of the prepared polymer in a variety of ways to a more or less disperse form (see Section 5.1 below). However, in general, the overall properties of the dispersions produced by these methods, such as particle size, dispersion stability and viscosity, have not been sufficiently satisfactory for their widespread application as surface coatings. Ideally, a process analogous to that of emulsion polymerization is required in which polymer dispersions of controlled particle size can be prepared directly by a heterogeneous process but in which the continuous phase is organic instead of aqueous.

Two basic types of heterogeneous polymerization process can be distinguished. In some, as in emulsion polymerization, the initial reaction mixture consists of two separate phases and the polymerization continues in a heterogeneous manner throughout. In others, as in dispersion polymerization in organic media, the reaction mixture is initially homogeneous but as polymerization proceeds, polymer soon separates out and the reaction then continues in a heterogeneous manner.

The characteristics of the various types of heterogeneous free-radical polymerization processes both in water and in organic liquids are summarized in Table 1.1. Precipitation polymerization can be carried out in a range of media—in water, e.g. aqueous solutions of acrylonitrile[4], or in organic precipitants, e.g. methyl methacrylate in cyclohexane, or in a monomer in which the polymer produced is insoluble, e.g. acrylonitrile or vinyl chloride[5]. An auto-accelerated polymerization is observed after the separation of polymer since the radical chain termination processes are restricted by the low mobility of the growing polymer radicals in the viscous reaction medium (see 'gel effect' below, Section 4.4.2). The polymer produced is in the form of an agglomerate or slurry. Emulsion polymerization in water is also characterized by a high rate of polymerization producing high molecular weight polymer, which in this case originates from the isolation of radicals in the separate growing polymer particles (Section 4.1). Here, the polymer produced is in the form of a stable latex of fine particle size.

Although the term 'dispersion polymerization' has been frequently used, particularly by Continental authors[6] to describe emulsion polymerization in general, it has been used by others[7] in a more restricted sense. This referred specifically to a method for preparing aqueous suspensions of polymer, particularly poly(vinyl acetate), which are stabilized by relatively high concentrations of water-soluble polymers, such as poly(vinyl alcohol) and it continues to be used by some authors in this sense to the present day[8]. The polymer particles produced are larger than those obtained by emulsion polymerization in its strictest sense and tend to settle out on further dilution of the aqueous phase.

Table 1.1. Characteristics of free-radical heterogeneous polymerization processes (Adapted from Schildknecht, *Polymer Processes*, Interscience Publishers Inc., New York, 1956)

| Type of polymerization | Continuous phase | Characteristics | Product |
|---|---|---|---|
| Precipitation | Water, organic liquids | Monomer and initiator soluble in continuous phase; auto-accelerated polymerization due to gel effect | Agglomerated polymer or slurry |
| Emulsion | Water | Low monomer solubility, initiator soluble in continuous phase; ionic/non-ionic surfactants; high rates due to radical isolation | Stable latex (0.1–0.3 μm) |
| Dispersion | Water | Low monomer solubility, initiator soluble in continuous phase; polymeric surfactants; gel effect | Coarse (0.5–1.0 μm) but stable 'emulsions' |
| Suspension (pearl, bead) | Water | Low monomer solubility, initiator soluble in monomer; low level of ionic surfactant; gel effect | Coarse (>5 μm) suspension in water |
| Dispersion | Organic liquids | Monomer and initiator soluble in continuous phase; graft copolymer dispersant; gel effect | Stable latex (0.1–0.5 μm); dispersions up to 5 μm possible. |

Suspension polymerization, involving the use of monomer droplets in water, is carried out with an initiator soluble in the monomer by an essentially micro-bulk type of polymerization[5].

In all of the heterogeneous dispersion polymerization processes described so far, in which the polymer is formed as a more or less stable dispersion of controlled particle size, an aqueous continuous phase has been used. Dispersion polymerization in organic liquids is a more recent development[9,10] and this is the sense in which the term 'dispersion polymerization' will be used throughout this book. It usually involves the polymerization of a monomer dissolved in an organic diluent to produce insoluble polymer dispersed in the continuous phase in the presence of an amphipathic graft or block copolymer as the dispersant (Chapter 3). In fact for most purposes, dispersion polymerization (in the sense defined above) can be regarded as a special type of precipitation polymerization in which flocculation is prevented and particle size controlled[11].

## 1.3. THE TERMINOLOGY OF DISPERSION POLYMERIZATION

The development of dispersion polymerization has utilized a blend of the disciplines of both polymer science and colloid science. Consequently, some of the terminology used for its description may be unfamiliar to some readers.

The term 'polymer colloids' is now frequently used to describe polymer dispersions of colloidal dimensions (that is, in the size range 0.01 to 10 μm) in any medium[12]. An abbreviated version, 'poloids', has also been used but in a more restricted sense[13] (see Section 3.7.2, below). Colloidal dispersions in organic diluents are often called 'organosols' with a corresponding term 'hydrosols' for similar dispersions of particles in water[14]. However, the term 'organosol' in the surface coating technology is usually restricted to polymer dispersions of a particular type in organic diluents (see Section 5.1). When plasticizers, like long-chain esters, are used as the organic diluent in order to aid subsequent film-formation processes, the dispersions are known as 'plastisols'[15]. The term NAD (non-aqueous dispersions) is also now frequently used in the literature on surface coatings to describe paint compositions which are based on polymer dispersions prepared in aliphatic hydrocarbons and similar diluents[16-23].

Polymer colloids can be considered as being of two basic types, lyophobic and lyophilic colloids. The interaction of lyophobic colloids with the continuous phase is so restricted that they are inherently unstable in the thermodynamic sense and the forces of attraction generated between such particles (see Section 2.1) rapidly lead to their aggregation. However, the interposition of a repulsive barrier having suitable strength and dimensions between the particles can so retard their aggregation that an indefinitely prolonged stability can be achieved for many practical systems. In contrast, lyophilic colloids interact strongly with the continuous phase and are thermodynamically stable. That is, the Gibbs free energy of mixing is negative and the dispersed condition is the preferred, lower-energy state.

Particles of polymer, such as poly(methyl methacrylate), dispersed in an aliphatic hydrocarbon are clearly lyophobic in character and rapidly aggregate. However, if the particles are each surrounded with a stabilizing barrier, such as an entire layer of solvated polymer chains of suitable dimensions, they become indefinitely stable provided that the protective layer remains intact. In a sense, such a stabilized particle can be regarded as a composite species whose core is lyophobic but whose outer layer is lyophilic in character. This is achieved in practice by the use of suitable amphipathic graft-copolymer dispersants, commonly called 'stabilizers' in dispersion polymerization.

The terms 'stabilizers' and 'stabilization', where used in this work, imply the use of a method for producing polymer dispersions which are stable towards aggregation processes. This is, of course, quite different from the sense in which these terms are frequently used in other branches of polymer science where they refer to processes and additives which confer on treated polymers an enhanced stability towards thermal and photolytic degradation processes[24].

In describing the structures of the various copolymers used as dispersants, the usual terminology is followed[25,26]. Thus, random copolymers derived from different monomer units, A and B, as represented by

$$\sim AAABAABBBBABBAABAABA \sim$$

are indicated by the prefix, -co-, as in poly(methyl methacrylate-co-styrene).

A linear block copolymer of the form

$$\sim\text{AAAAAAAAAABBBBBBBBAA}\sim$$

is indicated by the prefix, -b-, as in poly(methyl methacrylate-b-styrene). A graft copolymer of the type

$$\sim\text{AAAAAAAAAAAAAAAAAAAA}\sim$$
```
  B                   B
  B                   B
  B                   B
  B                   B
  B                   B
  B                   B
```

is represented by the prefix, -g-, as in poly(methyl methacrylate-g-styrene).

In an unstable dispersion, the primary colloidal particles rapidly form loose clusters, usually known as 'flocs', which can associate further to form loosely-bound aggregates. The term 'agglomerate' is sometimes used to denote more tightly-bound clusters of particles. However, in this work, the custom is followed in which terms such as 'flocculation', 'aggregation' and 'coagulation' are used synonymously to describe such processes unless reference is made to the detailed structure of the particular type of aggregate formed (for a general account of Colloid Science, see Reference 27).

## 1.4. REFERENCES

1. Solomon, D. H., *The Chemistry of Organic Film Formers*, John Wiley and Sons, Inc., New York, 1967.
2. Bovey, F. A., Kolthoff, I. M., Medalia, A. I. and Meehan, E. J., *Emulsion Polymerisation*, Interscience Publishers Inc., New York, 1955.
3. Warson, H., *The Applications of Synthetic Resin Emulsions*, Benn, London, 1972.
4. Dainton, F. S., Seaman, P. H., James, D. G. L. and Eaton, R. S., *J. Polym. Sci.*, **34**, 209 (1959).
5. Bamford, C. H., Barb, W. G., Jenkins, A. D. and Onyon, P. F., *The Kinetics of Vinyl Polymerisation by Radical Mechanisms*, Butterworths Scientific Publications, London, 1958.
6. Hölscher, F., 'Eigenschaften, Herstellung unt Prüfung', Part 1 of *Dispersionen Synthetischer Hochpolymerer*, Springer-Verlag, Berlin, 1969.
7. Schildknecht, C. E. (Ed.), *Polymer Processes*, Interscience Publishers Inc., New York, 1956.
8. Vanzo, E., *J. Appl. Polym. Sci.*, **16**, 1867 (1972).
9. Imperial Chemical Industries, *British Patent*, 893,429 (1962).
10. Rohm and Hass, *British Patent*, 934,038 (1963).
11. Barrett, K. E. J., *Br. Polym. J.*, **5**, 259 (1973).
12. Fitch, R. M. (Ed.), *Polymer Colloids*, Plenum Press, New York, 1971.
13. Bueche, F., *J. Colloid Interface Sci.*, **41**, 374 (1972).
14. Napper, D. H. and Hunter, R. J. in *Surface Chemistry and Colloids*, Vol. 7 (Ed. Kerker, M.), Butterworth, London, 1972.
15. Sarvetnick, H. A. (Ed.), *Plastisols and Organosols*, Van Nostrand Reichold Co., New York, 1972.
16. Berryman, D., *Austral. OCCA Proc. & News*, **7**, 4 (1970).
17. Thomas, H. R., *Farbe Lack*, **77**, 525 (1971).

18. Baylis, R. L., *Trans. Inst. Metal Finishing*, **50**, 80 (1972).
19. Derbin, G. M., *Amer. Paint J.*, 60 (1972).
20. Dowbenko, R. and Hart, D. P., *Ind. Eng. Chem. Prod. Res. Dev.*, **12**, 14 (1973).
21. Shibata, M., *Kobunshi Kako*, **21**, 331 (1972).
22. Waghorn, M. J., *Automotive and Other Ind. Finishing Symp.*, University of Warwick, 227 (1973).
23. Wigglesworth, D. J., *Trans. Inst. Metal Finishing*, **51**, 179 (1973).
24. Hawkins, W. L., *Polymer Stabilisation*, Wiley-Interscience, New York, 1972.
25. Ceresa, R. J., *Block and Graft Copolymers*, Butterworths, London, 1972.
26. Battaerd, H. A. J. and Tregar, G. W., *Graft Copolymers*, Interscience Publishers (John Wiley & Sons), New York, 1967.
27. Kruyt, H. R. (Ed.), *Colloid Science*, Elsevier Publishing Company, London, 1952.

CHAPTER 2

# The Theoretical Basis for the Steric Stabilization of Polymer Dispersions Prepared in Organic Media

D. W. J. OSMOND AND F. A. WAITE

| | |
|---|---|
| 2.1. THE FORCES OF ATTRACTION BETWEEN PARTICLES. | 10 |
| 2.1.1. The van der Waals–London attraction. | 10 |
| 2.1.2. The Hamaker integration. | 12 |
| 2.1.3. The retardation effect. | 14 |
| 2.1.4. The Hamaker method applied to liquid media. | 14 |
| 2.1.5. The Hamaker method applied to particles enclosed within rigid homogeneous layers. | 16 |
| 2.1.6. The continuum electrodynamic model. | 16 |
| 2.1.7. Implications for colloidal stability. | 18 |
| 2.2. THE FAILURE OF CHARGE STABILIZATION IN ORGANIC MEDIA OF LOW POLARITY. | 19 |
| 2.3. THE FORCES OF REPULSION GENERATED BY BARRIERS OF SOLUBLE POLYMER. | 22 |
| 2.3.1. The nature of the theoretical models. | 24 |
| 2.3.2. The volume restriction models. | 25 |
| 2.3.3. Models with interacting polymer chains. | 27 |
| 2.3.4. Stability in theta ($\theta$) solvents. | 33 |
| 2.3.5. The current view of stabilization theory. | 34 |
| 2.3.6. Polymer mobility and desorption | 36 |
| 2.3.7. Non-equilibrium situations. | 38 |
| 2.4. THE STERIC STABILIZATION OF POLYMERS DISPERSED IN ORGANIC MEDIA. | 39 |
| 2.5. REFERENCES. | 42 |

Colloidal stability is normally achieved by the interposition of repulsive forces of sufficient magnitude to overcome the inherent attractive forces of electromagnetic origin which arise as particles approach each other. In the absence of any repulsive forces, the particles rapidly aggregate. The repulsive electrostatic forces of the type generated in aqueous media are not generally available for stabilizing particles in organic liquids of low polarity such as the aliphatic hydrocarbons used as the continuous phase in dispersion polymerization. Consequently, in the latter case, stabilization is accomplished by the use of repulsive forces generated by the interaction of opposing dissolved polymer chains attached to the dispersed polymer particles—that is, by steric stabilization.

The quantitative theory of the stabilization of charged colloidal particles was based on an analysis of the net result of the interaction between forces of repulsion and forces of attraction as a function of the distance between the particle surfaces. A parallel approach is used in the present treatment of lyophobic polymer particles which are stabilized by attached polymer chains dissolved in organic liquids.

An outline only of the theories on the source and magnitude of the attraction between colloidal particles is given here since recent developments have been reviewed in detail elsewhere[1,2]. Charge stabilization is also only briefly described since the basic theory has been comprehensively dealt with in other publications[3,4]. Its application to non-aqueous media has also been considered[5]. Emphasis here is given to the reasons for the general inapplicability of charge stabilization to organic media of low polarity. Finally, the source and magnitude of the repulsive forces generated by soluble polymer chains in organic media is described in detail with special reference to the stabilization of polymer particles in hydrocarbon diluents of low polarity involved in dispersion polymerization. Although a number of the more recent reviews on colloidal stability[1,6,7] now include a consideration of steric modes of stabilization, this aspect has, in general, received much less attention than the corresponding electrostatically-stabilized systems.

## 2.1. THE FORCES OF ATTRACTION BETWEEN PARTICLES

### 2.1.1. The van der Waals–London attraction

The attractive force which operates between two adjacent particles, usually called the van der Waals force, originates in the interactions between the atoms and molecules of which the particles are composed. Interest in the nature of these intermolecular forces was first aroused by the study of the departure from ideal behaviour of real gases, as described by modified equations of state such as that due to van der Waals. As a result, the early work was mainly concerned with relatively simple systems of gases at low pressures in which the individual molecules were separated by distances much larger than the size of the molecules involved.

The first model used to describe the source of the attractive force was that due to Keesom[8] and was concerned with the interaction between permanent dipoles in separate molecules. A statistical analysis of a random arrangement of mobile dipoles showed that pairs of molecules would tend to orientate themselves so that their permanent dipoles would be aligned to produce the maximum attractive effect. Thermal perturbation causes some randomization in molecular configuration with a consequent departure from the ideal alignment. The Keesom attraction therefore tends to decrease as the temperature rises.

An additional source of attraction, recognized by Debye[9], is due to the interaction between a permanent dipole in one molecule with a dipole induced by it in a second molecule. This effect is quite independent of whether or not the second molecule has a permanent dipole. If the second molecule has a preferred direction of polarization, then the attractive force generated decreases as the temperature rises in the same way as the Keesom effect. However, for many types of molecules which are similarly polarizable in all directions, the induced dipole can follow the random motions of the permanent dipole with ease and the temperature dependence is normally quite small.

The observed deviation from the ideal gas laws in the case of gaseous helium still remained unexplained since the helium atom on a time–average basis has a spherically symmetrical electrostatic field. In 1930, London[10] showed that the attraction operating with this type of atom or molecule was essentially a quantum-mechanical effect and could occur between all molecules quite independently of the permanent or induced dipole effects. The quantum-mechanical attraction is essentially additive and is based on 'pair-wise' interactions of molecules in the gaseous state. This additive character of the quantum-mechanical attraction is unlike the attractions due to the Keesom and Debye effects. In the latter instance, it can be seen that if two dipoles are in an optimum alignment for attraction to one another, then they cannot both be in optimum alignment for attraction to a third dipole.

The physical basis of the London attraction can be envisaged as arising from the application of the Heisenberg uncertainty principle which requires that the electrical field of any atom or molecule is subject to random fluctuations. London considered that these fluctuations (taking place at frequencies corresponding to ultraviolet radiation) could result in the formation of a transient dipole able to induce dipoles in another atom or molecule. No actual dissipation of energy by the fluctuating field was envisaged as the total energy involved is less than one quantum. An alternative concept mentioned by London, and more recently elaborated by Krupp[11], regards the random fluctuations of all the molecules as forming a background radiation field with a finite value of the zero-point energy, $E_0$, of $\frac{1}{2}h\nu$. All the molecules are then in equilibrium with this radiation field, continually absorbing energy from it and re-emitting equal amounts of energy to it.

In either case, random fluctuations of the electrical fields of two adjacent molecules become coupled and oscillate together more nearly in phase than the fields of otherwise similar molecules in isolation. This coupling of adjacent molecules reduces the free energy of the system—in other words, there is an attraction between the molecules as they approach one another.

The multiplicity of the oscillatory modes associated with the random fluctuations of the electrical field of a molecule allow it to 'harmonize' its London interactions with several other molecules simultaneously without significant mutual interference. For example, a gas molecule can attract all of its neighbouring molecules simultaneously—not just the individual molecule which happens to be in its immediate vicinity at any given time. This behaviour is fundamentally different from that of the Keesom and Debye fields and is the basis for the concept of 'pair-wise' additivity for the London type of attraction.

It can be shown[10] that the attraction arising from the coupling of random fluctuations of the London type is a second-order perturbation effect and consequently the attractive potential energy ($V_A$) decreases with the distance of separation as follows:

$$V_A = -\lambda/r^6 \tag{2.1}$$

where the London interaction constant,

$$\lambda = (3/4)Z^{\frac{1}{2}}h\nu_v\alpha_0^2$$

where

$Z$ = number of electrons in the outer shell,

$\nu_v$ = the characteristic frequency,

and

$\alpha_0$ = the static polarizability of the molecules concerned[12].

### 2.1.2. The Hamaker integration

The magnitude of the attractive potential energy $(V_A)$ generated by the London interaction between condensed bodies in a vacuum (as opposed to individual gas molecules) has been calculated by a method initiated by Bradley[13] and de Boer[14] and developed by Hamaker[15]. The method is based on three assumptions:

(i) For condensed bodies separated by distances which are large compared to their molecular dimensions, the attractive and repulsive forces generated by permanent dipoles should cancel out.
(ii) The 'pair-wise' additivity concept used in calculating the London attraction between gas molecules can be applied to the corresponding interactions between atoms in different condensed bodies.
(iii) For bodies containing many molecules, the sum of all the 'pair-wise' interactions can be replaced by a double integral.

In this way, an integration of all possible interactions between the attracting elements of a pair of particles, results in an expression of the form:

$$V_A = A \cdot H \tag{2.2}$$

where $A$ is the Hamaker constant and $H$ the geometrical function.

The Hamaker constant has the dimensions of energy and is a function of the strength of the attraction between two elements, $\lambda$ (the London interaction constant) and their concentration, $q$ (the number of elements per unit volume) as follows:

$$A = \pi^2 q^2 \lambda \tag{2.3}$$

The methods available for obtaining values of the Hamaker constant for various materials have been reviewed[16,17]. The number and strength of the London oscillators in an element can be estimated directly from fundamental properties of the material concerned such as polarizability, magnetic susceptibility or optical data. An analysis of surface-wetting data, since surface free energy is related to the London attraction, has also been used to obtain a value for the Hamaker constant. Table 2.1 lists selected values for the Hamaker constants of typical solids and liquids from data given by Visser[17].

Table 2.1. Values of Hamaker constants for various solids and liquids (Based on data given by Visser[17]. (*Advan. Colloid Interface Sci.*, **3**, 331 (1972), with permission of Elsevier.)

| Solid | $A \times 10^{20}$ (J) | Liquid | $A \times 10^{20}$ (J) |
|---|---|---|---|
| Carbon (diamond) | 28.4 | Water | 4.38 |
| Carbon (graphite) | 47.0 | Cyclohexane | 4.64 |
| Silver | 40.0 | n-Decane | 4.6–5.8 |
| Aluminium oxide ($Al_2O_3$) | 15.5 | Benzene | 23–90.5 |
| Poly(styrene) | 6.15–6.6 | Toluene | 10–42 |
| Poly(methyl methacrylate) | 6.3 | | |

In general, the Hamaker constant tends to increase in value with the polarizability of the material concerned. Organic polymers and water show the lowest values, followed by ionic crystals and metal oxides, the highest values being obtained for metals and graphite. The reason for the large scatter of values often quoted for particular materials is ascribed to a combination of the nature of the approximations used in applying the underlying theory together with errors in experimental technique.

Nearly all the values of the Hamaker constant from optical data are based only on interactions at frequencies in the ultraviolet region corresponding to the principal excitation modes of the atoms concerned. More recent work on intermolecular forces[18] has shown that significant resonance can also occur in other regions, particularly the infrared, and under certain conditions these may also make a contribution to the total attractive force between molecules. In principle, the ordinary Hamaker constant could be replaced by a new constant $\bar{A}$ (equal to $A_1 + A_2 + A_3 \ldots$, where $A_1$, $A_2$ etc. are the appropriate constants for each frequency mode) in order to take account of these additional contributions.

The geometrical function, $H$, is an expression concerned with the solid geometry of the system under consideration and has the dimensions of a pure number. Although computers can now be used for the calculation of the actual numerical values for the complex functions involved in interactions between spheres and plates[19], the complexity of such expressions has often led in practice to the use of simplified expressions for ease of analysis. For example, in the case of spheres of equal size where the distance between their surfaces, $h$, is much less than their radius $a$, an approximate expression of the form

$$H = a/12h \qquad (2.4)$$

has been used. This gives a broad indication of how the geometrical function and hence the attractive potential varies with distance of separation. In general, these approximate calculations indicate that for typical colloidal systems the attraction may operate over distances of several tens of nanometres between the particles and not just within contact distances of less than 1 nm.

### 2.1.3. The retardation effect

The propagation of the electromagnetic coupling between any pair of fluctuating dipoles takes place at the speed of light in the intervening medium. Hence, when the fluctuating dipoles are separated by distances which are significant compared to the wavelength of the fluctuation frequency, the fluctuations move out of phase and there is a consequent reduction in the attractive potential generated. This effect was recognized by Overbeek[3] and subsequently quantified using fourth-order quantum mechanical theory by Casimir and Polder[20]. This latter work demonstrated that the inverse sixth-power relationship between distance and attraction between elements at close ranges (equation 2.1), changed to an inverse seventh-power relation at distances greater than the effective wavelength.

This effect can reduce significantly the attraction generated between medium-sized particles at separations down to 10 nm, especially for the correlation modes in the ultraviolet region and is even more pronounced when an intervening medium of a finite dielectric constant rather than a vacuum is present (see Section 2.1.4, below).

### 2.1.4. The Hamaker method applied to liquid media

Although it is now realized that the Hamaker method for calculating the attractive potential between particles is based on assumptions which are not fully valid (see Section 2.1.6, below), it is convenient at this point to review the various refinements which allow the method to be applied to actual colloidal systems.

The original calculations of Hamaker were based on the assumption that the two condensed bodies were separated by a vacuum. For actual colloidal particles which are suspended in a liquid medium, it is first necessary to consider the two distinct ways in which the presence of the medium can modify the forces of attraction. The *primary* medium effect is concerned with the influence of the liquid medium on the transmission of the London field between the elements of the macroscopic bodies. The *secondary* medium effect involves the finite attraction of the particles for the medium.

A recent account of the primary medium effect has been given by McLachlan[21], who has shown that the London field is modified by the medium in two ways. Firstly, the strength of the radiation of the electrical field is inversely proportional to the dielectric constant of the medium (an effect first described by Overbeek[4]). The second, and probably more important, effect is that the optical path-length between the particles is increased in proportion to the square root of the dielectric constant of the medium. This effect further reduces the attraction and also causes retardation to operate at a much closer range.

The secondary medium effect was discussed by Hamaker but has been overlooked in some subsequent studies. In the presence of a finite medium, the overall energy of aggregation is given by the difference between the energy required to separate two particle–medium pairs and the energy released by the formation of the corresponding particle–particle and medium–medium pairs. Consequently, the energy of aggregation is usually much smaller for particles suspended in a liquid medium than for the same particles in a vacuum. For particles which are not similar in chemical composition, the net energy of aggregation can be negative and the particles will repel each other. However, for similar particles which always attract each other, the secondary medium effect can be allowed for by the use of an expression which is derived for the effective Hamaker constant $A_{12}$:

$$A_{12} = (A_1^{\frac{1}{2}} - A_2^{\frac{1}{2}})^2 \qquad (2.5)$$

where $A_1$ is the Hamaker constant for the particles and $A_2$ that for the medium. The expression only strictly applies when the characteristic frequencies of the London oscillators of the elements of the particles and of the medium are identical, but it can provide an adequate estimate unless the respective frequencies are widely different[16].

The application of the medium correction (equation 2.5) to obtain an effective Hamaker constant can be illustrated by the case of two bodies of material 1, embedded in a medium of material 3, as in poly(styrene) dispersed in water. Although any errors in the values of the primary constants would tend to be multiplied up in the derived value, in general an effective Hamaker constant lower than either component is obtained, for example,

$A_{11}$—poly(styrene); $\quad A_{33}$—water $\rightarrow$ $A_{131}$—poly(styrene) dispersion in water

$\quad$ 6·15–6·6 $\quad ; \quad$ 4·38 $\quad \rightarrow \quad$ 0·15–0·23 ($\times 10^{20}$ J)

Clearly, for polymer colloids in aliphatic hydrocarbons, in which both phases are similar chemically, the application of the medium correction will also lead to a small value for the effective Hamaker constant, which will usually be at least an order of magnitude less than the separate values for the polymer and the diluent.

It is possible to conceive of an extreme case, where the values of the Hamaker constants for the particles and medium are identical, in which the London attraction is zero. Although permanent dipole effects may result in a finite attraction between the particles at contact distances, that is within a few tenths of a nanometre, a barrier consisting of a suitable soluble component with a corresponding thickness should be sufficient to counteract the residual attractive forces operating and to maintain a stable dispersion. Also if the chemical natures of the particles and diluent are similar, then even the short-range interactions at contact distances due to permanent dipole effects should be greatly reduced (to small values relative to $kT$), enabling a stable dispersion of stabilizer-free particles to be obtained. Strictly, such systems do not exist in practice. However, a close approximation to dispersions of this type can be prepared using polymer particles containing cross-links which prevent their complete dissolution but allow a limited degree of swelling to take place (see Section 5.1, below).

### 2.1.5. The Hamaker method applied to particles enclosed within rigid homogeneous layers

In 1961, a theoretical treatment by Vold considered the effect of enveloping spherical particles with layers of various materials of differing Hamaker constants on the interaction energies of the modified particles in a finite medium[22]. Surprisingly, it was found that the minimum attractive energy occurred when the value of the Hamaker constant for the enclosing layer was different from that of the medium—the so-called Vold effect. A later analysis of this approach[23] located an algebraic error in the original work and gave the correct value of the Hamaker constant for the enclosing layer at which the minimum is found. This new value suggests that the particle, together with its sheath of adsorbed material, actually behaves like a compound particle. With an appropriate Hamaker constant for the sheath, the average value obtained for the compound particle is closer to that of the dispersion medium than for the original bare particle, hence the net attraction is again reduced in the same manner as for the effective Hamaker constant. Unfortunately, the complex behaviour of the London attraction with increased distance, together with the layered structure of the compound particle, makes the averaging process very inefficient and reduces the corresponding corrections, unless the distance between the particles is large compared to their dimensions. As a result, at the smaller distances relevant to colloid stability, the correction is negligible when compared to the simple geometric or steric-spacing effect of the adsorbed layers.

### 2.1.6. The continuum electrodynamic model

The original derivation for the London attraction between molecules was concerned primarily with the problem of intermolecular forces in gases, where the mean distance of separation of the molecules is comparable to the wavelength of the principal coupling correlations. Only a few neighbouring molecules are subjected to attractive forces of any magnitude and the assumption of 'pair-wise' additivity of the attraction is thus quite reasonable. However, the Hamaker approach for the calculation of the attraction between condensed bodies considers the sum of the attractions between pairs of individual elements, one in each particle, as if each pair of elements were in complete isolation. In reality, each element is surrounded by many others which go to make up the particle and which are within the range of strong interactions. Hence the interaction between any pair of elements, each in a real particle, would be expected to be much modified when compared to that between the same elements in isolation. In addition, unless the pair of elements are located at the surface of the two bodies, they are actually attracting each other, not through a void, but through the intervening material of the particles themselves so that the actual interactions must be further modified by medium effects of the type described above. Clearly, therefore, the use of 'pair-wise' additivity can hardly be justified in calculating the attraction between condensed, macroscopic bodies.

Although, in principle, it should be possible to extend the work of Halicioğlu and Sinanoğlu[24] on multi-body interactions to provide a modification of the Hamaker procedure which would account for all the interactions involved in both the particles and the medium, so far this has yet to be carried out. However an alternative approach does exist, that of the continuum electrodynamic model, in which the complex effect of multi-body interactions is allowed for directly by using experimentally-determined macroscopic values for the electromagnetic constants of the bulk phases involved.

This alternative approach, introduced by Lifshitz[25] in 1955, is based on the concept that the bodies are ideal continua, with the same dielectric properties throughout. In some respects, Lifshitz and his co-workers[26] followed the quantum mechanical approach of Casimir and Polder[20] as used to derive expressions for the retardation correction (Section 2.1.3). The continuum approach has the advantage that it automatically incorporates the medium and retardation effects which have to be introduced as separate corrections in the Hamaker model. In addition, the full theory includes correlations at all frequencies, from zero to infinity, and not just those in the ultraviolet region originally considered by London. The predictions of the continuum theory, particularly in relation to the magnitude of the attractive forces generated and the retardation effect, are in good agreement with the results of direct measurement of attractive forces between solids *in vacuo*. Formerly, the available techniques only allowed measurements at distances of separation greater than 100 nm[16], that is, within the region of retardation. More recently, both the retarded and unretarded London attractions have been determined at distances of separation down to 5 nm[27], albeit in the absence of an intervening medium. These results indicated a transition from normal to retarded forces at a distance of about 15 nm, as predicted theoretically. Direct determinations across a liquid aliphatic hydrocarbon medium, a n-decane triple film[28], suggested a value for the effective Hamaker constant of $5.6 \times 10^{-21}$ J, which is very close the value predicted (5.5–6.1) by Parsegian and Ninham[18] using the modification of the continuum approach described below.

The work of Lifshitz and the Russian school[25,26] has been extended by other theoreticians in the field of molecular forces. McLachlan[29] has shown that many of the results from the quantum mechanical approach can also be achieved by simpler mathematical procedures using a quasi-classical method involving Maxwell's equations. In addition, Langbein[30] and Ninham and Parsegian[31–33] have applied the equations derived by Lifshitz to a number of systems relevant to colloid chemistry problems and have shown how approximate values for the required optical data may be obtained.

The way now appears to be open for a more general application of the continuum approach for the calculation of attractive forces between bodies in the condensed state. However, very complex mathematical calculations are still required although the development of interpolation procedures has facilitated the computations to some extent. It is likely therefore, that many workers in the field of colloid chemistry will continue to use the simpler Hamaker method for calculating the attractive forces between colloidal particles, despite a growing realization of its fundamental defects.

Since the completion of this manuscript, calculations for the van der Waals attraction between polymer latex particles in both aqueous and non-aqueous media have been reported (Smith, E. R., Mitchell, D. J. and Ninham, B. W., *J. Colloid Interface Sci.*, **45**, 55 (1973); Evans, R. and Napper, D. H., *ibid.*, 138). These demonstrate that provided the dielectric permittivities of the particles and medium are similar (as in the case of organic polymers dispersed in hydrocarbons), the predictions of the modified macroscopic continuum theory accord quite well with those of the microscopic Hamaker theory.

### 2.1.7. Implications for colloidal stability

The achievement of stability in sterically-stabilized colloidal dispersions depends on maintaining the particles at a sufficient distance from each other so that they are beyond the effective range of their inherent mutual attraction. Early calculations using the Hamaker method, especially in its simplified form, suggested that the attractive force would often still be large compared to $kT$, when the particles were separated even by distances of up to many tens of nanometres[34]. Subsequently, over the years, various refinements both in the corrections to be applied to the Hamaker method and in the experimental techniques used, have led to a position where both theory and experiment indicate that the attraction between colloidal particles, especially for polymer particles in organic media, is in many instances less than $kT$, even when their surfaces are separated by distances as small as 5–10 nm. However, the results from the more recent Hamaker techniques which incorporate frequencies additional to the ultraviolet and also those from the use of the continuum model, now suggest that not only should the absolute value of the attraction at a given distance be larger than previously believed, but also that the attraction should decrease more slowly with increasing distance of separation between the particles. This would imply that larger particle separations are required for stability than appear to be necessary in practical systems (see Section 3.4).

The recent theoretical advances in macroscopic (continuum) models have underlined the fundamental defects of microscopic (Hamaker) models. Essentially, in the Hamaker model no allowance is made for the effect of short-range correlations between adjacent elements in the same particle, thus underestimating the magnitude of the attractive force generated. At the same time however, no allowance is made either for the effect on the radiation from a particular element, as it travels past the many other elements of both its own particle and the attracted particle. The many successive interactions which are possible for this radiation as it traverses condensed bodies (of the same composition as the radiating element) to an element in another particle, greatly reduce the likelihood of direct phase correlation between the two primary elements. Hence, the attraction should be less than that predicted by the 'pair-wise' summation. The apparent success of the predictions of the original Hamaker method in many practical systems may well be due to the mutual cancellation of such opposing, but mechanistically-linked, effects so that their joint omission does not introduce first-order errors.

Quite a separate problem arises from the introduction of additional interaction frequencies in the infrared and microwave regions into the Hamaker method of calculation. In principle, there is an effective Hamaker constant for each frequency mode for particles dispersed in a liquid medium. Since there is a basic similarity in the values for the ultraviolet Hamaker constants for materials based on the lighter elements, the net effective value is usually significantly smaller than the value for the material of the particles alone. However, as Parsegian and Ninham[32] have pointed out, the infrared and microwave modes for the material of the particles and of the medium may occur at very different frequencies and with very different strengths. In these circumstances, the secondary medium effect involving the attraction of the intervening medium by the particles, results in only a trivial reduction in the value of the effective Hamaker constant.

Just this sort of situation would be expected to apply to polar polymers, such as poly(methyl methacrylate), dispersed in aliphatic hydrocarbons. Not only are large additional Hamaker constants introduced by these low-frequency coupling modes, but also their associated long wavelengths imply that attenuation due to retardation only becomes effective at distances much larger than those at which the attraction due to the ultraviolet coupling modes has fallen to very low values.

There still remains therefore, an apparent conflict between the predictions of the more recent continuum theories of attractive forces and the well-established experimental results for colloidal systems. However, even with the mathematical simplifications, the general solution for the attraction between small particles in a liquid medium is not yet available. Clearly, a satisfactory resolution of these differences must await further theoretical developments, together with additional experimental work on systems more representative of the phase dimensions of colloidal particles.

## 2.2. THE FAILURE OF CHARGE STABILIZATION IN ORGANIC MEDIA OF LOW POLARITY

The basic quantitative theory for the stabilization of charged colloidal particles, the so-called DLVO theory, was developed by the independent efforts of the Russian[35] and Dutch[3] colloid schools. In this treatment, colloidal stability results when the potential energy of repulsion arising from the approach of charged particles exceeds the inherent attractive energy between the particles over a certain distance of separation. The electrostatic stabilization of lyophobic colloidal dispersions in non-aqueous media has already been reviewed in terms of the DLVO theory[5].

Aqueous dispersions of colloidal polymer particles, virtually free from extraneous electrolyte and therefore containing only aqueous autoprotolytic ions and charging counter-ions, have been prepared in recent years, generally by exhaustive dialysis techniques[1,36]. These lattices are highly stable at all accessible concentrations, and indeed Krieger[37] has shown that long range repulsive forces, significantly greater than $kT$, can exist between the particles in dispersions of low concentration when the particle separation is large. As these aqueous lattices are stable, presumably as a result of simple coulombic repulsion, one must therefore seek for an explanation for the failure of this type of mechanism to stabilize polymer dispersions in organic liquids. The reason for this failure, however, appears to be quite straightforward.

The coulombic repulsive potential, $V_R$, between two such charged particles is given by the expression:

$$V_R = \frac{Q^2}{R\varepsilon} \tag{2.6}$$

where $Q$ is the charge on each particle, $R$ is the centre-to-centre distance between the particles and $\varepsilon$ is the dielectric constant of the liquid phase.

The upper limit of the equilibrium potential which can be reached in the absence of specific ionic interaction effects, is broadly similar for all systems, irrespective of the medium, since the work required to remove further ions would be large compared to $kT$. However, the charge required to produce this broadly constant potential may vary widely as it is directly proportional to the capacity of the particle (which behaves in the liquid medium as a spherical condenser). The capacity $C$, of a conducting sphere of radius $a$, in a dielectric is given by the expression:

$$C = a\varepsilon \tag{2.7}$$

By substituting in equation (2.6), the following expression for the repulsion results:

$$V_R = \frac{\zeta^2 a^2 \varepsilon^2}{R\varepsilon} = \frac{\zeta^2 a^2 \varepsilon}{R} \tag{2.8}$$

where $\zeta\ (=Q/C)$ is the surface potential of the particle.

It is therefore apparent, that the essential difference in the power of electrostatic stabilization in water and in typical organic media lies in the difference between the values of their respective dielectric constants. The large value for water results in a large capacity for the particles dispersed within it. Hence, at a potential in the range often found in practice (50–100 mV), the charge on the particles is much larger than for similar particles in an organic liquid. Indeed, the actual charge for a colloidal particle in a medium of low dielectric constant (say $\varepsilon = 2$) may be as small as 10 $e$, where $e$ is the charge of an electron.

Although the repulsive force or potential at a given charge is reduced in proportion to the dielectric constant, because the repulsion is proportional to the square of the charge, it is clear that the repulsion for similar particles is directly proportional to the dielectric constant of the medium. For water, this will be between one and two orders of magnitude greater than for typical organic liquids.

It is of course correct, as Albers and Overbeek[38] have pointed out, that since with coulombic repulsion there is a slow fall-off in repulsive potential with surface-to-surface distance, one has to take the difference between the repulsive potential at the required separation and that at the mean particle separation corresponding to the concentration of the dispersion. For media of high dielectric constant however, this difference is still so large compared with the van der Waals attraction of the particles at medium ranges, that highly stable dispersions result.

It has been suggested[34], that because the charge at a given surface potential and hence the repulsion increases with particle size, then it should be easier to stabilize coarse particles in low dielectric constant media. Returning to equation (2.8) above, it can be seen that the centre-to-centre separation, $R$, is equal to $2a + h$, where $h$ is the surface-to-surface separation. For close approach of the particles, $h$ is very small compared to $2a$, so that the actual repulsion at close approach is roughly proportional to the first power of the particle size, that is:

$$V_R \approx 1/2\zeta^2 a\varepsilon \qquad (2.9)$$

Similarly, since the London attraction between the particles at close approach also increases roughly in proportion to the first power of particle size, the ratio of the attractive and the repulsive potentials is approximately independent of the particle size. Thus, the absolute height of any potential barrier will be directly proportional to the particle size. As it is this barrier height, compared to $kT$, which determines the stability of the dispersion against thermal motion, stability increases with the exponential of the first power of particle size.

However, if dispersions are to be of any practical use, they must be stable not only towards flocculation caused by the normal Brownian motion of molecules but also against shear-induced collisions or orthokinetic flocculation[4]. With this type of flocculation, particles do not have the same average kinetic energy (distributed according to the Boltzmann exponential law) but instead have the same velocity, varying only with local shear gradient. As a result, the energy available to overcome the potential barrier is proportional to the mass of the particle, $M$, that is to $a^3$ for a given particle density. Hence the resistance to orthokinetic flocculation of particles stabilized by coulombic forces falls steeply with increasing particle size. Because of the narrow range of particle velocities for a given shear gradient, there tends to exist a critical particle size, for which the kinetic energy, $V_K$, given by the expression:

$$V_K = \tfrac{1}{2}M(\delta v)^2 \qquad (2.10)$$

(where $\delta v$ is the relative particle velocity in a collision) is just equal to $V_R$ the peak of the coulombic repulsive potential energy barrier.

All particles below this critical size are indefinitely stable at this shear rate; all those above, aggregate on every collision. Clearly, this aggregation will increase the particle size further, until a 'run-away' situation develops and the particles increase in size to become macroscopic lumps. It also follows from equation (2.10) above, that this critical size will fall sharply as the maximum shear gradient increases, since $V_K$ is also proportional to $(\delta v)^2$. It is thus evident that practical, as opposed to ideal or Brownian stability, cannot be improved by increasing the particle size. It would therefore appear that only relatively few liquids, such as water, with dielectric constants greater than about 50, are able to provide adequate dispersion stability as a result of electric charges on the particles.

Since the completion of this manuscript, a more extensive treatment involving the use of Maxwell's equations for the calculation of the electrostatic repulsion between charged conducting spheres in a medium of low dielectric constant has been reported (Parfitt, G. D., Wood, J. A. and Ball, R. T., *J. Chem. Soc. Faraday Trans. I*, **69**, 1908 (1973). These workers point out that the assumption that a charged conducting sphere can be represented by an equivalent charge at its centre is only valid when the sphere is effectively in isolation. The results obtained for charged spheres using Maxwell's equations were compared with those derived from the simple coulombic law for repulsion between point charges. This comparison indicates that the simple point charge assumption is a reasonable approximation for small particles, especially in transient encounters where the constant charge regime is likely to apply. For larger particles, i.e. greater than 1 μm in diameter, especially under equilibrium conditions where a constant potential regime is more likely, a point charge calculation will over-estimate significantly the strength of the repulsive potential. The errors introduced are still probably no larger than those in other factors involved in the determination of colloidal stability.

## 2.3. THE FORCES OF REPULSION GENERATED BY BARRIERS OF SOLUBLE POLYMER

As two surfaces, each covered by a layer of adsorbed soluble polymer chains, approach each other to within a distance less than the combined thickness of the adsorbed layers, an interaction between the polymer layers will occur. This interaction, which is the source of steric stabilization, will in most cases, generate a repulsive force between the opposing surfaces and many attempts have been made to describe its nature and magnitude. The problem has usually been treated in terms of the change in free energy which takes place when the two surfaces covered by adsorbed polymer are brought together from an infinite separation to some small, specified distance apart. It must be realized at the outset however, that although a calculation of the free energy change, particularly when restricted to specific regions of polymer perturbation in the liquid phase between the particles, can serve for the estimation of the magnitude of the interaction, it cannot provide information directly about the actual mechanism of repulsion. Repulsive forces can only be generated between the particles by events occurring immediately adjacent to their surfaces.

So far, the theories of steric stabilization have mainly been based on the basic model of an array of polymer chains each attached to a surface by one or at most a few links. The free energy changes accompanying the various interactions of the polymer chains, including interactions with each other, with solvent molecules, with similar chains on an opposing surface and with the opposing surface itself, have all been considered as contributing to the repulsive forces which give rise to steric stabilization. Most of the treatments have been based on theories actually derived for free polymer molecules and subsequently modified to allow for the condition that the chains are attached to a surface in some way.

The simplest model of a polymer molecule used is that of a freely-jointed chain in which each segment is of equal length but is effectively volumeless. Such a model can be treated by random flight statistics and in this way the entropy changes associated with the constraints imposed on the molecule calculated. A further version of this treatment is to generate the chain on a lattice by computer calculation. If more than one segment of polymer can occupy a site on the lattice at the same time, this again simulates the volumeless segment condition. However, a refinement giving a closer approximation to reality is to generate a chain on a lattice in which the choice of adding a segment to the chain is restricted (by incorporating a notional bond angle) and excluding the multiple occupancy of a single site (that is, the segments now have real volume). Such a model essentially simulates a polymer molecule dissolved in an athermal solvent in which the chain segments are equal in size to the solvent molecules. When the system is not athermal, the effect of the solvent can be simulated by including a weighting parameter in the selection process for the position of the added segment. Again, entropy changes accompanying constraints on the molecule can be calculated.

On the other hand, the application of statistical thermodynamics to 'real' polymers in free solution has given rise to a number of theories describing their behaviour, the name of Flory being associated with several of the treatments. Only a brief account of the more important features is given here. For further details the reader is referred to the excellent accounts given in the text-books of Flory[39] and Tanford[40] (see also Section 4.2.3, below).

In general, these treatments calculate the non-ideal contribution (that is, taking no account of the entropy of dilution of the polymer as a whole) to the free energy changes which take place when polymer molecules interact with solvent molecules or when polymer solutions are intermixed. The parameters, $\kappa_1$ and $\psi_1$, which are associated with the enthalpy and entropy of mixing respectively are of paramount importance in these calculations. These are related to $\chi_1$, a parameter expressing first neighbour interaction, by the expression

$$\kappa_1 - \psi_1 = \chi_1 - \tfrac{1}{2} \tag{2.11}$$

Alternatively, $\psi_1 - \kappa_1$ may be written as

$$\psi_1(1 - \theta/T) \tag{2.12}$$

where $\theta$ and $T$ are the theta temperature (see below) and absolute temperature respectively.

The equations describing the free energy of mixing polymer segments in solution all contain such expressions which tend to zero when $\theta = T$, $\kappa_1 = \psi_1$ or $\chi_1 = \tfrac{1}{2}$. Under such conditions the polymer solution is said to be 'ideal' or at its theta temperature. This implies that the only contribution to the free energy of mixing is that derived from the whole polymer molecule considered as a volumeless point in space and that the second virial coefficient in the expression for osmotic pressure is zero. Alternatively, since the enthalpy and entropy of dilution are in principle equal, two polymer molecules might be expected to interpenetrate freely and quite independently of the presence of the other (see Section 3.2, below).

It must be emphasized, however, that although these theories have been used to describe steric repulsion, they were originally derived to describe the behaviour of dilute polymer solutions. Since steric repulsion is concerned with the interaction of polymer chains on opposing surfaces, the conditions are not those which exist in dilute solutions. Nevertheless, the magnitude of the repulsive potentials predicted on the basis of such theories have proved to compare reasonably well with experimental determinations.

### 2.3.1. The nature of the theoretical models

Before reviewing the various theoretical treatments developed to describe the interaction of polymer chains in adsorbed layers, it is first necessary to appreciate the nature of the fundamental models on which the treatments are based. In describing such interactions it is convenient to consider one surface and its adsorbed layer as fixed and termed the *primary* surface whilst the other, the *secondary* surface approaches it.

As it is the simplest model for polymer adsorption, most authors have visualized the polymer chain as terminally attached or 'anchored' at a plane or spherical surface, so that no desorption or lateral displacement of the chain can occur. The chains are considered to be either so widely spaced on the surface that they can be treated as individual chains, or (more relevant to practical cases) to be so close together that the layer can be considered as uniform in composition parallel to the surface. In each case, the chains naturally adopt mean configurations which represent their lowest free energy states. Any constraint on the chains which alters this mean configuration must therefore increase the free energy of the system. If such a constraint arises from the approach of any obstruction from a direction having a component normal to the primary surface, the change in free energy will be manifested as a force of repulsion. This repulsion will act initially only upon those segments of the primary polymer chain which are prevented from occupying a site already filled, but then, because the whole configuration of the adsorbed chain is modified by this repulsion in the outer region, the concentration of segments at the surface increases, so that the surface itself is subjected to a repulsive force.

Such obstructions, which constrain the polymer chain, have been recognized as arising from two sources. The first is due to the presence of an impenetrable surface and the second to the presence of a real polymer molecule (i.e. one in which the segments have finite volume). Theories involving both types of interaction have been put forward. Unfortunately, there has been a tendency to make a sharp distinction between the perturbation of the primary chain caused by the two processes as is indicated by the nomenclature commonly used. Obstruction by a solid surface is commonly called the *volume restriction* term, whilst the changes in free energy associated with the obstruction by the segments of a real polymer molecule is usually described as the *free energy of mixing* term. This nomenclature has tended to obscure the essential similarity of the processes as far as the primary polymer chain is concerned. This in turn has led to entirely different methods for calculating the energy changes involved, and to confusion as to whether they interact mutually, or can be calculated separately and then added together. In some cases, there has been uncertainty as to whether one or other of the processes has any real existence at all. However, although for the sake of continuity the established terminology will still be used in the present review, these inherent defects must be borne in mind.

### 2.3.2. The volume restriction models

The perturbation of primary polymer chains by an impenetrable surface has usually been analysed along the following lines. When a secondary, plane parallel surface is brought to a distance $h$, from a primary plane surface carrying an adsorbed layer of polymer chains (where $h$ is less than the thickness, $\delta$, of the adsorbed layer), the layer is compressed (Figure 2.1). Such a compression reduces the number of possible configurations of each chain as a whole and results in an overall decrease in entropy. By applying the appropriate

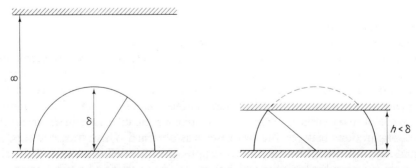

Figure 2.1. Schematic representation of the loss of configurational entropy on close approach of flat plates[41]. (After Mackor, *J. Colloid Sci.*, **6**, 492 (1951), with permission)

thermodynamic principles, the corresponding free energy change can be obtained. Since this volume restriction term results from a constraint on the configurations of the molecules as a whole, repulsion occurs even when the volumeless statistical chain equivalent rather than a 'real' molecule is considered. Indeed, most of the theoretical results have been obtained using statistical chain models in which the segments of the chain are volumeless, i.e. any number of chain segments may occupy one and the same position in space at any one time.

Mackor[41] made the first attempt to obtain an expression for the numerical value of the volume restriction term as a function of $h$. He pictured the adsorbed polymer chain to be a rigid rod, firmly attached but freely-jointed at one end to a flat surface, and assumed that all possible orientations of the rod at the surface had an equal probability of occurrence (Figure 2.1). His expression

$$\Delta G_v = kTN(1 - h/b) \tag{2.13}$$

(where $N$ = the number of chains per unit area and $b$ = the length of the rod; $b = \delta$ in this case), is quite simple, but the rigid rod used as a model or its subsequent refinement[42] scarcely simulated an adsorbed flexible polymer chain containing many segments.

If it is assumed that the rod is volumeless (i.e. only has length), there is no reduction in the number of possible orientations available to each rod, even if their points of attachment to the surface are separated by less than $2\delta$, i.e. there is no lateral interaction between the rods. Equally, as described above, if the secondary surface carries a similar adsorbed layer of rods there is no interaction between the primary and secondary rods. Thus, on the basis of this model, there is no repulsion between the surfaces even when both carry adsorbed chains, until the surface-to-surface separation is less than the length of one chain.

More sophisticated treatments of the volume restriction term for multi-segment chains have been carried out by Meier[43] and by Hesselink and co-workers[44,45]. The same basic assumptions are made as in the Mackor treatment, except that the polymer chains are considered to be long and flexible, although each uses a different approach to the random flight statistics involved. Both treatments, however, yield complex expressions which rely on unknown parameters and thus their application to actual systems is difficult. Equally, both expressions predict an enormous repulsive potential when the surface-to-surface separation, $h$, is less than $\delta/2$.

Yet a further treatment is due to Clayfield and Lumb[46,47] whose approach to the random flight statistics was similar to that of Hesselink. Both treatments involve the generation of chains on a cubic lattice from a point on a primary, impenetrable plane surface, both in isolation and in the presence of a secondary, impenetrable obstructing parallel plane surface placed at a specified distance from the primary surface. In this way, a value for the difference in the number of configurations between the two cases was obtained. There was, however, a significant difference between the two approaches. Hesselink generated chains on a six-choice cubic lattice in which there was no restriction on bond angle or occupation of a particular site. On the other hand, Clayfield and Lumb used a computer simulation to generate chains on a four-choice cubic lattice with the extra conditions in which the bond angle was restricted to 90° and a segment was not allowed to occupy a site already occupied by another. It is clear, that both initially and after obstruction by the secondary surface, the number of possible configurations available to the Clayfield chain was considerably less than that available to the Hesselink chain (i.e. at least $6^n$–$4^n$ times less, where $n$ is the number of segments in the chain). Thus, although the proportion of configurations for a given molecule lost on obstruction may be greater in the Clayfield model, the actual number lost will generally be much less than in the Hesselink model. As a result, the repulsion due to the volume restriction term calculated on the basis of the Clayfield model is considerably less than that calculated from the Hesselink model. Since a real molecule does contain actual segments which cannot occupy the same position in space at the same time (even under theta conditions in which in some senses a molecule behaves as if its segments were volumeless), the Clayfield model gives a much better simulation of a real system than either the models of Hesselink or of Meier.

To summarize the various volume restriction models used, the repulsion generated due to the volume restriction term was overestimated by both Meier and Hesselink, especially at small values of $h$. Again, on the basis of the Meier and Hesselink models of volumeless chains, no repulsive force will be generated, even if the secondary surface is covered by adsorbed layers of polymer, when the secondary surface is more than one layer-thickness away from the primary surface. In this case, the primary and secondary chains will interpenetrate without mutual obstruction or perturbation. Clayfield and Lumb on the other hand, assume no interpenetration of primary and secondary polymer chains, i.e. they virtually interpose a phantom impenetrable surface between the adsorbed layers. Hence, with their model, strong repulsions begin at $2\delta$, when both surfaces are covered with polymer chains.

### 2.3.3. Models with interacting polymer chains

In the treatments described above, in which the adsorbed chains on both the primary and secondary surfaces were assumed to be volumeless, there was no interaction between penetrating chains. In the Clayfield model, which implies a real volume for the segments of the polymer molecule, chain interpenetration was assumed not to occur. Clearly, actual polymer segments occupy a real volume and this is relevant not only to the number of configurations adopted by the unperturbed chains and to the number of configurations lost (and hence the repulsion produced) when a primary molecule is perturbed, but it also makes it possible for the segments of another polymer molecule to perturb the primary chains and produce a repulsion.

As early as 1942, Flory[48] laid the basis of his classical theory of macromolecular solutions in which he introduced the thermodynamic concept of the excluded volume of a polymer chain and from which properties such as the osmotic pressure and free energy of mixing were derived in thermodynamic terms. Rather than treat the perturbation of a primary polymer chain by the presence of segments of another polymer chain *ab initio*, using methods essentially similar to those described above for calculating the perturbation produced by an infinite, impenetrable surface, a number of workers have attempted to use the Flory treatment for calculating the repulsion produced by interactions between polymer chains in sterically-stabilized systems.

In 1958, Fischer[49] suggested that the repulsive energy of steric barriers was produced by the overlap of adsorbed polymer layers. The value of this repulsion was calculated by assuming that the adsorbed layer of polymer chains had a definite thickness and that the concentration of polymer segments was uniform within the layer. When such surfaces approached each other, the layers were considered as overlapping and, as no redistribution of polymer chains was assumed to occur, the concentration of polymer segments was doubled in the overlap volume (Figure 2.2). From the geometry of this system and the necessary

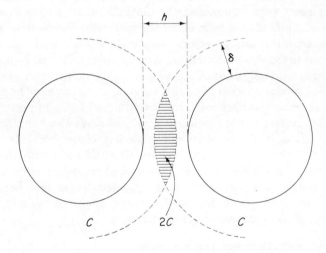

Figure 2.2. Model representing steric repulsion by overlap of dissolved polymer chains: $C$, concentration of polymer chains[50]. (After Ottewill and Walker, *Kolloid-Z.Z. Polymere*, **227**, 108 (1968), with permission of Dr Dietrich Steinkopf Verlag)

thermodynamic parameters derived from the Flory treatment, the change in the free energy of mixing ($\Delta G_m$) can be calculated as a function of the degree of overlap. This has been called the 'mixing' term and for the case of spherical particles, the following equivalent expressions have been used by various authors:

$$\Delta G_m = A' . B \quad \text{Fischer}^{49} \tag{2.14}$$

$$\Delta G_m = A' \frac{(\psi_1 - \kappa_1)}{V_1} \quad \text{Ottewill and Walker}^{50} \tag{2.15}$$

$$\Delta G_m = A' \frac{(1 - \theta/T)\psi_1}{V_1} \quad \text{Napper}[51,52] \tag{2.16}$$

where

$$A' = \tfrac{4}{3}\pi k T C^2 \left(\delta - \frac{h}{2}\right)^2 \left(3a + 2\delta + \frac{h}{2}\right)$$

$C$ = concentration of segments in adsorbed layer
$a$ = radius of particles
$h$ = surface-to-surface separation
$\delta$ = adsorbed layer thickness
$\psi_1$ = entropy parameter
$\kappa_1$ = enthalpy parameter
$V_1$ = partial molar volume of solvent
$\theta$ = theta temperature
$B$ = second virial coefficient

All of these expressions contain a common geometric term and an equivalent thermodynamic term which, depending upon the nature of the solvent, can be positive, zero or negative. Thus, in a good solvent, the free energy of mixing will be repulsive (i.e. work must be done on the system to overlap the adsorbed layers). In a theta solvent there will, in principle, be no change in the free energy of mixing. In a thermodynamically poor solvent, the free energy of mixing will be attractive and work must be done to separate the adsorbed layers.

The Fischer model, however, has two significant defects which result in the overestimation of the repulsive forces generated, particularly in the early stages of interaction. The first is the assumption that the concentration of polymer segments in the adsorbed layer is uniform. The second defect is that it does not take account of the redistribution of the polymer segments in the overlap zone. Consequently the model becomes meaningless beyond 'half overlap', i.e. when $h < \delta$; (half overlap is defined as the situation where the surface to surface distance between the particles, $h$, is equal to the thickness of one adsorbed polymer layer, $\delta$). A more realistic model would be one which allowed for the redistribution of polymer segments. This then would, to some extent, compensate for these defects and the entirely unreasonable step function in segment concentration at greater distances from the particle surface (i.e. at the overlap boundary) used in the Fischer model.

A model which allows for redistribution to take place has been proposed by Doroszkowski and Lambourne[53]. Here the polymer chains are considered to be irreversibly attached to the surface of the particles. As the particles approach each other, the polymer segments are redistributed to a new uniform value over the toroidal volume of interaction shown in Figure 2.3. The repulsion generated can then be calculated in two ways, either from the force produced or from the free energy of mixing.

In the first case, the segmental osmotic pressure in the zone of interaction and the total force on the contact plane was calculated and integrated with

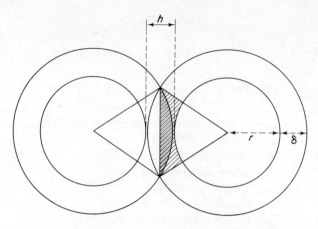

Figure 2.3. Model representing steric repulsion by overlap allowing redistribution of overlapped polymer segments ($V_R$, dotted area) to new volume ($V_H$, hatched area)[53]. (After Doroszkowski and Lambourne, *J. Polym. Sci. C*, **34**, 253 (1971), with permission)

respect to the distance of approach in order to obtain the total repulsive potential–distance relation[54]. In the second case, the basic method of Flory[48] was applied[55] to the more general model

$$V'_1(C'_1) + V'_2(C'_2) \rightarrow V'_3(C'_3)$$

where $V'$ represents volume and $C'$ represents segment concentration respectively. The expression obtained for the important condition

$$C'_1 = C'_2; \quad V'_1 = V'_2$$

was

$$\Delta G_m = C'_3(C'_3 - C'_1)V'_3\left(\frac{\psi_1 - \kappa_1}{V_1}\right)kT \qquad (2.17)$$

By incorporating the values for the relevant volumes and the associated concentrations at any degree of overlap for spherical particles (Figure 2.3), the following expression for the energy of repulsion is obtained:

$$\Delta G_M = 2kT\left(\frac{\psi_1 - \kappa_1}{V_1}\right)C_0^2\left(\frac{V_H + V_R}{V_H}\right)V_R \qquad (2.18)$$

where $C_0$ = initial average concentration of polymer in the layer, and in this case the values of the new ($V_H$) and initial ($V_R$) volumes are:

$$V_H = \pi/3(\delta - h/2)(a + h/2)(2a + \delta + h/2) - \left(\frac{2a^3}{a + \delta}\right),$$

and,

$$V_R = \pi/3(\delta - h/2)^2(3a + 2\delta + h/2)$$

In general, the results obtained gave good correlation with direct determinations of the repulsive force obtained by measuring the compression of monolayers of sterically-stabilized polymer dispersions on a surface balance[53]. In models with curved surfaces, as in Figure 2.4, it can be shown that larger interaction volumes are involved between polymer and polymer ($V_{pp}$) than between polymer and the particle surface ($V_{ps}$), i.e. $V_{pp}/V_{ps} \geqslant 4:1$. It is not surprising therefore, that agreement with experimental results is obtained even when the additional repulsion due to polymer–surface interactions is neglected in the treatment.

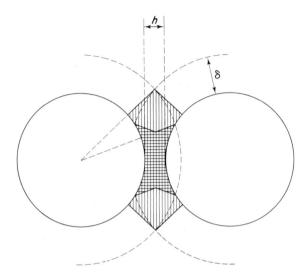

Figure 2.4. Model representing steric repulsion by overlap allowing redistribution of polymer segments at close approach: Polymer–polymer interaction volume, $V_{pp}$ (vertical shading); Polymer–particle surface interaction volume, $V_{ps}$ (crossed shading)

Haydon and co-workers have studied the interaction of adsorbed surface-active agents in thin hydrocarbon films supported between two bulk phases of a polar liquid such as water. Most of this work has been concerned with aliphatic chains of about 18 carbon atoms in length[28], but more recently the interactions of poly(12-hydroxy stearic acid) in thin films of heptane bounded by water have been examined[56]. The film was compressed by the application of a d.c. field and the resulting repulsive force generated between the adsorbed chains was determined. The corresponding changes in the thickness of the layers were estimated by optical or electrical capacitance methods. The experimental curve in Figure 2.5 shows the relation between film thickness and repulsive energy generated. The corresponding theoretical curves were obtained by inserting estimated values for the average polymer segment concentration and for the thickness of the layer into a planar version of the original expression

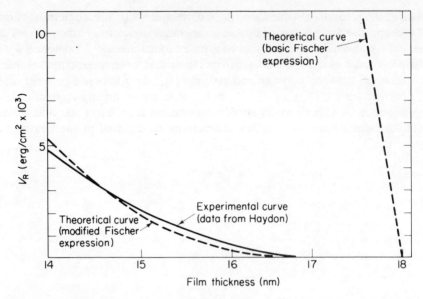

Figure 2.5. Theoretical and experimental curves relating repulsive energy ($V_R$) to film thickness

derived by Fischer[49]. Clearly, Figure 2.5 shows that the basic Fischer expression (here re-derived for flat plates) overestimates the repulsive potential at large film thicknesses. However, the incorporation of an analytic form for the segment density distribution which simulated that due to Hesselink[57], agrees well with the experimental curve. The largest compression applied to the film only reduced its thickness by about 4 nm, corresponding to about 20% of the maximum possible value. However, even for these small compressions, the value of the repulsive energy obtained ($V_R = 2 \times 10^{-3}$ erg/cm$^2$) corresponds to a value of $5\,kT$ for a cubic particle with sides of 100 nm in length.

To sum up, in the situation in which two particles carrying adsorbed layers of polymer chains approach each other, theories based wholly or partly on the mixing model predict that the layers will interact as soon as they come into contact with one another. Moreover, it is also predicted (and confirmed by experiment) that provided the diluent is thermodynamically a good solvent for the adsorbed polymer chain, the interaction of the polymer layers will give rise to large repulsive forces at small degrees of interaction. In thermodynamically poor solvents, of course, the magnitude of the repulsive force is less and in the limit a reversal from repulsion to attraction can take place. When sterically-stabilized spherical particles interact in a medium which is a good solvent for the stabilizing chains, the mixing term will be the dominant source of stabilizing repulsions at all ranges. This is a result, both of the nature of the geometry involved and the fact that most of the possible configurations are lost as a result of interactions between polymer chains before the stage of half-overlap is reached. For parallel surfaces covered with stabilizing chains, mixing models based on a realistic distribution of polymer segments, but not allowing for their redistribution, predict results which are only valid for the early stages of the interaction.

### 2.3.4. Stability in theta (θ) solvents

The change in the free energy of mixing ($\Delta G_m$), which arises when adsorbed layers of polymer chains overlap, is given by the form of the expressions (2.14–2.16), described above (Section 2.3.3). In a theta solvent, since the second virial coefficient (term $B$) is then zero, there is no resultant change in the free energy of mixing on overlap. Consequently, when the solvency of the diluent for the dispersion is reduced to that of a theta solvent, the repulsion generated falls to zero. Since such a change in the solvent environment will not at the same time alter significantly the attractive potential between the dispersed particles, it would be expected that flocculation would ensue when the diluent of a sterically-stabilized dispersion is converted to a theta solvent for the adsorbed polymer chains. This prediction, of course, does not take account of any effect of the volume restriction term which should still operate when the distance between the particles is less than the thickness of the adsorbed polymer layer, ($h < \delta$), even under theta conditions. In spite of this, Napper[51,52] who has made an extensive study of the effect of the solvency of the medium on the stability of sterically-stabilized polymer dispersions, found that flocculation took place under theta conditions, and in many cases slightly better than theta conditions. In addition, Doroszkowski and Lambourne[54] have shown that such dispersions at solvencies slightly worse than the true theta point, although flocculated, had a small but significant repulsion at $h \leq \delta$ which can be attributed to the volume restriction effect. Since these results were obtained with polymer dispersions with an inherent attraction which is still small when $h \simeq \delta$, it is surprising that flocculation does in fact take place under theta conditions where although $\Delta G_m = 0$, the volume restriction effect should still be operative.

For an explanation of these results, closer attention must be given to the original concept of the theta condition[39,58] and the experimental methods used for its determination[59-61]. On the basis of either approach, it is evident that the theta concept can only be strictly applied to polymer solutions of effectively infinite dilution. At higher concentrations, where the polymer chains are not in isolation, the concept breaks down. Hence, a condition which corresponds to a true theta point at infinite dilution becomes effectively worse than theta at high concentrations. Thus, although theta conditions could exist for the opposing polymer chains at infinite dilution, the actual conditions for the polymer chains in the adsorbed layer are worse than theta solvent conditions. Consequently, $\Delta G_m < 0$, that is, the chain segments now tend to associate together, contributing an overall attraction between the particles which leads to flocculation. In other words, phase separation of the adsorbed polymer layers tends to take place under theta conditions. In fact, a reappraisal of some of the earlier experimental work[55] indicated a close correlation between the incidence of flocculation and the onset of the phase separation of polymer from solution at a concentration corresponding to that involved in the barrier layers.

### 2.3.5. The current view of stabilization theory

The success of a statistical thermodynamic approach in providing a quantitative theory of steric stabilization depends on the validity of the model used for its computation. It will be evident from the previous discussion that there are fundamental differences in the nature and interpretation of the various models used and in the way in which they relate to the underlying physical processes involved in steric stabilization. This has led to the situation where, for example, some authors have ignored the contribution due to polymer-polymer interactions (the mixing term) and others have omitted the polymer-surface interaction (the volume restriction term). In addition, others have suggested that a term describing the degree of elasticity of the polymer layer, as put forward by Jäckel[62], should also be included.

Even when both terms are recognized, as in the case of the most elegant and elaborate treatments such as that originated by Meier[43] and more recently developed by Hesselink[45], perturbation of the adsorbed polymer molecule by a rigid surface (the volume restriction term) and the interaction of its segments with those of another polymer molecule (the mixing term) are treated as fundamentally different phenomena and the total repulsion is taken as their sum.

In these treatments, which are so important that some fairly full account of them must be given in addition to the references to the individual steps made above, the segment density of a polymer chain as a function of distance from a primary surface is calculated for an ideal volumeless chain and with the counter-surface at infinity. A new segment density distribution is then calculated when the counter-surface is at some finite distance from the primary surface, again on the assumption that the polymer chain is ideal. The volume restriction term is then derived from the difference in the number of configurations available to the chain in the two situations. A second polymer chain is then considered to have been adsorbed onto the counter-surface opposite the primary chain. The segment densities of the two chains are then added at all points, both when the surfaces are infinitely far apart and when they have been moved close together. The assumption is made that interpenetration of the chains occurs with no further perturbation due to the segment interactions, over and above that originally produced in the ideal chain by the presence of the solid counter-surfaces. A calculation of the free energy of mixing of the polymer segments (now assumed to have real volume) at each point is carried out for the surfaces both at infinity and close together. Finally, the sum of these differences is taken as the contribution of the free energy of mixing to the overall repulsion. The repulsions thus calculated for the volume-restriction term and the free-energy-of-mixing term are added to give a final value for the total repulsion between the surfaces. This arbitrary separation of the repulsion into a volume-restriction component and a free-energy-of-mixing component with the total repulsion taken as their sum and with no account of cross interaction effects, would seem to be intrinsically unsound in that it can lead to significant double counting of the repulsions, quite apart from the errors involved in using volumeless chains to calculate the volume restriction term.

For sterically-stabilized dispersions in non-aqueous media of low polarity, the principal sources of repulsion are entropic changes arising from the loss of available configuration to the polymer chains. The contribution to repulsion arising from enthalpic changes due to the rearrangement of solvent molecules is much less important in non-polar liquids than in aqueous or other polar media[2]. It would appear therefore that in essentially athermal systems, a true value for the total repulsion can be derived from the difference between the number of configurations of a real polymer chain of finite volume in its original situation, and the number of configurations remaining for this real chain after it has been perturbed by interaction with the other particle. This perturbation is due to the presence of segments of finite volume from the counter-polymer only until the inter-particle spacing is less than $\delta$; thereafter it is due to the simultaneous obstruction of the primary chain by segments of counter-polymer and the impenetrable counter-surface. Unfortunately, for a rigorous analysis, it is not clear how this simultaneous perturbation can be separated into individual terms, and, if so, how these should be combined.

Within its limits, the computer simulation method of Clayfield and Lumb[46] comes nearest to a realistic formulation of the problem. It deals with the case of a primary surface bearing polymer chains which is perturbed by an impenetrable surface and the ideal athermal environment is also considered. Further extensions are required however, to allow for the actual equilibrium degree of overlap and redistribution of the polymer chains taking place, caused by the presence of the opposing surface and the polymer chains attached to it. A weighting term, similar to that of Bluestone and Vold[63], is also needed to include cases in which the solvent does not constitute an athermal environment for the polymer segments. Such a treatment would provide a more realistic account of the configurational entropy contribution to the free energy changes associated with the interactions between adsorbed layers of polymer. In a non-athermal situation, a further term based on the derived segment density distributions to account for the enthalpic effects, must be included to obtain the total free energy change.

Fundamental theoretical studies of this type would apply to infinite plane surfaces with fixed stabilizing chains under equilibrium conditions. For the very different problem of making a practical estimate of the order of magnitude of the repulsions generated in real colloidal systems, it seems clear that the type of model based on free energies of mixing, either with a constant segment density within the barrier but with provision for the redistribution of segments, or of the type in which a simplified segment density distribution is assumed together with interpenetration without redistribution, are in fact adequate. The former model is particularly convenient when considering colloidal dispersions of spherical particles, since in practice the volume restriction terms can be neglected for such systems. On the other hand, the latter model is particularly useful in the case of large planar surfaces at moderate degrees of approach. These two models probably describe the change in repulsion with approach as accurately, where relevant, as the commonly used reciprocal approximation of the Hamaker integral for the attraction. At the same time, the absolute values of the repulsions calculated in this way probably contain no more error than those involved in current estimates of the numerical values of the Hamaker constant. Clearly, further theoretical developments of steric stabilization by polymer chains in solution must depend to a great extent on an increase in the understanding of the general behaviour of polymers in solution, particularly at more realistic concentrations than those of infinite dilution.

Since the completion of this manuscript, a further theoretical treatment of the stabilization of colloids by adsorbed polymer has been given (Dolan, A. K. and Edwards, S. F., *Proc. R. Soc. Lond.* A, **337**, 509, 1974).

### 2.3.6. Polymer mobility and desorption

So far, it has been tacitly assumed that the adsorption of polymer chains onto the surface of a particle is complete and that the chains are firmly anchored to the surface. However, situations exist in which any of the following possibilities can arise:

(i) The polymer chains are capable of lateral movement on the surface of the particle.
(ii) The adsorption of polymer chains is incomplete (that is, the surface is not saturated).
(iii) The stress generated on a saturated surface by interaction with another surface is sufficient to cause desorption of the polymer chains.

Such possibilities will give rise to differing effects, depending on whether the interacting surfaces carrying the adsorbed polymer chains are planar and effectively infinite, or are curved and effectively finite. Since the effects produced are much greater in the latter case and also because the primary concern here is with the stabilization of particulate dispersions, the discussion which follows is restricted to cases involving finite or curved surfaces.

The first case concerns the approach of two spherical particles, each carrying a complete, irreversibly adsorbed layer of polymer chains of uniform segment density, but in which the chains are free to move laterally on the surface. At first contact, the concentration of polymer segments in the interaction zone would either be doubled (Figure 2.2), or would be increased to some lower, but still substantial, level if the redistribution type of model were used (Figure 2.3). However, since the polymer chains are mobile, more extensive re-arrangements to a lower free energy state of more uniform segment density can take place by appropriate lateral movement throughout the whole of the volume of each adsorbed layer. This effect is over and above any of the usual conformational compression or re-distribution of the segments within a given chain. Clearly as the final concentration of polymer chains is greater than their initial concentration, a repulsion due to the free energy of mixing would still be expected.

The general expression for $\Delta G_m$ (equation 2.18, Section 2.3.3) has been used[64] to obtain repulsion values for cases in which the polymer chains are free to move over the surface of the particle and where the redistribution zone encompasses the whole of the particle surfaces. In this case, the repulsion generated depends on the number of particle–particle contacts involved, being less for a doublet, as in a Brownian encounter, than for particles with six contacts (close-packed, single layer)[54], or with 12 contacts (close-packed, ordered three-dimensional system)[50]. The results obtained, compared with those obtained from a redistribution over a limited interaction volume, indicated that the repulsion rises rather less steeply as the particles approach than in the case in which the chains are assumed to be immobile on the particle surfaces. Nevertheless, very substantial repulsions can still be generated.

For the situation where there is both lateral chain mobility on the surface together with an incompletely adsorbed layer of polymer chains, the chains can be displaced sideways without generating any significant mutual lateral repulsions by polymer–polymer interactions. Under these circumstances, particles can approach each other with negligible repulsive work. Although the displacement of the chains, confining them to a fraction of the total surface, results in a fall in their surface translational entropy, this is trivial in value compared to the segmental entropy changes which normally result from segment–segment or segment–surface interactions. As a result, systems of particles in which the stabilizer is capable of lateral motion, but in which the surface of the particles is not completely saturated with stabilizer, would not be expected to be stable and indeed, this is found to be so in practice (Section 3.3).

A small excess of stabilizer is usually always present in the continuous phase of stable polymer dispersions and the surface is effectively saturated. Further adsorption of stabilizer above this saturation level is prevented by the lateral repulsion of the stabilizing chains generated by the mixing term.

It is possible, by the use of a modified form of the Fischer expression which takes account of the lateral redistribution of the stabilizing chains, to calculate the actual desorption stress to which each chain is subjected in practical systems. Despite the height of the potential energy barriers involved, this value is not large compared with $kT$. Thus, for systems in which desorption plays a significant part, an equilibrium will always exist between adsorbed polymer and polymer in the continuous phase, even in the absence of any stress due to particle–particle interactions.

Under these circumstances, any equilibrium interaction between two particles with adsorbed polymer on their surface will tend to generate repulsion by a process of desorption rather than by compression of the polymer chains. This type of repulsion arises not only from the work required to desorb the polymer chains from the surface of the particles but also from the additional work arising from the increase of the activity of the polymer desorbed into solution in the continuous phase. However, in many cases, where amphipathic graft copolymers are used for stabilization, the excess polymer will be in the form of aggregates or micelles in the continuous phase (Section 6.4). Here there will be a negligible change in the activity of the desorbed polymer, and the work gained from micellization must be subtracted from the work of desorption. In practice, the work of micellization and of adsorption at the particle surface for each stabilizer molecule is often similar so that the net repulsion at equilibrium tends to zero. It follows, therefore, that where the work of micellization is the same as (or even exceeds) the work of adsorption, then the London attraction between the particles will far exceed the negligible (or even negative) strength of the barrier at equilibrium, and flocculation will ensue. However, the stabilizer has to desorb before it can micellize, so the uncorrected work of desorption provides an activation energy barrier which has to be overcome before the particles can touch. If the activation energy is sufficiently large, although such systems can be regarded in a sense as ultimately thermodynamically unstable, for all practical purposes the dispersions will be stable to Brownian collisions and also to the high energy collisions produced by large shear fields. In effect, the stabilizer is permanently attached to the surface of the particle and repulsion is generated by stabilizer compression.

It is also possible to conceive that, in some systems, the equilibrium work of stabilizer desorption can not only be finitely positive, but can actually exceed the London attractive potential, even at contact. Such dispersions are, therefore, like lyophilic colloids, intrinsically thermodynamically stable.

The case in which repulsion was due to stabilizer desorption was actually described by Mackor and van der Waals in 1952[42]. Although this paper has often been regarded as one of the first to consider steric repulsion, it does not in fact treat segmental interactions and the associated repulsive forces generated between permanently adsorbed polymer molecules, in the sense that they have been considered here.

### 2.3.7. Non-equilibrium situations

In the previous discussion, it has been assumed that the relaxation time of the adsorbed polymer molecules is short compared with the contact time of two particles in a Brownian collision. If in reality this were not so, then one would expect that when two particles collided, the adsorbed polymer layers would be compressed with little or no interpenetration of the chains. Equally, it is possible that in many encounters, lateral displacement with or without desorption, although energetically favoured may not have time to occur, quite apart from the effect of the activation energy barriers discussed above.

Such a compression must lead to a higher energy of repulsion as soon as the layers touch because, for a given segment concentration, the changes in free energy due to self-mixing and cross-mixing effects are identical, but the number of configurations otherwise available to the molecules are fewer (i.e. the volume restriction term is greater).

The bulk modulus required in the treatment of Jäckel[62], may be derived in principle by substituting the instantaneous segment distribution in such a collision into the appropriate expressions for the total work of interaction of the polymer molecules.

## 2.4. THE STERIC STABILIZATION OF POLYMERS DISPERSED IN ORGANIC MEDIA

According to the DLVO model developed for charge-stabilized colloidal systems[3,4,35], colloidal stability is achieved by the generation of a repulsive force of a sufficient magnitude to overcome the inherent attraction manifested between particles as they approach one another. The effect of the superimposition of given patterns of repulsive and attractive energy with respect to the distance between the surface of the particles is conventionally represented in the form of potential energy–particle separation curves.

The behaviour of electrostatically-stabilized dispersions in aqueous media, where the repulsive and attractive energies are of the same order of magnitude, is represented by a net curve of the type shown in Figure 2.6(a). This curve results from the imposition of a repulsive potential energy larger than the attractive energy at intermediate distances but which fails to increase at the rate at which the particles approach to one another. As a consequence, coagulation (that is, the formation of closely-packed aggregates) can take place at these shorter distances. Such a disperse system is thermodynamically unstable in that the coagulated state is preferred on energetic grounds. However, provided that the potential energy barrier of the primary maximum is large enough compared with the thermal kinetic energy of the particles, the rate of coagulation can be extremely slow and many systems of this type can exhibit a *quasi*-stability of considerable duration in practice. At larger distances of separation at the limit of the range of electrostatic stabilization, a secondary minimum may be observed with coarse particles where flocculation (that is, the formation of more loosely-bound aggregates) takes place.

The behaviour of a disperse system stabilized by a sheath of soluble polymer chains in water or in organic media, in which the repulsive energy is generally much larger than the attractive energy, is represented by a curve of the type shown in Figure 2.6(b). Here the potential energy of repulsion exceeds that of attraction by an ever-increasing amount as the particles approach each other. Consequently, the resulting interaction is always repulsive and rapidly increases in magnitude with decreasing particle separation. Provided therefore that an entire barrier of soluble molecular chains of sufficient dimensions is maintained irreversibly at the surface of the particles, then steric stabilization produces an unconditionally stable disperse system. The attractive forces generated between dispersions of organic polymers are small compared with dispersions of more polarizable species (such as metal sols or graphite, Table 2.1, Section 2.1.2) and make virtually no contribution to the resultant potential energy curve provided that steric stabilization still operates. Evans and Napper[65] have calculated the

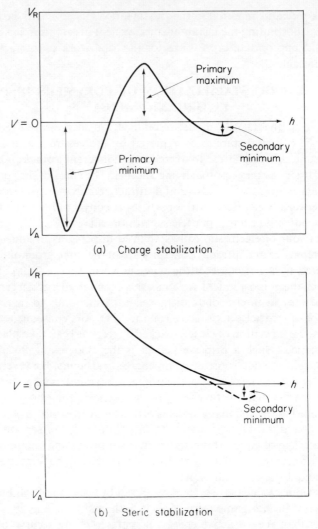

Figure 2.6. Form of net potential energy ($V$) curves as function of particle surface separation ($h$) for charge stabilization (a) and steric stabilization (b)

total potential energy curves for two sterically-stabilized spherical particles representing an aqueous dispersion of poly(vinyl acetate) stabilized with poly(ethylene oxide), [Figure 2.7(a)]. They show that provided the Flory solvency coefficient $\alpha$ is $>1\cdot000$ (where the value $1\cdot000$ corresponds to the $\theta$-point), the omission of the attractive component makes no significant difference to the overall repulsive potential [Figure 2.7(b)]. It was argued that stabilization failure due to a worsening of solvent conditions for the soluble polymer chains can be regarded as due to an attraction between the interposing

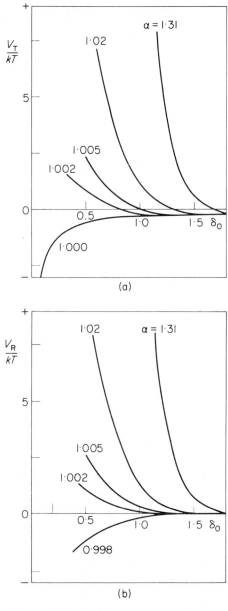

Figure 2.7. Potential energy curves for two sterically-stabilized spherical particles including (a) and excluding (b) the attractive component. (Radii $10^3$ Å, effective Hamaker constant $3 \times 10^{-21}$ J, $T = 303$ K)[65]. (From Evans and Napper, *Kolloid-Z.Z. Polymere*, **251**, 329 (1973), with permission of Dr Dietrich Steinkopf Verlag)

polymer chains themselves rather than invoking more fundamental sources of attraction.

Unlike electrostatically-stabilized systems, the steric barrier is of a finite dimension so that the very large repulsive energy generated by soluble polymer falls to zero beyond the effective range of the interacting soluble chains [Figure 2.6(b)]. It is thus conceivable that for certain combinations of polymer-layer thickness and particle size, a significant attractive trough might exist in this region [Figure 2.6(b), secondary minimum; compare also with Figure 2.7(a)]. Such systems would then show the type of behaviour corresponding to the weak flocculation at the secondary minimum which is observed with charge-stabilized systems in water, shown in Figure 2.6(a). The effect has yet to be unambiguously demonstrated with sterically-stabilized systems in non-aqueous media although it has been shown to occur with aqueous dispersions of poly-(styrene) stabilized with short poly(ethylene oxide) chains[66].

Clearly, the behaviour of sterically-stabilized dispersions of polymers in organic media can in general be represented by the essentially stable system shown in Figure 2.6(b). For this to be realized in practice however, it is essential that the very large repulsive forces involved are not dissipated by other stress-relieving effects during particle collision, such as by desorption of the chains from the particle surface as discussed in the previous Section. The energy of adsorption of the stabilizing chains should therefore be high. The concentration and dimensions of the polymer chains on the particle surface should also be large enough to provide an entire protective sheath during particle collision so that even transient contact between unprotected areas is avoided. Finally, in order for steric stabilization to operate at its maximum effect, the liquid medium used in a dispersion polymerization should be a thermodynamically good solvent for the stabilizing polymer chains.

The various methods developed for the practical operation of these criteria for successful dispersion polymerization in organic media with amphipathic graft and block copolymers as dispersants are described in detail in the following Chapter.

## 2.5. REFERENCES

1. Napper, D. H. and Hunter, R. J., in *Surface Chemistry and Colloids*, Vol. 7 (Ed. Kerker, M.), Butterworth, London, 1972.
2. Ottewill, R. H., in *Colloid Science*, Vol. I, Specialist Periodical Reports, The Chemical Society, London, 1973.
3. Vervey, E. J. W. and Overbeek, J. Th. G., *Theory of the Stability of Lyophobic Colloids*, Elsevier, Amsterdam, 1948.
4. Overbeek, J. Th. G., in *Colloid Science*, Vol. I (Ed. Kruyt, H. R.), Elsevier, Amsterdam, 1952.
5. Lyklema, J., *Advan. Colloid Interface Sci.*, **2**, 65 (1968).
6. Napper, D. H., *Ind. Eng. Chem. Prod. Res. Develop.*, **9**, 467 (1970).
7. Vincent, B., in *Colloid Science*, Vol. I, Specialist Periodical Reports, The Chemical Society, London, 1973.
8. Keesom, W. H., *Physik. Z.*, **22**, 129, 643 (1921).
9. Debye, P., *Physik. Z.*, **21**, 178 (1920); **22**, 302 (1921).

10. London, F., *Z. Physik.*, **63**, 245 (1930); *Trans. Faraday Soc.*, **38**, 8 (1937).
11. Krupp, H., *Advan. Colloid Interface Sci.*, **1**, 111 (1967).
12. Moelwyn-Hughes, E. A., *Physical Chemistry*, 2nd edition, Pergamon, London, 1961.
13. Bradley, R. S., *Phil. Mag.*, **13**, 853 (1932).
14. Boer, J. H. de, *Trans. Faraday Soc.*, **32**, 10 (1936).
15. Hamaker, H. C., *Rec. Trav. Chim.*, **55**, 1015 (1936); **56**, 3, 727 (1937).
16. Gregory, J., *Advan. Colloid Interface Sci.*, **2**, 396 (1969).
17. Visser, J., *Advan. Colloid Interface Sci.*, **3**, 331 (1972).
18. Parsegian, V. A. and Ninham, B. W., *J. Colloid Interface Sci.*, **37**, 332 (1971).
19. Vincent, B., *J. Colloid Interface Sci.*, **42**, 270 (1973).
20. Casimir, H. B. G. and Polder, D., *Phys. Rev.*, **73**, 360 (1948).
21. McLachlan, A. D., *Discuss. Faraday Soc.*, **40**, 239 (1965).
22. Vold, M. J., *J. Colloid Sci.*, **16**, 1 (1961).
23. Osmond, D. W. J., Vincent, B. and Waite, F. A., *J. Colloid Interface Sci.*, **42**, 262 (1973).
24. Halicioğlu, T. and Sinanoğlu, O., *Phys. Rev.*, **A5**, 2223 (1972).
25. Lifshitz, E. M., *J. Exper. Theor. Phys. USSR*, **29**, 94 (1955); *Soviet Physics, JETP*, **2**, 73 (1956).
26. Dzyaloshinskii, I. E., Lifshitz, E. M. and Pitaevskii, L. P., *Uspekhi Fig. Nauk*, **73**, 381 (1961); *Advan. Phys.*, **10**, 165 (1961).
27. Tabor, D. and Winterton, R. H. S., *Proc. Roy. Soc. A*, **312**, 435 (1969).
28. Haydon, D. A. and Taylor, J. L., *Nature (London)*, **217**, 739 (1968).
29. McLachlan, A. D., *Proc. Roy. Soc. A*, **271**, 387; **287**, 80 (1963); *Mol. Physics*, **7**, 381 (1964).
30. Langbein, D., *J. Adhesion*, **1**, 237 (1969).
31. Ninham, B. W. and Parsegian, V. A., *Biophys. J.*, **10**, 646 (1970); *J. Chem. Phys.*, **52**, 4578 (1970).
32. Parsegian, V. A. and Ninham, B. W., *Biophys. J.*, **10**, 664 (1970).
33. Mitchell, D. J. and Ninham, B. W., *J. Chem. Phys.*, **56**, 1117 (1972).
34. Koelmans, H. and Overbeek, J. Th. G., *Discuss. Faraday Soc.*, **18**, 52 (1954).
35. Deryaguin, B. V. and Landau, L. D., *Acta Physicochim., URSS*, **14**, 633 (1941).
36. Ottewill, R. H. and Shaw, J. N., *Kolloid-Z. Z. Polymere*, **215**, 161 (1967).
37. Krieger, I. M., private communication, 1969.
38. Albers, W. and Overbeek, J. Th. G., *J. Colloid Sci.*, **14**, 501 (1959).
39. Flory, P. J., *Principles of Polymer Chemistry*, Cornell University Press, Ithaca, New York, 1953 (Chapters 10, 12, 13, 14).
40. Tanford, C., *The Physical Chemistry of Macromolecules*, John Wiley & Sons, New York, 1963 (Chapters 2–3).
41. Mackor, E. L., *J. Colloid Sci.*, **6**, 492 (1951).
42. Mackor, E. L. and van der Waals, J. H., *J. Colloid Sci.*, **7**, 535 (1952).
43. Meier, D. J., *J. Phys. Chem.*, **71**, 1861 (1967).
44. Hesselink, F. Th., *J. Phys. Chem.*, **75**, 65 (1971).
45. Hesselink, F. Th., Vrij, A. and Overbeek, J. Th. G., *J. Phys. Chem.*, **75**, 2094 (1971).
46. Clayfield, E. J. and Lumb, E. C., *J. Colloid Interface Sci.*, **22**, 269, 285 (1966).
47. Clayfield, E. J. and Lumb, E. C., *Macromolecules*, **1**, 133 (1968).
48. Flory, P. J., *J. Chem. Phys.*, **10**, 51 (1942).
49. Fischer, E. W., *Kolloid-Z.*, **160**, 120 (1958).
50. Ottewill, R. H. and Walker, T., *Kolloid-Z. Z. Polymere*, **227**, 108 (1968).
51. Napper, D. H., *Trans. Faraday Soc.*, **64**, 1701 (1968).
52. Napper, D. H., *J. Colloid Interface Sci.*, **29**, 168 (1969).
53. Doroszkowski, A. and Lambourne, R., *J. Polym. Sci. C*, **34**, 253 (1971).
54. Doroszkowski, A. and Lambourne, R., *J. Colloid Interface Sci.*; **43**, 97 (1973).
55. Waite, F. A., ICI Paints Division, unpublished results, 1971.
56. Haydon, D. A., private communication, 1972.

57. Hesselink, F. Th., *J. Phys. Chem.*, **73**, 3488 (1969).
58. Flory, P. J., *Statistical Mechanics of Chain Molecules*, Interscience, New York, 1969.
59. Elias, H.-G., *Makromol. Chem.*, **33**, 140 (1959); **50**, 1 (1961).
60. Napper, D. H., *Makromol. Chem.*, **120**, 231 (1968).
61. Napper, D. H., *Polymer*, **10**, 181 (1969).
62. Jäckel, K., *Kolloid-Z. Z. Polymere*, **197**, 143 (1964).
63. Bluestone, S. and Vold, M. J., *J. Polym. Sci. A*, **2**, 289 (1964).
64. Osmond, D. W. J., ICI Paints Division, unpublished results, 1971.
65. Evans, R. and Napper, D. H., *Kolloid-Z. Z. Polymere*, **251**, 329, 409 (1973).
66. Long, J., Osmond, D. W. J. and Vincent, B., *J. Colloid Interface Sci.*, **42**, 545 (1973).

# CHAPTER 3

# *The Design and Synthesis of Dispersants for Dispersion Polymerization in Organic Media*

D. J. WALBRIDGE

| | |
|---|---:|
| 3.1. THE ROLE OF THE DISPERSANT | 45 |
| 3.2. THE REQUIREMENTS OF THE SOLUBLE COMPONENT | 49 |
| 3.3. THE SURFACE CONCENTRATION OF THE ADSORBED SOLUBLE COMPONENT | 54 |
| 3.4. THE DIMENSIONS AND STRENGTH OF THE STERIC BARRIER | 60 |
| 3.5. THE SELECTION OF THE ANCHORING COMPONENT | 63 |
|    3.5.1. Anchoring by physical adsorption processes | 64 |
|    3.5.2. Anchoring in emulsions | 67 |
|    3.5.3. Anchoring by covalent links | 68 |
|    3.5.4. Anchoring by acid: base interactions | 70 |
| 3.6. THE BEHAVIOUR OF GRAFT COPOLYMERS IN DISPERSION POLYMERIZATION | 74 |
| 3.7. THE DESIGN AND SYNTHESIS OF GRAFT COPOLYMERS AS DISPERSANTS | 79 |
|    3.7.1. Random grafting on to soluble polymer | 80 |
|    3.7.2. Transfer grafting on to specific sites | 84 |
|    3.7.3. Grafting through randomly distributed copolymerizable groups | 88 |
|    3.7.4. Formation of graft copolymer from soluble polymers carrying terminal polymerizable groups | 96 |
| 3.8. THE SYNTHESIS OF BLOCK COPOLYMERS AS DISPERSANTS | 103 |
|    3.8.1. Formation by coupling with condensation reactions | 104 |
|    3.8.2. Formation by ionically-initiated polymerization | 104 |
| 3.9. SELECTED RECIPES FOR THE PREPARATION OF GRAFT COPOLYMER DISPERSANTS | 106 |
|    3.9.1. Preparation of graft dispersant precursor based on poly(lauryl methacrylate) | 107 |
|    3.9.2. Preparation of preformed 'comb' graft copolymer dispersant from the copolymerization of the poly(hydroxy-stearic acid)/glycidyl methacrylate adduct with methyl methacrylate/methacrylic acid | 108 |
|    3.9.3. Preparation of dispersant precursor based on poly(lauryl methacrylate-co-methacrylic acid) reacted with allyl glycidyl ether | 109 |
|    3.9.4. Preparation of block copolymer dispersant by anionic polymerization of t-butyl styrene and styrene | 110 |
| 3.10. REFERENCES | 110 |

## 3.1. THE ROLE OF THE DISPERSANT

The function of the dispersant in a sterically-stabilized colloidal dispersion is to provide a layer of material solvated by the dispersion medium on each particle surface. Every particle is thus surrounded by a tenuous cloud of freely-moving polymer chains which are, in effect, in solution in the continuous phase.

This layer prevents the particles from coming into direct contact and also ensures that at the distance of closest approach of the two particles, the attraction between them is so small that thermal energy renders contact reversible. The theory of stabilization with soluble polymer chains has been fully discussed in Chapter 2, so the present discussion is limited to those considerations which are important for the actual preparation of polymer dispersions in organic media.

In a sub-micron colloidal dispersion in a non-aqueous medium, the particle collision frequency even at low concentrations is such that the number of free particles is reduced by flocculation almost to vanishing point in a few seconds, unless some mechanism for reducing the attraction between the particles is present. This is readily demonstrated, for example, by attempting to disperse titanium dioxide in an organic liquid or poly(vinyl chloride) powder in an aliphatic hydrocarbon in the absence of any surface active agents. One method of achieving some measure of dispersion stability for polymer particles is by the technique of 'organosol' preparation, in which the surface layers of the polymer are swollen by the addition of controlled amounts of solvent or plasticizer to the non-solvent medium (for further details, see Section 5.1). This is most easily attained when the interiors of the polymer particles are cross-linked either covalently or by microcrystalline regions. However, it is not a method which can be readily applied to heterogeneous polymerization processes. In contrast, the method of steric stabilization, in which polymeric chains soluble in the dispersion medium are attached to the surface of the particles, is more general and may be applied to any type of particle and has long been employed in the manufacture of paints and inks and other similar suspensions[1].

The nature of the stabilization process can be explained in simple terms in the following way. When two particles which have polymer molecules firmly attached to their surfaces approach one another in a medium in which the polymer molecules are soluble, a change in free energy ($\Delta G_R$) occurs as the polymer molecules interpenetrate or are compressed. The resulting increase in the concentration of the polymer segments in the layers of adsorbed polymer generates an osmotic pressure, i.e. $\Delta G_R$ is positive. To counteract this effect, solvent diffuses into the regions of higher polymer concentration forcing the particles apart until the steric barriers are no longer in contact. An effective stabilizer must therefore be able to maintain a complete coverage of the particle surface. In addition, the solvated cloud surrounding the particles must be concentrated enough to generate sufficient osmotic pressure to resist the inherent attractive forces, yet show no tendency to form a separate phase. The solvated sheath is localized at the surface of the particle only because it is adsorbed or attached in some other way to the surface.

It follows that since a positive value for $\Delta G_R$ ($= \Delta H_R - T\Delta S_R$) is necessary to achieve stability, this can arise from the separate enthalpic ($\Delta H_R$) and entropic ($\Delta S_R$) components. Napper[1] has classified the contributions to steric stabilization in the following way:

|  |  | Combined enthalpic– |
| --- | --- | --- |
| *Enthalpic stabilization* | *Entropic stabilization* | *entropic stabilization* |
| $\Delta H_R$, $\Delta S_R$ positive; $\Delta H_R > T\Delta S_R$ | $\Delta H_R$, $\Delta S_R$ negative; $T\Delta S_R > \Delta H_R$ | $\Delta H_R$ positive, $\Delta S_R$ negative |
| Dispersions flocculate on heating | Dispersions flocculate on cooling | Dispersions stable throughout normal temperature range |

In non-aqueous media, the entropic mode of stabilization is of primary concern since the heat of solution of a polymer is generally positive and hence $\Delta H_R$ is negative. However, examples of enthalpic stabilization can also be envisaged in non-aqueous media where the stabilizing polymer is maintained in solution by hydrogen bonding or by acid–base interactions between the polymer and solvent components.

The thickness of the solvated layer normal to the particle surface is not a critical factor in practice. In the early stages of the development of the dispersion polymerization process, the prevailing theories of colloid stability predicted that polymer chains of molecular weights of 10,000 upward in value would be necessary to provide adequate stability[2]. It has since been found however, that solvated chains of about 18 carbon atoms in length are sufficient to prevent flocculation and it may be that much shorter chains would suffice if satisfactory and practical methods of assembling such solvated layers at the particle surface could be found.

Each solvated moiety must be strongly attached or anchored to the surface of the particle so that it is neither desorbed from the surface nor displaced laterally when two particles collide. This is an important consideration when high reaction temperatures or strong solvent conditions are used in the dispersion polymerization. Homopolymers and random copolymers soluble in organic media are, for example, too weakly and reversibly adsorbed on the low energy polymer surfaces and are ineffective as stabilizers. Attempts to disperse polymer particles in aliphatic hydrocarbons in the presence of such polymers have generally been unsuccessful, e.g. poly(vinyl chloride) particles, when milled in a petroleum spirit solution of poly(isobutene), gave highly flocculated, stiff suspensions[3]. Similarly, when methyl methacrylate is polymerized with an azo initiator in a dilute solution of poly(isobutene) or poly(lauryl methacrylate) in n-heptane, the polymer formed precipitates out and adheres to the walls of the reaction flask or rapidly forms large aggregates and ultimately a separate swollen polymer phase[3]. The presence of the soluble polymer has little or no effect on the state of dispersion of the poly(methyl methacrylate) produced. It might be supposed that materials which are known[4] to be good dispersants for pigments and other fine particles in non-aqueous media, such as alkyd resins, copolymers of alkyl methacrylates with amino acrylates, metal soaps and sorbitan oleates, would also act as dispersants for polymer particles. However, many such dispersants have been examined in the course of the development of dispersion polymerization processes and almost invariably the results are just as unsuccessful as those with soluble polymers. The polar groups present on the surface of inorganic pigments or easily polarizable organic pigments are absent in the case of most of the commercially useful vinyl, acrylic and condensation polymers. Consequently, the dispersants are only weakly adsorbed and do not provide an efficient barrier to flocculation. It is possible to provide sufficiently strong interactions between polymer particles and dispersants by incorporating suitable functional groups, as will be described later in this Chapter (Section 3.5.4), but although stable polymer dispersions can be prepared in this way, the method has been of little practical importance so far.

The most successful type of dispersant devised for use in dispersion polymerization has been based on a block or graft copolymer which consists of two essential polymeric components—one soluble and one insoluble in the continuous phase. The insoluble component, or anchor group as it is often called, associates with the disperse-phase polymer. In some cases, it may become physically absorbed into the polymer particle and can be designed so that it reacts chemically with the disperse phase after absorption. Even without such absorption processes, these types of stabilizers are extremely effective since, by virtue of the insolubility of one of their components, they are strongly adsorbed onto a particle surface. In this way, the soluble component is firmly attached at the surface and so provides a lyophilic layer completely covering the surface of the particle. It is not necessary for the adsorbed polymer to be identical in composition with the disperse phase polymer produced in the polymerization process. For example, poly(vinyl acetate) can be employed as the insoluble component of the dispersant in the dispersion polymerization of an acrylic monomer[5]. The main requirement for the anchor group is that it be insoluble in the dispersion medium, but its effectiveness may be greatly enhanced if it has some specific affinity for the dispersed polymer. As one might expect from the solution behaviour of graft dispersants (see Section 3.6), they can form aggregates or micelles in the dispersion medium, as a result of the self-association of the anchoring components. This effect will influence the activity of the dispersant in a dispersion polymerization in a manner resembling the behaviour of aqueous surfactants in emulsion polymerization[6] (see also Section 4.3.3). The extent of the micellar association will depend largely on the degree of the insolubility of the anchor group in the dispersion medium and also the relative sizes of the soluble and insoluble components of the stabilizer. The criterion of insolubility of the anchor group also defines, in practice, the minimum size of the soluble group. For a polymer to be sufficiently insoluble in the dispersion medium, the molecular weight usually has to be of the order of 1000 or greater. The soluble chain attached to such an anchor must be at least of similar molecular weight, otherwise a stable micellar solution of dispersant cannot be formed in the dispersion medium and precipitation occurs. The minimum molecular weight of the soluble component must therefore be at least 500 to 1000, which is considerably greater than the minimum requirement for an effective steric barrier.

Examples of block and graft copolymers used as dispersants for dispersion polymerization in various organic media are illustrated in Table 3.1.

Table 3.1. Examples of block and graft copolymers used as dispersants in dispersion polymerization. (For further details of patents, see Appendix and Tables 8.2 and 8.1.)

| Dispersant | Continuous phase | Disperse polymer | Patent |
|---|---|---|---|
| Poly(2-ethylhexyl-acrylate-g-vinyl acetate) | Aliphatic hydrocarbon | Poly(methyl methacrylate) | British 1,101,984 |
| Poly(12-hydroxystearic acid-g-methyl methacrylate) | Aliphatic hydrocarbon | Poly(methyl methacrylate-co-methacrylic acid) | British 1,122,397 |
| Poly(lauryl methacrylate-b-methacrylic acid) | Ethyl alcohol | Poly(lauryl methacrylate) | British 941,305 |
| Poly(styrene-b-dimethyl siloxane) | Hexane | Poly(styrene) | German 2,142,598 |
| Poly(2-ethylhexyl-acrylate-g-methyl methacrylate) | Aliphatic hydrocarbon | Poly(styrene) | British 1,101,984 |
| Poly(styrene-b-methacrylic acid) | Ethyl alcohol | Poly(styrene) | British 941,305 |
| Poly(butadiene-g-methacrylic acid) | Ethyl alcohol | Poly(styrene) | British 941,305 |
| Poly(styrene-b-t-butyl styrene) | Aliphatic hydrocarbon | Poly(styrene) | Netherlands 72/06366 |

## 3.2. THE REQUIREMENTS OF THE SOLUBLE COMPONENT

The functions of the soluble component and the anchor component of graft copolymer dispersants used in dispersion polymerization are so distinct that they can usefully be considered separately before going on to discuss the design and preparation of the graft copolymers themselves.

The soluble component, in providing a layer of polymer in dilute solution on the surface of each particle, is responsible for the stabilization of the dispersion against flocculation. The chemical nature of the dispersed particle is of little significance since the steric barrier formed is so effective that differences in the magnitude of the attractive forces produced between particles are largely irrelevant. For example, poly(hydroxystearic acid) of molecular weight in the range 1500–1800 has been used successfully to stabilize dispersions of materials as diverse as poly(methyl methacrylate) and poly(acrylonitrile)[7], nylon and poly(ethylene terephthalate)[8] and titanium dioxide and copper phthalocyanine[9], some of them, over a wide range of particle diameters up to 20 μm or more in size. The same effect is probably also produced by soluble polymer of even lower molecular weight although the range of materials which has been dispersed by such species is more limited, being confined mainly to acrylic polymers and titanium dioxide (see Table 3.13, Section 3.4). With soluble polymer of higher molecular weight, the steric barrier formed will be thicker but it has not been established with certainty that the resulting polymer dispersion will be more stable to flocculation. In practice, where flocculation of a non-aqueous dispersion has been observed, its cause has invariably been traced either to an incorrect choice of the soluble polymer for a given continuous phase or to incomplete coverage of the particle surface by the soluble polymer attached to it (Section 2.4). Sterically-stabilized particles adsorb strongly on to unstabilized surfaces, as can be observed when an uncoated glass slide is immersed in a non-aqueous dispersion[10]. It is therefore essential for stability against flocculation that the surface of every particle is fully covered.

The principal consideration in selecting an appropriate soluble component for the dispersant is that it should be freely soluble in the dispersion medium to be used. This requirement had been recognized empirically for some years before any quantitative studies were undertaken and it was known, for example, that the addition of a small amount of alcohol to a sterically-stabilized polymer or pigment dispersion in petroleum hydrocarbons would cause the dispersion to flocculate, even though the amount of alcohol added was not sufficient to cause actual precipitation of the stabilizing polymer from the dispersion medium[10]. Subsequently, systematic studies were carried out on this region of incipient instability in non-aqueous dispersions in order to provide information about the practical criteria of dispersancy and to enable the current theories of stabilization to be tested[5]. These studies demonstrated that in order to achieve indefinite stability, the dispersion medium must be 'better than a theta solvent' for the stabilizing soluble polymer moieties.

In order to illustrate the relevance of the theta point to the practical aspects of steric stabilization, it is first necessary to consider in a little more detail some aspects of the solution properties of polymers. For sterically-stabilized dispersions in media of low molecular weight such as organic solvents or water, $\Delta G_R$, the free energy of repulsion, is derived principally from the non-ideal behaviour of the soluble polymer in the stabilizing layer. The segments of a polymer molecule in solution undergo interactions with other segments of the extended chain. The effect of these interactions is to exclude some other possible chain configurations and to cause an expansion of the molecule as a whole. It is thus the 'excluded volume' effect which in a good solvent results in the positive deviation of the osmotic pressure from ideality. This effect is in some ways comparable to the non-ideal behaviour of gases and has its analogue in the $b$ term of the van der Waals equation[11]. The osmotic pressure produced by dilute solutions of small molecules is given by the van't Hoff analogue of the ideal gas equation, i.e. for $n$ moles of solute

$$\Pi V = nRT \quad \text{or} \quad \Pi/C_s = RT/M \tag{3.1}$$

where $\Pi$ is the osmotic pressure and $C_s$ the weight concentration of a solute of molecular weight $M$. For polymer molecules of molecular weight $M_n$, the relationship may, as with a non-ideal gas, be expressed as a virial expansion of the form

$$\Pi/C_s = A + BC_s + \cdots \tag{3.2}$$

where $A = RT/M_n$ and $B$ is the second virial coefficient for the polymer solution. In a good solvent, where there is a large excluded volume effect, $B$ is large and positive. However, by the appropriate selection of less favourable conditions of solvency, the finite volume of the chain unit can be compensated for exactly by the attractions between the chain units. Under these conditions, the excluded volume effect no longer operates and $B$ is then equal to zero. This is the theta point for a polymer solution which is analogous to the Boyle point for a gas. At the theta point, the osmotic pressure of the solution obeys the van't Hoff equation. Under worse than theta conditions, negative deviations from the ideal osmotic pressure are observed, $\Delta G_R$ becomes negative and flocculation of a sterically-stabilized dispersion should ensue.

Napper[5] found, that for a variety of polymer dispersions stabilized by graft copolymers in hydrocarbon and hydrocarbon–alcohol mixtures, incipient flocculation was observed at or near the theta point. In one series of experiments, the theta point was approached by the addition of alcohol to the polymer dispersion in a hydrocarbon medium until a critical flocculation volume (c.f.v.) of alcohol (wt-%) was found at which the turbidity of the dispersion increased rapidly. This type of experiment was repeated over a range of temperatures and correlated with the theta solvent compositions for the solvated stabilizing moieties in free solution at these temperatures (Table 3.2).

Table 3.2. Theta compositions and critical flocculation volume (c.f.v.) for dispersions of poly(methyl methacrylate) in n-heptane[5]. (From Napper, *Trans. Faraday Soc.*, **64**, 1701 (1968), with permission.)

| Stabilizing moiety | Temp (K) | Flocculant | c.f.v. (%) | Theta composition (%) |
|---|---|---|---|---|
| PSA[a] | 274 | Ethanol | 39.5 | 39 |
| PSA | 297 | Ethanol | 50.5 | 51 |
| PSA | 313 | Ethanol | 58 | 59 |
| PSA | 274 | n-Propanol | 61 | 65 |
| PSA | 297 | n-Propanol | 78 | 84 |
| PSA | 274 | n-Butanol | 74 | 81 |
| PLM[b] | 249 | Ethanol | 31 | 33 |
| PLM | 274 | Ethanol | 38 | 39 |
| PLM | 297 | Ethanol | 44 | 45 |
| PLM | 313 | Ethanol | 51 | 51 |
| PLM | 274 | n-Propanol | 56 | 52 |
| PLM | 297 | n-Propanol | 70 | 67 |

[a] PSA = poly(12-hydroxystearic acid).
[b] PLM = poly(lauryl methacrylate).

The correspondence between the c.f.v. and the theta composition was very close when ethanol was used as the flocculant. With the other alcohols, the dispersions flocculated in media which were significantly better solvents than the measured theta compositions. Fair correspondence was also observed between the theta temperature and the critical flocculation temperature (c.f.t.) in single solvent systems, as shown in Table 3.3.

Table 3.3. Theta temperature and critical flocculation temperature (c.f.t.)[5]. (From Napper, *Trans. Faraday Soc.*, **64**, 1701 (1968), with permission.)

| Dispersion | Dispersant | Dispersion medium | c.f.t. (K) | Theta temp. (K) |
|---|---|---|---|---|
| PMMA | PMMA/PSA | n-Heptane | 260 | 240 |
| PMMA | PMMA/PSA | n-Hexanol | 265 | 245 |
| PMMA | EK/PSA | n-Heptane | 249 | 240 |
| PMMA | PVA/PSA | n-Heptane | 250 | 240 |
| PMMA | PMMA/PLM | n-Heptane | 215 | 175 |
| PVA | PVA/PSA | n-Heptane | 250 | 240 |
| PVA | PMMA/PSA | n-Heptane | 245 | 240 |

Key to symbols: PMMA = poly(methyl methacrylate); PVA = poly(vinyl acetate); PSA = poly(12-hydroxystearic acid); EK = epoxy resin, 'Epikote' 1004 (Shell Chemicals, U.K. Limited); PLM = poly(lauryl methacrylate).

The critical flocculation temperature almost invariably occurred some 10 to 20 K above the theta temperature—that is, as with the mixed solvent compositions above, flocculation was observed at better than theta conditions.

Despite these discrepancies, which may have their origin in difficulties in estimating the actual theta conditions for the soluble polymer moieties when incorporated in a graft copolymer and adsorbed on a particle surface, the correspondence of flocculation with theta point is sufficiently close to provide an excellent practical limiting criterion for the selection of the stabilizing group for a particular dispersion medium.

It follows from the theory of high polymer solutions that the theta point, and consequently the critical flocculation conditions, should be independent of the molecular weight of the polymer[12]. Although there appears to be no practical evidence to support this point for dispersions in organic media, studies have been carried out on dispersions of poly(vinyl acetate) in water stabilized with poly(ethylene oxide) chains[13] and with graft copolymers containing poly(acrylic acid) as the soluble component[14]. In the first case, incipient flocculation was produced by warming the latex and in the second, the flocculation temperature was approached by cooling the latex.

In both cases (Table 3.4), there was a close correspondence between critical flocculation temperature and theta temperature, and the values obtained were virtually independent of the molecular weight of the soluble component. Since similar flocculation behaviour would be expected in organic media, it would therefore seem reasonable to conclude that the critical flocculation conditions will also be independent of the molecular weight of the stabilizing polymer in dispersion in organic liquids. It must be borne in mind however, that the separate contributions to stability from the enthalpic and entropic effects in aqueous media are likely to be very different from those in non-aqueous media.

Table 3.4. Dependence of the critical flocculation temperature (c.f.t.) on the molecular weight of the stabilizing polymer for aqueous dispersions of poly(vinyl acetate)[13,14]. (From Napper, *J. Colloid Interface Sci.*, **32**, 106 (1970); *Polymer Letters*, **10**, 449 (1972), with permission.)

| Stabilizing polymer | Mol. wt. | c.f.t. (K) | Theta temperature (K) |
|---|---|---|---|
| Poly(ethylene oxide) | $1.0 \times 10^6$ | 317 | 315 |
| | $8.0 \times 10^5$ | 316 | 316 |
| | $3.2 \times 10^5$ | 315 | 314 |
| | $9.6 \times 10^4$ | 316 | 315 |
| | $4.9 \times 10^4$ | 316 | 314 |
| | $2.3 \times 10^4$ | 314 | 315 |
| | $1.0 \times 10^4$ | 318 | 319 |
| Poly(acrylic acid) | $1.0 \times 10^4$ | 287 | 287 |
| | $1.9 \times 10^4$ | 289 | 287 |
| | $5.2 \times 10^4$ | 283 | 287 |
| | $9.0 \times 10^4$ | 281 | 287 |

The choice of the dispersion medium in a dispersion polymerization is primarily dictated by the constraint that the polymer to be prepared in it should be insoluble. It is then necessary to choose a soluble stabilizing polymer for which the polymerization medium is a good solvent in the sense described above. Table 3.5 summarizes the theta conditions for a range of the more

Table 3.5. Theta conditions for some common polymers[15]. (Based on data of Elias *et al. Polymer Handbook*, 1966, pp. IV-167 *et seq.* with permission of Interscience Publishers, New York.)

| Polymer | Theta solvent | Theta temperature (°C) |
|---|---|---|
| Poly(ethylene) | Diphenyl ether | 161.4 |
| Poly(isobutene) | Benzene | 24 |
| Poly(propylene)[a] | Cyclohexanone | 92 |
| Poly(styrene)[a] | Cyclohexane | 35 |
| Poly(vinyl acetate)[a] | 3-Heptanone | 29 |
| Poly(acrylic acid)[a] | Dioxane | 30 |
| Poly(methyl methacrylate)[a] | Butyl acetate | −20 |
| Poly(1,4-butadiene) (90% *cis*) | Hexane/heptane (50/50) | 5 |
| Poly(butadiene-co-styrene) (70/30) | n-Octane | 21 |
| Poly(adipic acid–ethylene glycol) | Cyclohexanol | 114.5 |

[a] Atactic polymer.

common polymers. This Table gives selected values from a more complete compilation[15] which the reader should consult for fuller details. Unfortunately, the data available on theta conditions for polymers, even in pure solvents, is somewhat limited and in practice it is desirable to have an alternative method as a basis for selecting the soluble component of the stabilizer. Fortunately, a large body of experimental data which fulfils this requirement is available in the form of the solubility parameters of solvents and polymers. Since this concept is described in detail in a later Chapter (Section 4.2) in connection with the formation of polymer particles by dispersion polymerization in organic liquids, it will not be described further here, but it will be evident that the data available can also be used to select a given polymer as the soluble group for a particular dispersion medium. However, although theta conditions and solubility parameters can be used to guide the choice of a suitable soluble group, general experience or trial and error methods must sometimes be used.

## 3.3. THE SURFACE CONCENTRATION OF THE ADSORBED SOLUBLE COMPONENT

There is a close relationship between the total surface area of the particles in a sterically-stabilized dispersion and the quantity of soluble stabilizing group adsorbed on the particle surfaces. To understand this relationship it is necessary first to consider the process of particle formation and growth (for a more detailed discussion see Chapter 4).

At the onset of a dispersion polymerization, the initial number of polymer particles precipitated is determined by a variety of factors, including the amount of graft copolymer dispersant in solution in the dispersion medium, the tendency of the anchor group of the dispersant to associate with the precipitating polymer, the relative insolubility of the precipitating polymer and the rate at which this polymer is being formed. However, once the initial crop of particles precipitates, the process conditions can usually be controlled so that the particles already formed grow steadily without fresh nucleation. Excess stabilizer present at any stage of the polymerization can cause further nucleation to occur, whilst a deficiency of stabilizer leads either to flocculation or to particle coalescence, depending on the nature of the dispersed particle. It is therefore important, that once the initial precipitation conditions have been adjusted to produce the correct number of particles for obtaining the final particle size required, the quantity of graft copolymer dispersant present should only be sufficient to cover the additional surface formed during subsequent particle growth and should not exceed this amount significantly.

The degree of adsorption of graft copolymers onto the surfaces of polymer particles under conditions used in polymerization has been measured[7]. In these experiments, the amount of graft dispersant associated with the polymer particles was determined as the difference between the total amount in the latex preparation and the amount left unconsumed after the completion of the dispersion polymerization, as measured by the non-volatile content of the supernatant liquor obtained after high-speed centrifugation of the latex. The polymer dispersions were prepared in aliphatic hydrocarbons at 80 °C. At this temperature, in the presence of free monomer, adsorption of the graft copolymer is probably an equilibrium process. At the lower temperatures employed for centrifugation, with no monomer remaining, the adsorption was substantially irreversible and, therefore, it is likely that only small errors can arise through re-equilibration of graft dispersant on dilution of the latex for centrifugation.

The average area occupied by each soluble polymer moiety was calculated from these adsorption results and the particle surface areas determined from electron micrographs. The square root of this area was taken as the linear spacing between the points of attachment of the soluble polymer chains. The dispersant used in the preparation was a poly(hydroxystearic acid-g-methyl methacrylate-co-methacrylic acid) copolymer. In this graft copolymer, the soluble group, poly(hydroxystearic acid), is attached to the anchor backbone via its terminal carboxyl group. The dispersion polymers prepared were either poly(methyl methacrylate) or copolymers of methyl methacrylate with minor amounts of other acrylic monomers, and the polymerization medium was an aliphatic hydrocarbon. A seed stage was formed by polymerizing a small amount of the acrylic monomer in a solution of graft copolymer in hydrocarbon. The remaining monomer and dispersion stabilizer were then added slowly to produce controlled growth of the particles.

The results determined for a range of different particle sizes and compositions are summarized in Table 3.6.

Table 3.6. Spacing of poly(12-hydroxystearic acid) chains on surface of polymer particles prepared by dispersion polymerization[7]. (From Osmond and Walbridge, *J. Polym. Sci. C*, No. 30, 388 (1970), with permission.)

| Dispersion polymer | Mean particle diameter (μm) | Mol. weight of soluble chain, $M_n$ | R.m.s. end-to-end distance in free solution (nm) | Surface coverage per soluble moiety ($nm^2$) | Av. spacing per soluble moiety (nm) |
|---|---|---|---|---|---|
| PMMA | 0.035 | 1500 | 3.2 | — | — |
| PMMA | 0.106 | 1500 | 3.2 | 3.0 | 1.7 |
| PMMA | 0.138 | 1500 | 3.2 | 3.5 | 1.9 |
| PMMA | 0.214 | 1500 | 3.2 | — | — |
| PMMA | 0.5 | 1500 | 3.2 | 3.1 | 1.8 |
| PMMA | 2.24 | 1500 | 3.2 | — | — |
| PMMA | 0.14 | 1600 | 3.3 | 3.0 | 1.7 |
| PMMA | 0.22 | 1600 | 3.3 | 3.8 | 2.0 |
| PMMA/MA (98:2) | 0.05 | 1600 | 3.3 | 4.9 | 2.2 |
| PMMA/MA (98:2) | 0.1 | 1600 | 3.3 | 5.1 | 2.3 |
| PMMA/MA (98:2) | 0.1 | 1600 | 3.3 | 5.6 | 2.4 |
| PMMA/MA (98:2) | 0.15 | 1600 | 3.3 | 4.3 | 2.1 |
| PMMA/MA (98:2) | 0.2 | 1600 | 3.3 | 5.6 | 2.4 |
| PEA/MMA (50:50) | 0.11 | 1600 | 3.3 | 3.6 | 1.9 |
| PEA/MMA (50:50) | 0.22 | 1600 | 3.3 | 3.9 | 2.0 |
| PEA/MMA (50:50) | 0.32 | 1600 | 3.3 | 4.5 | 2.1 |
| PEA/MMA (50:50) | 0.46 | 1600 | 3.3 | 3.3 | 1.8 |
| PMMA | 0.2 | 600 | — | 1.3–1.8 | 1.1–1.4[a] |

[a] Additional data from unpublished results of J. M. King and F. A. Waite, ICI Paints Division, 1964.
PMMA = poly(methyl methacrylate);
PMMA/MA = poly(methyl methacrylate-co-methacrylic acid);
PEA/MMA = poly(ethyl acrylate-co-methyl methacrylate).

It will be evident that the surface coverage per soluble moiety was independent of the particle diameter over a wide range. The distance between the points of attachment, even at the lowest coverage, was somewhat less than the root mean square (r.m.s.) end-to-end distance for the unattached molecule, calculated from intrinsic viscosity measurements, which suggests that these relatively low molecular weight soluble groups are packed close together, perhaps occupying a cylindrical rather than a spherical volume. It is of interest to note in this context that similar values have also been obtained for the spacing of poly(12-hydroxystearic acid) chains adsorbed on a very different substrate, namely titanium dioxide[16].

An indication of the practical limits of dispersant surface coverage for poly(hydroxystearic acid) groups was given by a subsequent set of experiments (Table 3.7) in which a common poly(methyl methacrylate) seed stage was prepared so that the final particle diameter was 0.2 μm, but in which the amount of graft copolymer dispersant added during the particle growth stage was varied[17].

Table 3.7. Variation of surface coverage with dispersant concentration[17].

| Relative dispersant concentration in feed stage | Surface coverage per soluble moiety (nm$^2$) |
|---|---|
| 0.58 | 5.55 |
| 0.71 | 4.00 |
| 0.80 | 4.42 |
| 0.88 | 3.39 |
| 1.0 | 2.56 |

At the low level of dispersant used, the first signs of flocculation became evident as foaming during the last stages of polymerization and subsequent difficulty in filtration. Above the highest concentration used (given in Table 3.7), electron micrographs revealed the presence of a second crop of very fine particles. The value of 2.56 nm$^2$ must therefore represent the closest packing of soluble chains that can be achieved in this dynamic situation. These results are in good agreement with those obtained with a latex similar in both polymer and stabilizer composition to those already described[5]. In this latter work, the surface coverage of poly(hydroxystearic acid) in the original latex was defined as unity. The size of the polymer particles in the latex was then increased by the addition of methyl methacrylate and initiator without further addition of graft copolymer dispersants, thus distributing the dispersant originally present over a larger surface area. The critical flocculation temperatures (c.f.t., see Section 3.2 above) were determined for a series of samples of fractional surface coverage relative to the starting sample. A steady increase in c.f.t. was observed (Table 3.8) until the latex preparation finally flocculated when the relative surface coverage was 0.57. This value is in remarkable agreement with that (0.58) obtained

Table 3.8. Variation of critical flocculation temperature (c.f.t.) with surface coverage[5]. (Based on data from Napper, *Trans. Faraday Soc.*, **64**, 1701 (1968), with permission.)

| Relative surface coverage | c.f.t. (K) |
|---|---|
| 1·00 | 258 |
| 0·89 | 259 |
| 0·80 | 266 |
| 0·72 | 270 |
| 0·68 | 272 |
| 0·65 | 286 |
| 0·57 | 323 (reaction temperature) |

in the previous experiment (Table 3.7). Even at this coverage, the average distance between the points of attachment is less than the r.m.s. end-to-end distance.

It has been shown[10] that by continued washing of the latex with weak solvents at room temperature it is possible to remove much of the graft dispersant and yet still retain some stability to flocculation. Surface coverages of 27 nm$^2$ per soluble moiety were obtained for poly(hydroxystearic acid). The limits of surface coverage for stability under static conditions at room temperature would be expected to be quite different from those obtaining in the mobile surface conditions during polymerization at 80 °C.

Surface coverage experiments have also been carried out with poly(lauryl methacrylate) as the stabilizing soluble group[7]. The dispersant was prepared as a graft copolymer precursor, by attaching methacrylate residues randomly to the poly(lauryl methacrylate) chains (see Section 3.7.1). A seed stage, in which stabilizing graft copolymer was formed concurrently with insoluble polymer, was prepared by polymerizing methyl methacrylate and all the graft precursor in hydrocarbon at 80 °C. The remainder of the acrylic monomer was then fed in slowly. The results of the surface coverage measurements using poly(lauryl methacrylate) soluble groups are listed in Table 3.9.

Although the experimental results were few in number and the measurements were less precise than those obtained with poly(hydroxystearic acid), the average spacing of the soluble moieties obtained is once again independent of particle size. For these linear polymers of higher molecular weight, anchored at random points along their length, the chain-spacing figures are much closer to the r.m.s. dimensions of the polymer chains in free solution than was the case with the low molecular weight polymers. The steric barrier for these polymers may, therefore, be visualized as equivalent to close-packed spheres of soluble polymer covering the surface, compared with the cylindrical packing of the regularly-branched poly(12-hydroxystearic acid) molecules.

The same two soluble groups have also been studied in dispersion polymerizations at the lower temperature of 30 °C[18]. Both soluble groups were converted to graft copolymers by copolymerization of unsaturated precursors with methyl

Table 3.9. Spacing of poly(lauryl methacrylate) chains on surface of polymer particles prepared by dispersion polymerization[7]. (From Osmond and Walbridge, *J. Polym. Sci. C*, No. 30, 388 (1970), with permission.)

| Dispersion polymer | Mean particle diameter (μm) | Mol. weight of soluble chain | R.m.s. end-to-end distance in free solution (nm) | Surface coverage per soluble moiety (nm$^2$) | Average spacing of soluble moieties (nm) |
| --- | --- | --- | --- | --- | --- |
| PMMA/MA (98:2) | 0·02 | 30,000 | 11 | 120 | 11·0 |
| PMMA/MA (98:2) | 0·03 | 30,000 | 11 | 90 | 9·5 |
| PMMA/MA (98:2) | 0·1 | 30,000 | 11 | 270 | 16·5 |
| PMMA/MA (98:2) | 0·18 | 30,000 | 11 | 120 | 11·0 |
| PMMA/MA (98:2) | 0·2 | 30,000 | 11 | 140 | 11·8 |
| PMMA/MA (98:2) | 0·3 | 30,000 | 11 | 90 | 9·5 |
| PMMA/MA (90:10) | 0·2 | 430,000 | 37 | 460 | 21·2 |
| PMMA/MA (60:40) | 0·6 | 430,000 | 37 | 1200 | 34·6 |

PMMA/MA = poly(methyl methacrylate-co-methacrylic acid).

methacrylate and methacrylic acid. The graft dispersants were then dissolved with methyl methacrylate in n-heptane and the monomer polymerized at 30 °C to give fine particle dispersions. From an analysis of the products, the area occupied per soluble moiety was computed. The results are summarized in Table 3.10. The main findings of Osmond and Walbridge[7] relating to the absence of a particle size effect and to the configuration of the soluble polymers were confirmed, but for both stabilizers the area stabilized per soluble moiety was only one third of the value found by the previous workers. This difference was attributed to the lower temperature of polymerization permitting a closer

Table 3.10. Surface coverage of poly(methyl methacrylate) particles by polymeric stabilizers (gravimetric results)[18]. (From Fitch and Kamath, *J. Indian Chem. Soc.*, **49**, 1209 (1972), with permission.)

| Soluble polymer | Particle diameter (μm) | Particle No. (per $1 \times 10^{-15}$) | Area stabilized per soluble moiety (nm$^2$) |
| --- | --- | --- | --- |
| PLMA | 0·115 | 7·6 | 34 |
| PLMA | 0·118 | 6·0 | 32 |
| PLMA | 0·113 | 4·7 | 29 |
| PSA | 0·033 | 48·8 | 0·8 |
| PSA | 0·035 | 41·0 | 1·0 |
| PSA | 0·027 | 92·4 | 0·8 |
| PSA | 0·023 | 137·0 | 1·0 |

PLMA = Poly(lauryl methacrylate), $\overline{M}_n = 37,500$;
PSA  = Poly(12-hydroxystearic acid), $\overline{M}_n = 1500$.

packing of the soluble groups. The poly(lauryl methacrylate) moieties were found to be more efficient than those of poly(12-hydroxystearic acid) in that the area stabilized per unit weight of polymer adsorbed was somewhat greater. The 'feather-like' structure of poly(lauryl methacrylate), consisting as it does of $C_{18}$ chains attached at regular intervals to a backbone, was considered to allow a higher segment density in the solvated barrier layer.

All the results quoted above were derived from dispersion polymerizations initiated by free radicals. The soluble stabilizing polymers which were prepared either by condensation or free radical-initiated addition polymerization, had a wide distribution in molecular weight. An alternative method of dispersion polymerization, which is attractive for the purpose of quantitative study of dispersant behaviour, utilizes anionic initiation[19].

In this method, the soluble component of the dispersant, e.g. poly(t-butyl styrene), was prepared by the use of lithium butyl in n-heptane at 25 °C to give a narrow molecular weight distribution, the average molecular weight being determined by the relative concentrations of monomer and initiator. Each of the soluble chains had one 'living' end from which the insoluble poly(styrene) component was grown. This copolymer then precipitated in the form of a stable dispersion, in which each particle consisted entirely of aggregates of block copolymer. Samples were taken for analysis at various ratios of poly-(styrene) to soluble polymer component. From a measurement of the particle diameter of these samples and subsequent calculation using known values for the molecular weight and the mass ratios of soluble and insoluble polymer, it was possible to obtain a value for the surface area stabilized by a given soluble moiety, as the particle size increased during the dispersion polymerization.

Table 3.11. Surface area coverage for poly(t-butyl styrene) soluble groups attached to poly(styrene) particles in n-heptane[19].

| Mol. weight poly(TBS)[a] ($M_n$) | Poly(styrene): poly(TBS) mass ratio | Particle diameter[b] (μm) | Surface coverage per soluble moiety[c] (nm$^2$) |
|---|---|---|---|
| $1 \times 10^4$ | 2.0 | 0.038 | 5.8 |
| $1 \times 10^4$ | 4.4 | 0.047 | 9.2 |
| $1 \times 10^4$ | 6.5 | 0.058 | 11.0 |
| $2 \times 10^4$ | 2.4 | 0.033 | 15.0 |
| $2 \times 10^4$ | 4.9 | 0.047 | 21.0 |
| $2 \times 10^4$ | 7.7 | 0.058 | 25.0 |
| $2 \times 10^4$ | 10.0 | 0.066 | 28.0 |
| $6.3 \times 10^4$ | 2.2 | 0.034 | 4.2 |
| $6.3 \times 10^4$ | 4.4 | 0.047 | 5.8 |
| $6.3 \times 10^4$ | 6.7 | 0.072 | 5.6 |
| $6.3 \times 10^4$ | 9.2 | 0.094 | 5.8 |

[a] Polydispersity $M_w/M_n = 1.3 - 1.4$.
[b] Values calculated for poly(styrene) particle core from electron micrographs; estimated random error $c. \pm 10\%$.
[c] Values quoted $\pm 15\%$.

The values obtained for surface coverage (Table 3.11) were low compared to those determined for poly(lauryl methacrylate) of comparable molecular weight (Table 3.9) in dispersion polymerizations carried out at 90 °C. However, they were in reasonable agreement with those determined at 30 °C (Table 3.10). Ideally, more information is required on the effect of temperature on the packing of adsorbed molecules on particle surfaces before any firm conclusions can be drawn, but the present evidence suggests that the spacing between the soluble groups increases rapidly with increasing temperature. This effect may well have implications for other dispersion processes, such as the dispersion of pigments, which are performed at temperatures above those at which the products are ultimately used. The results also showed that the area of the particle surface covered by each soluble moiety became larger as the particle diameter increased.

For the purpose of designing dispersants for latex preparations in organic media, it would be useful to have an empirical relationship between the weight of stabilizer employed and the latex particle diameter. Unfortunately, the evidence available is insufficient for any general relationship to be deduced so far. The greatest number of results are available for the low molecular weight poly(hydroxystearic acid) as the soluble component and in this case, the value for the area stabilized by a given weight of adsorbed soluble polymer ranges from between 1 and $2 \, m^2/mg$ soluble polymer at 90 °C[7] to $0.36 \, m^2/mg$ at 30 °C[18]. For poly(lauryl methacrylate) of molecular weight 30,000–40,000, the value is less accurately known but is within the range $1.75$–$2.5 \, m^2/mg$ at 90 °C[7] to $0.54 \, m^2/mg$ at 30 °C[18]. For poly(t-butyl styrene) in the same molecular weight range, but in a very different polymerization system at room temperature, the values were in the region of $0.7 \, m^2/mg$[19]. At higher molecular weights, quantitative results are even more scanty, the only data available being those for poly(lauryl methacrylate), which has a molecular weight of approximately 500,000, which were in the range $0.6$–$1.7 \, m^2/mg$ at 90 °C[7]. These values can therefore be used as a rough guide to determine the amount of the dispersant soluble group required for the preparation of a latex of a required particle size. It is of interest to note that the amounts of soluble polymer required for the stabilization of dispersions of titanium dioxide in organic liquids are very close to those required for polymer dispersions and are also, in general, insensitive to the value of the molecular weight of the adsorbed polymer (Table 3.13, Section 3.4, below).

## 3.4. THE DIMENSIONS AND STRENGTH OF THE STERIC BARRIER

The hydrodynamic thickness of the surface layer of adsorbed soluble polymer on particles of both pigments and polymers has been determined from the viscosities of the particle suspensions by a number of workers[10,16,20,21]. Although the methods used differ in detail, each of them depends on the principle of measuring the apparent disperse phase volume from the relative viscosity of the suspension and the continuous phase and then calculating the thickness of the surface layer from the difference between the apparent and actual disperse phase volumes. By the use of adsorption isotherms or surface-coverage data of the type described in the previous Section (3.3), it is then possible to calculate the actual polymer concentration in the surface layer.

Although it is not possible to stabilize dispersions of polymer with soluble random copolymers during dispersion polymerization, they can provide some degree of stabilization for titanium dioxide dispersions. In Table 3.12,

Table 3.12. Adsorption of random copolymers on dispersions of titanium dioxide and poly(methyl methacrylate) in aliphatic hydrocarbon[21]. (From Doroszkowski and Lambourne, in *Chimie, Physique et Applications Pratiques des Agents de Surface*, Vol. II, Ediciones S. A., Barcelona, 1969, with permission.)

| Substrate | Adsorbed polymer composition (molar) | Surface layer thickness (nm) | Weight of soluble polymer adsorbed (g/1000 m$^2$) | Equilibrium polymer concentration (g/100 ml) | Soluble polymer concentration in surface layer (g/100 ml) |
|---|---|---|---|---|---|
| Titanium dioxide | LM/MM/1·0/1·0 | 21·0 | 1·08 | 7·96 | 5·1 |
| Titanium dioxide | CM/SM/MM 0·2:0·8:1·0 | 16·0 | 1·26 | 7·88 | 7·8 |
| Titanium dioxide | HSA-g-(MM/GM) 1·0:15:0·6 | 10·0 | 0·95 | 3·4 | 12·0 |
| Poly(methyl methacrylate)[a] | HSA-g-(MM/MA) 1·0:14·7:0·26 | 6·2 | 0·83 | — | 13·2 |

Key to monomers: LM = Lauryl methacrylate; MM = Methyl methacrylate; GM = Glycidyl methacrylate; MA = Methacrylic acid; HSA = 12-Hydroxystearic acid; CM = Cetyl methacrylate; SM = Stearyl methacrylate.
[a] Data from Ref. 20.

the surface layer thicknesses and the concentration of various soluble polymeric species, including random copolymers, are compared for dispersions of both titanium dioxide and poly(methyl methacrylate) in aliphatic hydrocarbon media. It is evident that the random copolymers form diffuse layers, in which the concentration of polymer on the barrier is not appreciably greater than in solution. Graft copolymers, in contrast to the random copolymers, give much more concentrated surface layers and the equilibrium concentration of polymer in solution is very low in comparison.

Precise data for the values of the hydrodynamic barrier thickness of graft copolymers adsorbed on particles are only available for poly(hydroxystearic acid) as the soluble component (Table 3.13)[22]. The hydrodynamic values determined are of considerable practical interest since they allow a calculation of the apparent phase volumes of dispersions of any particle size to be made For example, a dispersion of particles 0·05 μm in diameter stabilized with poly(hydroxystearic acid), would have an apparent phase volume approximately

Table 3.13. Hydrodynamic barrier thickness of graft copolymers containing poly(hydroxystearic acid) adsorbed on dispersions of titanium dioxide and poly(methyl methacrylate) in aliphatic hydrocarbon[22].

| Substrate | Adsorbed soluble material | $\overline{M}_n$ | Hydrodynamic barrier thickness (nm) |
|---|---|---|---|
| Titanium dioxide | HSA dimer[a] | c. 600 | 5.0[b] |
| Titanium dioxide | HSA trimer | c. 900 | 6.1 |
| Titanium dioxide | HSA pentamer | c. 1500 | 8.0 |
| Titanium dioxide | Poly(HSA) | 2000 | 7.0 ⎱ 10.0 ⎰ |
| Titanium dioxide | Poly(HSA) | 4700 | 7.6 |
| Poly(methyl methacrylate)[c] | Poly(HSA) | 1600 | 6.2 |

[a] HSA = hydroxystearic acid.
[b] The values for polymer adsorbed on titanium dioxide were determined by the method of Ref. 21.
[c] Data from Ref. 20.

double the disperse phase volume and this would clearly limit the concentration at which such a dispersion could be prepared.

The values for the thickness of the hydrodynamic barriers given (Table 3.13), with the exception of the polymer of highest molecular weight, are in good agreement with the values calculated for 'stretched' end-to-end distances based on the number-average molecular weights of the poly(hydroxystearic acid) samples used. For the pure oligomers, it is probable that the actual steric barrier thickness, i.e. the thickness at which significant repulsion is generated when two particles come into contact, is similar to the hydrodynamic thickness. However, for unfractionated molecules with a broad molecular weight distribution, there is some indication that the steric barrier thicknesses are considerably larger than the hydrodynamic values[20].

A technique for the determination of both the thickness and the strength of the steric barrier has been developed, in which the forces of repulsion generated by the steric barriers of a series of monodisperse polymer particles of varying diameter were measured with a surface balance[23]. The particles used were stabilized with graft copolymer consisting of a poly(methyl methacrylate) anchor group to which was attached soluble poly(hydroxystearic acid) side chains of $\overline{M}_n$ 1600. The particles were supported at a water/heptane interface so that the stabilizing polymer was in the heptane phase in the plane of compression. From pressure–area measurements, a steric barrier thickness of 13 nm was calculated which was independent of particle size. This value is considerably in excess of the hydrodynamic thickness of 6.2 nm for the same polymer. The value for $\overline{M}_w$ (c. 3800), however, indicated the presence of molecules with 'stretched' lengths of 15–20 nm. It therefore seems plausible to suppose that substantial repulsion can be generated when these longer molecules overlap and are forced into contact whereas in relatively dilute suspensions the effective hydrodynamic thickness is more closely related to the number-average dimensions.

## 3.5. THE SELECTION OF THE ANCHORING COMPONENT

We have already shown in the preceding Sections, that the criteria for the selection of the soluble component of a polymeric dispersant can be defined in reasonably precise terms. For a given dispersion medium therefore, it is usually possible, using solubility parameters or other related quantities, to select from a number of types of polymer to act as the soluble component. The next problem to be considered is how to anchor the soluble groups selected to the surface of the polymer particle. It is essential for effective stabilization that the soluble groups are not readily displaced or desorbed from the surface of the polymer, either during its preparation or in subsequent applications of the dispersion. Since the soluble polymers alone are not strongly enough adsorbed onto the surface of the insoluble polymer particle, it is necessary to provide a second component in the dispersant to function as an anchor. The method most frequently employed has been to provide a second polymeric component insoluble in the diluent, which is attached to the soluble polymeric component in the form of a block or graft copolymer. The forces of attraction between the insoluble components of the dispersant and the polymer particle are then essentially similar to those which lead to the aggregation of unstabilized polymer particles in liquid media (see Section 2.1).

It is of interest to compare anchoring phenomena in non-aqueous media with the behaviour of surfactants used in the preparation of aqueous colloidal dispersions. In aqueous dispersions, the associations between the anchoring group of a surfactant and the surface to which it is adsorbed are often very weak and the anchoring component and the dispersed particle may be very different in composition. This is because of the gain in energy arising from the rejection of the aqueous surfactant from the hydrogen-bonded structure of water. In most organic solvents a much smaller 'rejection' energy is involved and consequently in this case large energies of association between the particle and the anchor group are required.

The degree to which the anchoring of the dispersant to the particle is irreversible depends on the amount of entanglement of the anchor group within or on the disperse phase polymer. At the elevated temperatures of the dispersion polymerization process, the monomer-swollen particle is likely to be a viscous, semi-liquid medium in which lateral movement of the dispersant is more probable than its actual desorption.

In some applications of the dispersions, the adsorption type of anchoring has proved inadequate, particularly when the latex produced has to withstand the subsequent addition of strong solvents and the dispersant can be desorbed from the particle surface. Anchoring must then be reinforced by the introduction of covalent links between the dispersed polymer and the insoluble polymeric component of the dispersant. Alternatively, the dispersant molecules can be linked together chemically after adsorption to form a cross-linked network enclosing the particle surface.

A different form of anchoring which does not involve graft copolymers and is analogous to the adsorption of dispersants on mineral or pigment surfaces in non-aqueous media, can be achieved by the incorporation of complementary acidic and basic groups into the dispersed polymer particle and the dispersant respectively.

Most of the information relating to the anchoring of polymeric dispersants is still of a qualitative nature since very little systematic work has been carried out on this subject so far. Consequently, the discussion which follows only attempts to provide some general, empirical rules for the selection of anchor groups for dispersants in non-aqueous dispersion polymerization.

### 3.5.1. Anchoring by physical adsorption processes

For the convenience of our classification, physical adsorption is understood to cover the processes of adsorption due to van der Waals forces, dipolar interactions and weak, easily-dissociated hydrogen bonds. It excludes strong specific bonds of an electrostatic or covalent nature. Physical adsorption in this context is therefore typified by the attractions between polymer molecules such as exist in amorphous polymers like acrylics, polyesters, poly(styrene), etc.

In the two main processes employed for the steric stabilization of such polymer dispersions in non-aqueous media, the stabilizer is a block or graft copolymer which is either produced concurrently with the disperse phase polymer by a process of grafting on to a soluble polymer dissolved in the continuous phase, or it is made separately and added as a preformed block or graft polymer to the dispersion reaction medium. In the former process, a graft stabilizer precursor is present in solution at the start of the dispersion polymerization. This precursor is the soluble component of the stabilizer, so modified that it contains one or more groups which take part in the dispersion polymerization by copolymerization or transfer reactions. A typical example is natural rubber, which in the presence of a peroxide initiator and an acrylic monomer, forms graft copolymer very readily by the growth of acrylic polymer from the radical sites generated by hydrogen abstraction[24].

In a process of this type, shortly after polymerization has commenced, the reaction mixture contains disperse phase polymer, graft polymer and soluble polymer. When precipitation of polymer occurs, the graft copolymer is associated with the disperse phase polymer and a stable dispersion of seed particles results. The number and size of the particles formed is determined by a combination of the ratio of the amounts of the disperse phase polymer and the graft copolymer, the strength of the association between them, and their insolubility in the continuous phase.

In grafting processes which take place during dispersion polymerization, the anchor group must inevitably be closely similar in composition, if not identical to the dispersed polymer, because it is being formed simultaneously. As it is mainly produced by polymerization in solution, the anchor group will usually be of a lower molecular weight than the polymer formed within the particles. Relatively little control can be exercised over the anchor group in this type of process, therefore, without at the same time also changing the composition or molecular weight of the dispersed polymer. In practice, however, this does not constitute a serious drawback and indeed it may be turned to advantage. For example, the introduction of a small proportion of a highly polar monomer at the beginning of the dispersion polymerization reduces the solubility of both the disperse polymer and the dispersant anchor group and also increases the attraction between them. The polymer therefore precipitates out more rapidly and strong adsorption or co-precipitation of the anchor group of the graft copolymer is also encouraged. This results in a much finer initial precipitate being formed. In this way, a very effective control of particle size can be achieved by relatively minor changes in the initial monomer concentration. If the remainder of the monomer is then added as a separate feed, the overall change in polymer composition need only be very small. A similar cooperative effect is of course observed when the solvency of the medium is reduced; conversely the particle size may be coarsened by the addition of relatively small amounts of strong solvent or, in those cases where the monomer is a solvent for its own polymer, by increasing the amount of monomer present in solution at the beginning of the dispersion polymerization. A further minor advantage of the identical nature of anchor and disperse polymer in an *in-situ* process is the absence of any compatibility problem which might adversely affect the degree of association of the anchor group with the surface of the polymer particle.

However, that this is not a serious problem in most polymerization processes is illustrated by the surprisingly wide range of anchor groups which can be employed successfully in preformed block or graft stabilizers. Most non-crystalline polymers and copolymers, particularly addition polymers, will adsorb or co-precipitate with almost any anchor group of relatively low molecular weight in a dispersion polymerization. Thus, acrylic polymers have been stabilized with graft copolymers in which the anchor group was an identical or similar acrylic polymer, or a quite different polymer, e.g. poly(vinyl acetate)[5]. Alternatively, block copolymers in which the anchor group was an epoxide resin (based on a diphenylolpropane–epichlorhydrin condensate) have been used to stabilize acrylic polymer dispersions[25]. Dispersions of copolymers based on vinyl acetate have been prepared using dispersants having acrylic anchors[5]. In all of these examples, the dispersed polymer was one which would have been swollen by the monomers present during the polymerization (see Section 4.4).

Since most vinyl and acrylic monomers are good solvents for their own and many other polymers, the swelling of the particle by monomer during the polymerization process may increase the compatibility of the particle with dissimilar anchor groups and increase the anchoring energy. This probably is the reason why such a wide variety of anchor groups can be employed. Even a polymer such as poly(acrylonitrile) which is insoluble in its own monomer is heavily swollen with acrylonitrile during the initial stages of the dispersion polymerization (Section 4.4.4). Poly(vinyl chloride) is an exception in that under the conditions of dispersion polymerization, it is not extensively swollen by monomer. In this case, particle growth takes place by continuous aggregation of precipitating polymer and attempts to stabilize the polymer dispersion using preformed graft dispersants have been relatively unsuccessful[26]. The versatility of many types of block or graft copolymers in stabilizing dispersions can also be attributed in some degree to the fact that the majority of the pre-formed dispersants employed have been composed either of relatively low molecular weight polymer blocks, or are graft copolymers with short runs of anchor polymer. Such polymer groups of low molecular weight have a much wider range of compatibilities with other polymers than their higher molecular weight analogues.

There is however, a lower limit to the molecular weight of the anchor group below which it ceases to become effective. This lower limit is best considered in conjunction with the effect of the number of anchor groups in the block or graft copolymer dispersant. If the dispersant has the configuration

$$S-A-S-A-S-A-S-A-S$$

(where S = soluble group and A = anchor group) it is likely to be adsorbed at the particle surface as shown in Figure 3.1(a).

The total anchoring energy will thus be the product of the total number of separate anchoring groups and the energy of adsorption for each group. The compound stabilizer in Figure 3.1(a) will be more effectively anchored than the assembly of separate amphipathic molecules in Figure 3.1(b), even though the total energy of adsorption is approximately the same. This is because the probability of detaching all the anchor groups simultaneously in 3.1(a) is much less than the chance that a proportion of the AS molecules in 3.1(b) will

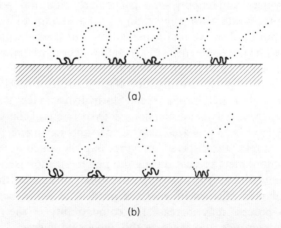

Figure 3.1. Schematic representation of adsorption of multiple-anchored (a) and single-anchored (b) graft copolymer dispersants at a particle surface: (····) soluble chain; (——) insoluble chain

be desorbed, which would leave unstabilized surface and could thus lead to flocculation. Unfortunately, there is very little quantitative experimental evidence to support this general picture so far as polymeric anchors are concerned, although it has been clearly demonstrated with anchoring of stabilizers by acid–base interactions (Section 3.5.4). Such evidence as there is suggests that, for the S—A—S—A—S type of structure, the minimum molecular weight of each anchor segment should be approximately 500, depending on the exact nature and insolubility of the anchor group[27]. Below this value, the dispersant behaves increasingly like a random copolymer.

As an illustration of this effect, 1 mol of stearic acid was reacted with a 2:1 molar copolymer of methyl methacrylate and glycidyl methacrylate, to yield an approximately 1:1 methacrylate–stearyl copolymer, with pendant soluble groups with a molecular weight of 300. The behaviour of this copolymer was compared with that of a copolymer of stearyl methacrylate and methyl methacrylate as a dispersant in a dispersion polymerization. Neither was very effective as a dispersion stabilizer, but it was at least possible to prepare coarse particle dispersions with the first copolymer. The anchoring of this copolymer was no doubt marginally more effective because of the presence of the hydroxyl group generated by the reaction between the stearic acid and the pendant glycidyl groups. The area of the particle surface stabilized by each soluble group was considerably below the calculated value, suggesting that the stabilizer was adsorbed as loops rather than as flat molecules on the surface.

### 3.5.2. Anchoring in emulsions

In the preparation of some types of polymer dispersions, it is necessary to pre-emulsify monomers which are insoluble in the continuous phase. The stabilization of such oil-in-oil emulsions is clearly closely related to the stabilization of polymer dispersions. The soluble component of the dispersant acts in the same general fashion so that the same rules apply for its selection. The criteria for the selection of the anchor group however, are somewhat more severe than those used for physical adsorption anchoring on to polymer particles.

It is well known that block and graft copolymers are very effective stabilizers for the preparation of oil-in-oil emulsions[28–30]. Most of the emulsifiers described have been well-defined block copolymers prepared by anionic polymerization. These have been used to emulsify one solvent in another, one polymer solution in another polymer solution, and to improve the mutual compatibility of two polymers. For the purposes of dispersion polymerization, we are concerned with liquid-in-liquid emulsions in which the continuous phase is a liquid of low molecular weight and the disperse phase is a second, immiscible liquid of low viscosity which may, for example, be a mixture of monomers or a reactive polymer of low molecular weight. The anchor group of the dispersant should be insoluble in the continuous phase, just as for the dispersion of solid particles in a liquid medium. However, Nicks and Osmond have demonstrated that there is an additional requirement in that the anchor group should be fully soluble in the disperse phase, or at least compatible with it, in order to form stable dispersions of fine particle size[31]. If the anchor group is not soluble, it is easily rejected from the mobile droplet surface and coalescence will readily ensue. However, if the anchor group dissolves in the droplet there is a large gain in energy due to the entropy of solution. Since coalescence would result in an overall reduction in surface area, it requires the reversal of this solvation process. It is found in practice that provided the anchor group is soluble, coalescence is not normally observed and indeed spontaneous emulsification often takes place. The condition that the anchor group should be soluble can be fulfilled fairly easily when both the anchor group and the disperse phase are of a low molecular weight. For example, poly(hydroxystearic acid-g-methyl methacrylate) will emulsify liquids which are solvents for poly(methyl methacrylate), such as, di(ethoxyethyl) phthalate, liquid epoxide resins or urea formaldehyde resins in aliphatic hydrocarbons. It is not effective for the emulsification of water or glycols, which are not, of course, solvents for the anchor groups. Moreover as the molecular weight of the disperse phase is increased the degree of mismatch which can be tolerated between it and the anchor group of the dispersant is rapidly reduced.

In a situation where the two phases to be emulsified are of low viscosity and are good solvents for the respective components of the amphipathic dispersant, spontaneous emulsification in both phases is observed. Under these conditions, it is advantageous therefore to dissolve the dispersant in the continuous phase. With an appropriate choice of components, emulsions of phase volume up to 70 or 80% and which are very resistant to coalescence under shear are readily obtained.

For most vinyl polymerizations carried out in organic media, the monomer is usually completely soluble in the reaction medium and the problem of emulsifying the monomers prior to polymerization does not arise. This is not the case however with condensation polymerizations, with reactants such as glycols, diacids or amine salts which are insoluble in hydrocarbon media (see Section 5.5.2). In this case, the block or graft copolymer dispersant may be employed initially as an emulsifier by observing the general rules given above. As polymerization proceeds however the dispersant should, preferably, also be capable of maintaining the stability of the polymer dispersion[8]. Since the nature of the disperse phase is changing throughout the course of the polymerization, it is unlikely that the dispersant will remain completely effective throughout unless the components of the dispersant anchor group take part to a limited extent in the polymerization reaction. In this way, the dispersant is chemically bound to the particle surface and remains effective. This type of process is discussed in more detail in the following Section (3.5.3).

### 3.5.3. Anchoring by covalent links

When a strong solvent is added to a latex which is stabilized by a block or graft copolymer dispersant, there is always a tendency for the dispersant to be dissolved off if it is anchored only by physical forces. This process results in the thickening and eventual gelation of the latex. It is also possible, by slowly increasing the solvency of the continuous phase, to remove a dispersant of this type from the particle surface without swelling the particles. When this operation is carried out in successive stages, separating the particles from the medium at each stage, the latex becomes less and less stable until it eventually flocculates. A variant of the effect of strong solvent has also been observed during the preparation of dispersions of polymers with low glass transition temperatures, such as poly(vinyl acetate) or poly(acrylates). Such polymer particles are almost fluid at reaction temperatures above 70 °C. Towards the end of their preparation, a dramatic increase in the number of large particles is frequently observed, which under the electron microscope can be seen to be coalesced aggregates of many smaller particles. The most plausible explanation for their formation is that the plasticizing action of the monomer at the elevated temperature has caused the anchor group of the stabilizer to desorb or at least to become mobile on the particle surface, thus permitting the lateral displacement of the soluble chain. In either eventuality, a collision with another particle in a similar state readily leads to aggregation and coalescence, thereby reducing the number of particles and the area of the surface to be stabilized and hence restoring the equilibrium.

Both of these effects have disadvantages under certain conditions. The aggregation of latex particles during their preparation is one obvious drawback. The flocculating action of strong solvent has also proved a serious difficulty in the formulation of surface coatings. It is often necessary to incorporate a proportion of relatively non-volatile coalescing solvent so that after the coating has been applied and the diluent evaporates, the residual strong solvent causes the particles to soften and adhere to one another. To overcome the tendency of physically-anchored dispersants to be displaced by the action of strong solvents, covalent links have been used to attach the anchoring component of the dispersant more firmly to the polymer particle. In one method[32], complementary reactive groups are incorporated into the stabilizer and dispersed polymer. After the polymerization is complete, the dispersion formed is kept at an elevated temperature (or alternatively a catalyst is added) to accelerate the subsequent reaction between the dispersant and the disperse phase. Examples of reactions used include that of maleic anhydride in the particle with hydroxyethyl methacrylate in the dispersant; the combination of glycidyl methacrylate and acrylic acid in the presence of a tertiary amine catalyst has also been utilized. A variant of this process employed a di-isocyanate, which linked hydroxyl groups in the particle with those in the dispersant.

In a further general method[33], polymerizable groups were attached to the anchor group of the dispersant. These groups copolymerized with the monomer which formed the disperse phase polymer when the latex was prepared. If more than one polymerizable group was attached to the anchor, the effect was to produce a cross-linked network on the surface of the particle. This reduced the mobility of the dispersant during the dispersion polymerization and was therefore more effective than the post-polymerization reaction used above in reducing the aggregation and coalescence of polymers prepared well above their glass transition temperatures. It was also equally effective in preventing flocculation after the addition of a strong solvent to a latex. However, unless the process is properly controlled it can interfere with the subsequent coalescence processes of the particles when used as paint films. Examples of this type of graft copolymer dispersant have been prepared by reacting the pendant glycidyl groups in the anchor group with methacrylic acid (and *vice versa*) to produce dispersants with the correct functionality with respect to polymerizable bonds (see later, Section 3.7). If the number of polymerizable groups is small, inadequate resistance to strong solvent is obtained: too many polymerizable groups can lead to the subsequent problems in the film formation of paints described above. On average, between two and four polymerizable groups attached to each graft dispersant molecule is usually close to the optimum but the exact functionality will of course depend to some extent on the particular end use.

A further variation of this method has been described[34] which utilizes *N*(hydroxy-ethyl)ethylene imine to attach methacrylic acid to the anchor group. This type of reaction has the added advantage of promoting adhesion of the film-forming polymer towards the substrate on which it is used.

### 3.5.4. Anchoring by acid: base interactions

#### 3.5.4.1. Multi-point anchoring

The lower limit to the molecular weight of the soluble stabilizing group of a polymeric dispersant is primarily determined by the anchoring power of the insoluble group which can be associated with it, whilst still retaining the overall solubility of the dispersant in the continuous phase of the dispersion. In practice, as described in the preceding Sections, the practical limit for the soluble group is a molecular weight of a value in the region of 500. However, it is possible to utilize stabilizing chains as short as eighteen carbon atoms in length by providing acid–base interactions between particle and dispersant to increase the energy of adsorption of the dispersant[35].

The most generally effective method for producing stabilization by this principle is to provide a multiplicity of acidic or basic groups along the polymeric backbone to which the soluble components are attached (Figure 3.2) where the X, Y groups may be entirely acidic or entirely basic. This dispersant is then used to prepare a polymer dispersion in which a small amount of a complementary acidic or basic monomer (1–2%) is incorporated in the starting monomers. The original example of this mode of anchoring employed a dispersant which was a styrene–maleic anhydride alternating copolymer converted to the acid amide by reaction with dilaurylamine. This was then used to stabilize a dispersion of a methyl methacrylate/t-butylaminoethyl methacrylate (99:1 molar) copolymer in an aliphatic hydrocarbon medium. In another example, a copolymer of cetyl/stearyl methacrylate with methacrylic acid was used as a dispersant for the preparation of a copolymer of methyl methacrylate with dimethylaminoethyl methacrylate in aliphatic hydrocarbon. The average surface

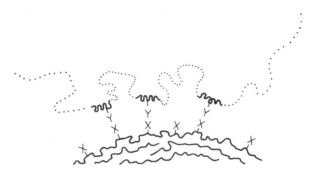

Figure 3.2. Schematic representation of multi-point anchoring of a graft copolymer dispersant on the surface of a polymer particle by acid/base interactions (groups X, Y): (····) soluble chain; (———) insoluble chain

area occupied by each octadecyl chain in this type of preparation, assuming all such groups in the absorbed stabilizer to be anchored to the surface, was found by experiment to be 0.45 nm$^2$ [27]. This value is in fair agreement with the average value, 0.21 nm$^2$, determined for a monolayer of close-packed stearic acid molecules on the surface balance[36–38], bearing in mind the more elevated temperatures used in dispersion polymerization.

Attempts to stabilize dispersions with similar copolymers using dodecyl groups in place of octadecyl groups were not successful. However, this may well have been due to the difficulty of anchoring the dodecyl chains sufficiently close to one another on the particle surface, rather than to any failure of the smaller barrier to keep the particles apart.

*3.5.4.2. Single-point anchoring*

Fatty acids, such as stearic acid and oleic acid, can be used to aid the dispersion of some pigments in hydrocarbon media but they are quite ineffective as dispersants in non-aqueous dispersion polymerizations. From the results with the multiple acid/base interactions described above, it might be supposed that by the incorporation of a sufficient number of amino-groups in the disperse phase it should be possible to produce stable dispersions using fatty acids. However, the dispersion polymerization of mixtures of methyl methacrylate and dimethylaminoethyl methacrylate in aliphatic hydrocarbon at 80 °C showed no signs of stabilization with oleic or stearic acid even when as much as 25 % of the amino monomer was used in the monomer mixture. A partial explanation for these observations may be found in a consideration of the association between carboxylic acids and tertiary amines in hydrocarbon media. The degree of association is dependent on temperature, because the interaction involves the formation of easily dissociated hydrogen bonds[39] rather than ion-pairs with formal charge separation, and there is little or no stabilization of ions by solvation. This is illustrated by the change with temperature of the heat of neutralization of oleic acid with dimethylaminoethyl methacrylate, as shown in

Table 3.14[27]. Although these results were obtained with a relatively simple calorimetric procedure[40] and should be considered as approximate values only, they clearly illustrate the general trends.

Table 3.14. Heat of neutralization of various acids with dimethylaminoethyl methacrylate in aliphatic hydrocarbon[27].

| Acid | Temperature (°C) | Heat evolved (kcal/mol) |
| --- | --- | --- |
| Oleic acid | 6 | 4 |
| Oleic acid | 20 | 3.2 |
| Oleic acid | 30 | 2.5 |
| Oleic acid | 40 | 2 |
| Oleic acid | 50 | 1.8 |
| Oleic acid | 60 | 1.4 |
| Oleic acid | 70 | 1.0 |
| Oleic acid | 80 | 0.4 |
| Oleyl acid phosphate | 20 | 6 |
| Oleyl acid phosphate | 80 | 6.5 |
| Dodecylbenzene sulphonic acid | 20 | 16 |

With increase in temperature, the heat of association falls indicating a corresponding shift in the equilibrium towards the dissociated state. At 80 °C therefore, the proportion of associated molecules is probably only about a tenth of those at 6 °C. A particle surface which was adequately covered at the lower temperature would therefore be virtually unprotected at the higher temperature. Polymerizations at 20 °C did in fact show some signs of stability and it was possible to prepare dilute latexes of copolymers of methyl methacrylate and dimethylaminoethyl methacrylate (75:25 molar) with oleic acid (using between 5 and 20% by weight on the dispersed polymer). At this level of basic monomer distributed evenly throughout the particle, the spacing of the amino groups on the surface of the particle should be in the region of one per square nanometre. There may of course be a slightly higher concentration of amine in the surface as there is excess acid in the continuous phase.

Extending the thesis that the acid–base association equilibrium at the operating temperature is important, it would then be expected that stronger acids would be more effective as dispersants. The calorimetric results for oleyl phosphate, a commerical 1:1 mixture of mono and di-oleyl phosphates, show little change with temperature and in fact oleyl acid phosphate or stearyl acid phosphate were found to be very good stabilizers for the preparation of fine particle methyl methacrylate/dimethylaminoethyl methacrylate copolymer dispersions, even when the proportion of amine used was as low as 2%. Unless some mechanism is operating which concentrates all the amine in the surface of the particle, the amine groups must be spaced much further apart than the adsorbed alkyl phosphate molecules, each of which occupied an experimentally determined area of 0.5–1.0 nm$^2$. One possible explanation for the surprising efficiency of the phosphate stabilizers[27] is that hydrogen bonding between the phosphate molecules can give the structure shown in Figure 3.3, which reinforces the adsorption by forming a two-dimensional surface network.

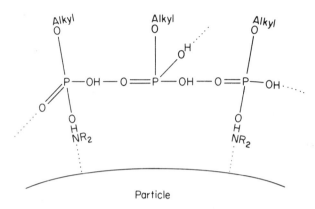

Figure 3.3. Postulated structure of two-dimensional network formed by phosphate stabilizers at the particle surface

Alternatively, the high efficiency may be due to the association of alkyl phosphate molecules which extend from the surface into the continuous phase, thereby raising the molecular weight of the soluble group and increasing its surface coverage. Commercial alkyl phosphates are approximately 1:1 mixtures of the mono and di-alkyl phosphate esters. The mono-alkyl ester has a greater potential for hydrogen bonding and should more easily form surface networks or extended associations. Separate evaluations of the pure mono and di-alkyl esters demonstrated that only the mono-alkyl ester was an effective stabilizer, the di-alkyl ester showing little sign of dispersant activity.

When even stronger acids were examined the pattern of results changed once again. Octadecane sulphonic acid was relatively ineffective as a stabilizer for dimethylaminoethyl methacrylate copolymer dispersions even at high levels of acid and base. The large heat of neutralization of dodecylbenzene sulphonic acid with the amine (Table 3.14) indicates a particularly strong association between the two components, most probably resulting in the formation of an ion-pair. This complex may in fact behave more like stearyl methacrylate and form random copolymers with methyl methacrylate which would have little stabilizing capability and would be co-precipitated and trapped within the insoluble polymer. The weaker association of the phosphates would tend to favour continual desorption of the soluble stearyl component followed by re-adsorption on a fresh surface, thus avoiding permanent burying of the alkyl group. However, if the strong acid residue is attached to a polymeric soluble group which is large enough to act as a dispersant precursor in the polymerization, then the acid–base interaction should provide a means of bonding polymerizable groups to the large soluble group. To test this thesis, a copolymer of lauryl methacrylate and glycidyl methacrylate was reacted with sulphuric acid to introduce a sulphate ester group on the polymer chain. When this was used in conjunction with an amino-methacrylate monomer, it was just as effective in a dispersion polymerization as a stabilizer precursor in which the methacrylate double bond had been attached covalently[27].

## 3.6. THE BEHAVIOUR OF GRAFT COPOLYMERS IN DISPERSION POLYMERIZATION

It is well established that amphipathic polymers, in both solution and under bulk conditions, generally exist in the form of aggregates, somewhat analogous to the micellar structures observed in aqueous soap solutions[41-45]. The aggregates formed can adopt a variety of configurations depending on their concentration, the size and composition of the polymer, the solvent environment and the temperature. At very low concentrations, a conventional monomolecular solution is formed, which quickly changes to an aggregated state when the concentration increases to a few per cent. At higher concentrations ($>20\%$), the aggregates coalesce into regular and periodic structures of three main types: spheres, rods or cylinders and lamellae (Figure 6.9; for a more extended account of micellization behaviour see Section 6.4 below). In these aggregate structures, the core is composed of the insoluble or less soluble component, and is surrounded by an outer layer of soluble polymer. On theoretical grounds[42], it is predicted that in solution the equilibrium size of the micelle increases as the cube root of the degree of polymerization, $n$ (where $n = n_A + n_B$ for the two polymeric components A and B). It also follows that the size of the micelle will increase as the interfacial contact energy per unit area between the core and the solvated outer layer of the micelle becomes larger. In practical terms, this implies that the more incompatible the components of a graft copolymer, the larger the micelles which are formed. Much of the published information on micelle formation relates to relatively monodisperse AB or ABA block copolymers prepared by anionic polymerization. However, random graft copolymers appear to behave in the same fashion and stable, concentrated dispersions of spherical aggregates of micellar dimensions have been prepared in hemi-solvents, i.e. a solvent for one component of the graft copolymer[46]. These were transformed to cylindrical or lamellar structures either by heating or by the addition of a swellant for the particle core.

In dispersion polymerization, the polymeric dispersant is normally employed at concentrations which do not exceed a few per cent by weight. From the foregoing discussion, it will be apparent that under these conditions the dispersant will tend to exist as loose aggregates in equilibrium with single molecules. Dilute solutions of polymeric dispersants do in fact exhibit light scattering effects characteristic of micellar dispersions. Compared with micellar solutions of aqueous soaps, it is probable that in non-aqueous media the equilibrium will be less dynamic because of the higher molecular weight of the associated polymeric components and the consequent reduction in the rate of desorption or release from the aggregate.

In dispersion polymerization, during or immediately following the precipitation of the insoluble disperse phase polymer, the dispersant becomes adsorbed on the polymer surface. The equilibrium between the adsorbed dispersant, single dispersant molecules and dispersant micelles may be represented as shown in Figure 3.4. It is therefore clear that a careful balance between the

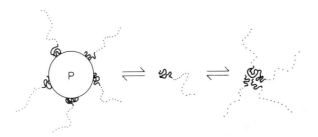

Figure 3.4. Dispersant equilibria during dispersion polymerization: (····) soluble group; (———) insoluble or anchor group; P, growing polymer particle

insoluble and the soluble components is necessary if the dispersant is to function efficiently. When the insoluble (or anchor) group is either too small or does not have some specific interaction with the particle, it will not adsorb to form a monolayer except at high equilibrium concentrations of dispersant in solution. If on the other hand, the anchor/soluble group balance leans too heavily towards the anchor side, the dispersant will exist predominantly in the form of micellar aggregates which do not readily dissociate. Under these conditions, the initial nucleation process will be followed rapidly by flocculation because of the insufficiency of free dispersant molecules. In the limit, the micelle may even become the actual locus of the polymerization, provided that separate precipitation of unstabilized polymer formed in solution does not occur. At even higher anchor/soluble component ratios, the dispersant molecule may not be able to form a stable spherical micelle (i.e. a monolayer of soluble component will not be formed around the aggregate of dispersant molecules) and the dispersant will then precipitate as a flocculated mass. Although it is difficult to lay down precise limits for the anchor/soluble balance required, it is a general rule that for polymeric anchors and soluble groups, a mass ratio of 1:1 is the best starting point for practical formulations with 1:3 and 3:1 as the extreme limits.

There are obvious links between the anchor/soluble balance concept and the hydrophile/lipophile balance (HLB) system for emulsifiers[47]. Obviously the majority of the dispersants employed in non-aqueous dispersion polymerization would have a low HLB value, but the constraints already mentioned for non-aqueous polymer dispersants, such as the minimum anchor size, prevent the straightforward application of the simple HLB system.

For a given anchor/soluble group mass balance, it is possible to vary the composition of the anchor group and thereby alter the equilibrium between the adsorbed dispersant molecules with the associated and single molecules in solution.

For example, the effect of dispersant concentration on particle size has been examined for four dispersants of similar general structure but with different anchor group compositions[48]. The graft copolymer dispersants used were of the 'comb' type (Section 3.7.4) with pendant soluble poly(hydroxystearic acid) chains attached to an acrylic backbone of varying composition at an anchor/soluble mass ratio of 1:1. The acrylic backbones used had the following compositions:

I. poly(methyl methacrylate-co-methacrylic acid) (98:2, w/w).
II. poly(methyl methacrylate).
III. poly(methyl methacrylate-co-ethyl acrylate) (1:1, w/w).
IV. poly(ethyl acrylate).

The effect of the use of these dispersants on the size of poly(methyl methacrylate) particles formed under otherwise identical conditions of dispersion polymerization at 80 °C is shown in Figure 3.5. With each of the dispersants I–III, an increase in the dispersant concentration led to a decrease in particle size, and a minimum particle size was reached. For dispersants I and II, the concen-

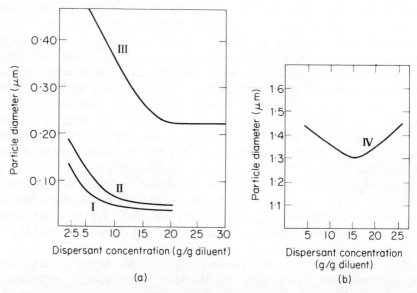

Figure 3.5. Effect of concentration of 'comb'-type dispersant of variable anchor group composition on particle size[48]. Anchor group composition: (a) Poly(methyl methacrylate-co-methacrylic acid), 98:2, w/w, I; Poly(methyl methacrylate) II; Poly-(methyl methacrylate-co-ethyl acrylate), 1:1, w/w, III. (b) Poly(ethyl acrylate) IV

tration required to reach minimum particle size was 10–15 × $10^{-3}$, and for III c. 20 × $10^{-3}$ (g dispersant/g diluent). These results can be interpreted in the following way. The number of particles formed, given a reasonably constant rate of precipitation, will depend on the rate of adsorption of dispersant. This in turn, depends on the strength of anchoring and the concentration of monomolecular dispersant. The constancy of particle size above a given concentration for each dispersant then corresponds to the point at which the monomolecular dispersant concentration reaches an upper limiting value, possibly the equivalent of the critical micelle concentration of an aqueous surfactant. The stronger the anchoring group, the greater the rate of adsorption of dispersant onto the precipitated polymer, the larger the particle number and the lower the 'critical micelle concentration'. Dispersant IV exhibited extreme behaviour—at comparable concentrations, the particle size achieved was very much larger than was obtained with the other dispersants and did not decrease further as the dispersant concentration was increased. At low concentrations, a stable latex could not be prepared with dispersant IV. It was therefore concluded that poly(ethylacrylate) although probably fully dissociated had very little anchoring power under the reaction conditions used.

It was also predicted that the dispersity of particle size should depend on the same factors; the greater the rate of adsorption of dispersant, the more continuous and extended the nucleation stage and hence the wider the particle size distribution obtained. Figure 3.6 illustrates the particle size dispersities of the latexes prepared with the same four dispersants. Dispersants I–III showed

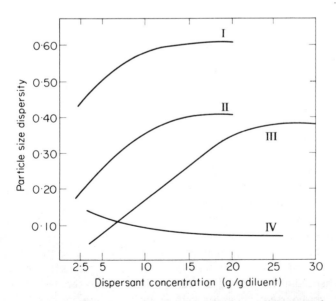

Figure 3.6. Effect of concentration of 'comb'-type dispersant of variable anchor group composition on particle size dispersity. Anchor group compositions as in Figure 3.5

a rise in particle size dispersity with concentration until a limiting value was reached at approximately the same concentration as the minimum particle size was observed. As expected, the dispersity ($= \sigma/R$, where $\sigma$ is the standard deviation and $R$ the mean particle diameter) is in the order I > II > III > IV. As the rate of dispersant adsorption decreases, the particle size and dispersity is increasingly controlled by the rate of aggregation of the precipitating polymer until, with Dispersant IV, the anchor is so weak that the particle size and dispersity remain constant and independent of dispersant concentration.

Additional evidence that stabilization is strongly influenced by the degree of self-association of the dispersant also comes from a particularly illuminating series of experiments[49] in which the effect of temperature, solvency of the medium and dispersant composition on the dispersion polymerization process was determined. The dispersants used were again of the poly(hydroxystearic acid) 'comb'-type, in which a 1:1 ratio of anchor to soluble group was maintained but the anchor component was varied in polarity and flexibility. The dispersion polymerizations were conducted in aliphatic hydrocarbon with either vinyl acetate or methyl methacrylate, to give a polymer content of 50 % at full conversion. Table 3.15 summarizes the results of these experiments.

Table 3.15. Effect of anchor group composition on latex stability[49].

| Disperse phase monomer | Dispersant anchor group | Reaction temperature (°C) | Particle diameter (μm) |
|---|---|---|---|
| VA | MMA:MA 98:2 | 80 | 0.2 |
| VA | MMA:MA 98:2 | 50 | Eventually flocculated |
| VA | VA | 50 | 0.1–0.2 |
| VA | EA:MMA 50:50 | 50 | 0.8–1.0 |
| MMA | MMA:MA 98:2 | 80 | 0.1 |
| MMA | MMA:MA 98:2 | 50 | Eventually flocculated |
| MMA | MMA | 50 | Eventually flocculated |
| MMA | EA:MA 50:50 | 50 | 0.2–0.3 |

VA = vinyl acetate; MMA = methyl methacrylate; EA = ethyl acrylate; MA = methacrylic acid.

A general conclusion which emerged from this study was that the higher the glass transition temperature of the polymeric anchor component of the dispersant (or the more strongly it associates with itself, e.g. by hydrogen bonding), the higher the reaction temperature necessary for the dispersant to be effective in a dispersion polymerization. This led in turn to the concept that the reaction temperature or the solvency of the medium must be sufficiently high to enable the micellar aggregates of dispersant molecules to dissociate freely. On the other hand, if either the temperature or the solvency is too high then the dispersant may be adsorbed too weakly onto the particle and consequently either coarse particles or even an unstable latex results. The chemical nature of the particle itself also plays a part as can be seen by comparing the particle diameters of poly(vinyl acetate) and poly(methyl methacrylate) formed with the same dispersant and at the same reaction temperatures. This behaviour is probably associated with differences in the rate of precipitation of the two polymers concerned.

The principles outlined above also apply to the dispersion of preformed polymer or pigment particles[50]. Attempts to disperse poly(vinyl chloride) by ball milling in a hydrocarbon diluent at room temperature using a dispersant with a poly(methyl methacrylate) anchor were unsuccessful even after many hours of milling. The product was flocculated and contained many unbroken aggregates of primary particles. By raising the milling temperature to 50 °C (or by adding a solvent or a plasticizer such as a phthalate ester to the mill) dispersion to a fine particle, fluid product was rapidly achieved. Alternatively, the same improvement could be obtained by lowering the glass transition temperature of the anchor group by the copolymerization of ethyl acrylate in the dispersant backbone.

## 3.7. THE DESIGN AND SYNTHESIS OF GRAFT COPOLYMERS AS DISPERSANTS

There are a number of possible ways in which the insoluble or anchor groups and the soluble groups of a polymeric dispersant can be assembled. Some of the more obvious configurations are shown in Figure 3.7. Examples a, b and c in Figure 3.7 were used extensively in the early development of acrylic dispersion polymers, whereas d and e were later developments, which for various reasons proved easier to control and to have less undesirable polymeric species liable to interfere with the dispersion polymerization process. Block copolymers of

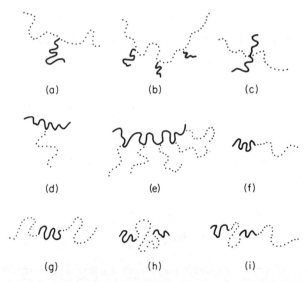

Figure 3.7. Possible configurations of copolymer dispersants: Graft copolymers (a)–(e); block copolymers (f)–(i); (····) soluble chain; (——) insoluble chain

types f, g and h have been widely reported in the literature as interesting materials in their own right, both in bulk and as micellar dispersions, but they have been less extensively utilized as dispersion stabilizers, although the evidence which is available indicates that they are extremely effective (Section 3.8). The synthesis of the various types of dispersants and their behaviour in dispersion polymerization will now be discussed. Experimental details for the synthesis of the main types of dispersant are given at the end of this chapter (Section 3.9).

The term 'graft copolymer' implies the attachment of one or more branches or side chains composed of one polymer, to a backbone or stem of a second polymer, to yield structures of the type shown in Figure 3.7. In the discussion which follows, the term is employed to describe the structure of the graft copolymer formed rather than its method of assembly.

### 3.7.1. Random grafting on to soluble polymer

This is the simplest method for preparing a graft copolymer and it has found wide application in the preparation of polymer dispersions. In the method most commonly employed, the polymer which is to become the soluble component of the stabilizer is first dissolved in the reaction medium. Then monomer (M) to form the disperse phase polymer, together with an organic peroxide ($R-O-O-R$) are dissolved in the reaction mixture and heated to initiate polymerization. The series of reactions given below illustrates the course of the reaction, which results finally in the formation of a stable polymer dispersion. The process is schematically represented in Figure 3.8.

$$R-O-O-R \rightarrow 2RO\cdot \qquad (3.3)$$

$$RO\cdot + M \rightarrow RO-M\cdot \xrightarrow{nM} P_{disp} \text{ (precipitated)} \qquad (3.4)$$

$$RO\cdot + P_{soln} \rightarrow ROH + P\cdot_{soln} \text{ (or} \rightarrow ROP\cdot_{soln}) \qquad (3.5)$$

$$P_{soln} + RO-M_n\cdot \rightarrow P\cdot_{soln} + P_{disp} \qquad (3.6)$$

$$P\cdot_{soln} + nM \rightarrow P_{soln}\text{-g-}P_{disp} \qquad (3.7)$$

$$P\cdot_{soln} + RO-M_n\cdot \rightarrow P_{soln}\text{-g-}P_{disp} \qquad (3.8)$$

$$P_{soln}\text{-g-}P_{disp} + P_{disp} \rightarrow \text{stabilized dispersed polymer} \qquad (3.9)$$

where

$$P_{disp} = \text{disperse phase polymer and}$$

$$P_{soln} = \text{soluble polymer}$$

The formation of graft copolymer (equations 3.7 and 3.8) is dependent on the generation of a reactive radical site on the soluble polymer molecule by hydrogen abstraction or by radical addition to a double bond (equation 3.5). Although nearly all polymers take part in such reactions to some extent, the most suitable

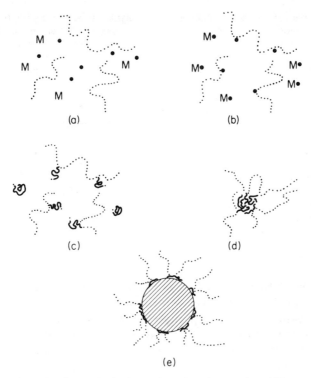

Figure 3.8. Stages in the formation of random graft stabilizers in dispersion polymerization: (a) Radical formation (initiation); (b) radical transfer; (c) random grafting; (d) aggregation; (e) particle growth; (····) soluble chain; (——) insoluble chain; •, free radicals; M, monomer

soluble polymers are those which have particularly reactive groups distributed along the polymer chain. Examples of such polymers which have been used for the formation of random graft stabilizers are listed in Table 3.16.

Natural rubber, which was used very successfully by a number of workers in some of the earliest examples of polymer dispersion in hydrocarbons[51–54], has a multiplicity of reactive groups which can undergo hydrogen abstraction to produce grafting sites. (An account of the early work by the British Rubber Producers' Research Association and others on the grafting of polymer on to natural rubber, both as latex and in bulk, is given in Reference 55.) In these systems, grafting often occurred at a number of sites on the same rubber molecule which resulted either in gelation of the reaction mixture, or thickening of the dispersion due to the multifunctional graft molecules acting as bridges between the polymer particles formed. To reduce such undesirable side effects, it was necessary to use degraded natural rubber which because of its reduced molecular weight had a lower potential functionality. Further control was achieved by adding transfer agents such as alkyl mercaptans or admitting traces

Table 3.16. Some examples of soluble polymers used for the formation of random graft dispersants during dispersion polymerization (see Tables 8.2 and 8.1, Appendix, for further details of patents).

| Soluble polymer | Patent reference |
|---|---|
| Degraded natural rubber | British 893,429; 990,154; 992,635; 1,017,931 |
| Poly(butadiene) | British 990,154; 1,008,001; 1,017,931; 1,243,274 |
| Poly(ethylene-co-dicyclopentadiene) | British 1,156,235 |
| Unsaturated oil-modified alkyd resins | British 1,007,723 |
| Poly(ethylene-co-vinyl acetate) | United States 3,645,959 |
| Poly(vinyl methyl ether) | British 1,009,004 |
| Poly(lauryl methacrylate) | United States 3,232,903 |
| Poly(vinyl oxazoline) | United States 3,654,201 |
| Poly(siloxanes) | German 2,142,598 |

of oxygen, which by competing for radicals effectively reduced the number of grafting sites on the rubber molecule. Natural rubber is the extreme example of a molecule with a great multiplicity of grafting sites and it is an excellent stabilizer for the preparation of fine particle dispersions because it produces a high ratio of graft copolymer to ungrafted disperse phase polymer during the critical precipitation stage at the beginning of the dispersion polymerization. Many other soluble polymers, although less reactive, form sufficient graft copolymer to stabilize relatively coarse but still sub-micron dispersions. In fact, it is very difficult in practice to eliminate random grafting completely, particularly when peroxide initiators are used[24]. Consequently virtually all non-aqueous dispersions are in part stabilized by graft copolymer formed via hydrogen abstraction. Variation in performance of supposedly identical stabilizers has often been due to an ignorance of the contribution of the random grafting reaction. If incipient grafting produces a larger proportion of the total stabilizer than is expected then small changes in the concentration of chain transfer agent or in the amount of terminal unsaturation on the soluble polymer chain arising from disproportionation reactions can then cause unexpected variations in particle size from one latex preparation to another. Clearly such a process could be quite difficult to control successfully on a commercial scale.

One of the mechanisms for the generation of radical sites on the soluble polymer is by transfer from the growing polymer radicals (equation 3.6). The ability of the terminal radical of a propagating polymer chain to abstract hydrogen atoms varies so much from one polymer to another that it can be a dominating factor in determining the fineness and stability of the latex. This can be illustrated by the behaviour of two soluble polymers, natural rubber and poly(lauryl methacrylate) and two disperse polymers, poly(methyl methacrylate) and poly(vinyl chloride), in aliphatic hydrocarbon media under similar reaction conditions[56] (Table 3.17).

Table 3.17. Effect of nature of soluble polymer on properties of polymer dispersion formed[56].

| Soluble polymer | Disperse polymer | Dispersion |
|---|---|---|
| Natural rubber | Poly(methyl methacrylate) | Sub-micron; low viscosity |
| Natural rubber | Poly(vinyl chloride) | $\ll 0.1$ µm; rapidly gels |
| Poly(lauryl methacrylate) | Poly(methyl methacrylate) | $> 1$ µm; coarse, flocculated precipitate |
| Poly(lauryl methacrylate) | Poly(vinyl chloride) | 0.2–0.4 µm; low viscosity |

Poly(vinyl chloride), in the presence of natural rubber, forms large amounts of graft polymer (much of which is probably polyfunctional in character) and produces an extremely fine dispersion. In contrast, poly(methyl methacrylate) in the presence of poly(lauryl methacrylate) produces so little graft copolymer that the amount is often insufficient to form a stable seed even at the commencement of polymerization. These examples represent the extremes of behaviour. Most other common monomers lie somewhere in between these extremes in their tendency to form graft copolymer by random hydrogen abstraction in free radical polymerizations.

Although frequently convenient, it is of course not necessary to form the graft copolymer simultaneously with the dispersion polymer. An alternative method is to prepare the graft copolymer in a bulk or solution polymerization process and to add it in the required amount to the dispersion medium. The random grafting method when operated in solution inevitably produces a mixture of ungrafted soluble and anchor components in addition to the graft copolymer required. When this mixture is added to the dispersion medium the ungrafted anchor component precipitates, usually as a coarse dispersion, which can in some processes act as a nucleus or seed for subsequent polymer growth. In many cases, this may be overcome by the addition of a sufficient amount of monomer in order to raise the solvency of the medium above the level required to dissolve the seed polymer. When the consumption of monomer during polymerization has proceeded far enough to lower the solvency of the medium sufficiently, then precipitation takes place in a more controllable manner[57].

Many polymers are not very efficient graft copolymer precursors since they have relatively few readily graftable sites on their polymer chains. In these cases therefore, it is often necessary to use excessive amounts of soluble polymers relative to the area of particle surface to be sterically protected. It is therefore an advantage if the soluble component can perform some further function related to the storage or end-use of the composition. For example, drying oil-modified alkyd resins have been used as dispersant precursors which also functioned as cross-linkable film formers in coating compositions[58]. In another application, thermosetting coatings have been developed in which etherified melamine–formaldehyde resins can apparently serve first as the dispersant precursor and subsequently as a means of cross-linking an acrylic polymer disperse phase during the film-baking operation[59,60].

A related process employs a soluble copolymer incorporating N-butoxy methyl methacrylamide and a hydroxymethacrylate both of which cross-link with each other or with co-reactive groups in the disperse phase on baking films of the dispersion[61].

Attempts have also been made to use the soluble polymer as a means of promoting adhesion of coatings based on the dispersion. This is a logical approach since it is inevitable that the continuous phase of the dispersion or the outside of the particle makes the first contact with the substrate to be coated. In one example, a fatty acid-modified vinyl oxazoline polymer was used for this purpose[62].

The non-aqueous dispersion process employing grafting on to rubbery polymers has been explored as a route to high impact plastics. Most established processes for high impact plastics use poly(butadiene) as a modifying rubber, but these products have poor resistance to outdoor degradation[63]. Terpolymers of ethylene, propylene and a polyene (EPDM rubbers) were known to be superior in this respect but were not sufficiently reactive to form the required level of graft in a conventional ABS process. However, good results were obtained by the polymerization of styrene and acrylonitrile in a solution of a rubbery copolymer of ethylene, propylene and 5-ethylidene-2-norbornene in a mixture of benzene and hexane. A grafting initiator, such as benzoyl peroxide, was essential[64]. Precipitation occurred after 15–20 minutes of reaction and the final product was a stable, low viscosity latex. In a development of this process a seed stage was first produced, followed by the addition of the remaining reactants and the polymerization was carried to completion[65].

This account of the preparation of random graft copolymers has so far been limited to free radical processes and in fact the majority of examples in the literature are of this type. However, another class of random graft copolymer is that formed by anionic initiation. Rubber again is a useful soluble polymer and its use has been claimed in stabilizing ionically-initiated dispersions of vinyl aromatic polymers and polylactams[66].

The preparation of stable dispersions of poly(α-methylstyrene) by anionic dispersion polymerization in heptane has also been described[67]. Poly(vinylethyl ether), of molecular weight approximately one million, was found to be effective as an interfacial agent when present at levels greater than 0.1 % by weight of monomer. Lower molecular weight polymer was ineffective. The mechanism of dispersion stabilization was believed to involve the attachment of the poly(vinyl ether) to the particle by solvation of the terminal carbanion of the poly(α-methylstyrene) chain but the possibility of graft formation by metallation of the soluble chain or its scission cannot be excluded.

### 3.7.2. Transfer grafting on to specific sites

The use of the random grafting process in dispersion polymerization described above (Section 3.7.1) has several disadvantages. For example, the particle size of the dispersed polymer is difficult to control and the formation of polyfunctional graft copolymer can lead to thickening and eventual gelation of the dispersion. Furthermore, in practice, the choice of readily-grafted soluble polymers is rather limited and more often than not the soluble polymer which is most suitable for stabilization efficiency may not be acceptable in the final composition in which the dispersion is to be used. In particular, natural rubbers are undesirable as components of surface coatings because of their poor resistance to solar radiation and oxygen.

A natural extension of the random graft process is provided by the abstraction of hydrogen atoms from a relatively small number of very active transfer sites, which have been introduced during, or subsequent to, the polymerization of the soluble component. The number of such grafting sites required on each molecule of soluble component must depend on the efficiency of the transfer group in generating graft sites under the reaction conditions of the dispersion polymerization. The ideal molecule would have only one anchor moiety of the appropriate size attached to each soluble group. However, the number of possible reactions in each system and the inevitable distribution of polymeric species is such that a fine control of structure is not possible in practice. Once a transfer group has been selected, which is known to be effective as a transfer agent in the polymerization of the disperse phase monomer, it then becomes necessary to use an empirical optimization of the soluble group composition in order to obtain the required results.

Examples of the types of transfer groups used in the dispersion polymerization of various monomers are listed in Table 3.18. There is, of course, no sharp distinction between the behaviour of the transfer groups included in this Section and the transfer reactions involved in the random grafting of ethylene/propylene/

Table 3.18. Transfer groups used for the formation of graft dispersant during the dispersion polymerization of acrylic monomers. (For further details of patents, see Appendix, Tables 8.2 and 8.1.)

| Soluble component polymers | Transfer group | Patent reference |
|---|---|---|
| Poly[alkyl (meth)acrylate] | Vinyl pyridine | British 956,453; British 956,454; United States 3,218,302 |
| Poly[alkyl (meth)acrylate] | $N$-Vinyl pyrrolidone | British 956,543; United States 3,218,302 |
| Poly[alkyl (meth)acrylate] | Dimethylaminoethyl methacrylate | British 956,453; British 1,223,343 |
| Poly(alkyl acrylate) | Allyl methacrylate | British 1,223,343; United States 3,660,537 |
| Poly(dimethyl siloxane) | Mercaptan | German 2,142,598 |
| Poly(alkyl acrylate) | Tetrahydrofurfuryl methacrylate | British 1,223,343 |
| Poly(alkyl acrylate) | Isopropyl styrene | British 1,223,343 |
| Poly(alkyl methacrylate) | Undecenyl methacrylate | British 1,174,391 |

diene terpolymers, drying-oil modified resins, etc. described in Section 3.7.1 above. It must also be recognized that some groups, e.g. allyl esters, may take part in copolymerization reactions in addition to acting as transfer agents (see Section 3.7.3, below). The examples quoted in this section have, in the main, been deliberately synthesized to exploit the transfer groups present, whereas those in the previous section were selected from the range of polymers already available.

The tertiary-amino methacrylate monomers and allyl methacrylate have proved particularly suitable for incorporating transfer groups. Such monomers can be readily copolymerized with other methacrylate monomers using azo initiators and under these conditions very little chain transfer takes place. The grafting reaction can be conducted subsequently, either with peroxide initiators or with monomers which form radicals which are much more active in transfer reactions. The use of a number of transfer groups which have been described, such as vinyl pyridine or $N$-vinyl pyrrolidone, would not be expected to be very efficient, but the dispersion polymerization processes in which they were employed were often designed to produce granular slurries rather than fine particle dispersions. For this purpose, of course, it is sufficient if gross flocculation is avoided and the reaction mixture remains fluid. Strangely, relatively little use has been made so far of the mercaptan group, which is most efficient in transfer reactions, but one example deserves special mention[68]. Here, the soluble component of the graft copolymer was a poly(dimethylsiloxane) with a terminal mercapto-ethyl grouping. Polymerization of a vinyl monomer in a solution of the siloxane in hexane yielded fine particle dispersions. The siloxane was thus able to play the dual role of stabilizing the dispersion and also modifying the properties of the final polymer dispersion.

A detailed study has been reported of the preparation of dispersions of poly(methyl methacrylate) in heptane stabilized with a commercial butyl rubber [i.e. a poly(isobutylene) copolymer] with a low unsaturation value, which was deliberately introduced for the purpose of vulcanizing the rubber[69]. The name 'poloid' has been coined for these types of colloidal polymeric suspensions. Examples of similar preparations have been described previously but in this work a systematic examination was made of the relationship between the stabilizer/disperse polymer ratio and the disperse phase volume (Table 3.19).

Table 3.19. Variation in particle diameter of poly(methyl methacrylate) suspensions in heptane stabilized by butyl rubber[69]. (Based on data from Figure 1 in Bueche, *J. Colloid Interface Sci.*, **41**, 376 (1972), with permission.)

| Methyl methacrylate in disperse phase (vol.-%) | 5 | | | 10 | | | $c$, 18 | | |
|---|---|---|---|---|---|---|---|---|---|
| Butyl rubber/methyl methacrylate ratio | 0.05 | 0.6 | 0.95 | 0.05 | 0.6 | 0.975 | 0.15 | 0.4 | 0.6 |
| Particle diameter (μm) | 0.24 | 0.136 | 0.098 | 0.28 | 0.15 | 0.124 | 0.3 | | 0.3 |

Stable suspensions above 0·3 μm diameter could not be prepared. The importance of the presence of unsaturation in promoting the formation of graft copolymers was illustrated by the fact that between two and five times as much pure poly(isobutylene) was required to prepare dispersions equivalent to those obtained using commercial butyl rubber. Some doubt was cast on whether the formation of graft copolymer was responsible for the stabilization of these dispersions and the suggestion was made that the rubber chains were held at the particle surface by entanglement. The difference between the behaviour of butyl rubber and poly(isobutylene) however would appear difficult to explain on this hypothesis. A decrease in the molecular weight of the poly(methyl methacrylate) prepared in the presence of the butyl rubber was also obtained, suggesting that chain transfer to butyl rubber was in fact taking place.

In the majority of examples in which transfer grafting is employed, the dispersion polymerization is carried out in a solution of the polymeric soluble component which carries the active transfer groups. However, this sequence is not essential and this is illustrated by a series of patents which relate to the preparation of acrylic lacquers for automotive finishes[70-72]. The acrylic polymers described in these patents consisted of graft copolymers, which on the addition of a non-solvent formed a 'self-stable' organosol. This could then either act as a 'seed' for a subsequent dispersion polymerization process or could be used without further modification as the basis for a paint. In the first step in the reaction sequence, the anchor component which is ultimately the insoluble portion of the graft, was prepared in solution in a good solvent. The anchor used was a copolymer of methyl methacrylate with either a small proportion of allyl methacrylate, or a tertiary-amino methacrylate or a combination of the two. The initiator was carefully selected to avoid hydrogen abstraction side-reactions. In a typical example, the amount of allyl methacrylate was 0·8% by weight of the methyl methacrylate which, if it was uniformly copolymerized, would correspond to between five and ten potential grafting sites per anchor molecule. When this first stage was completed, the grafting monomer and further initiator was added and the polymerization continued. The grafting monomer used, e.g. 2-ethylhexyl acrylate, was free from active grafting sites. The initiator for this second stage was deliberately chosen to be a hydrogen abstraction promotor, such as t-butyl peroxypivalate, to encourage the grafting reaction and an approximately 3:1 ratio of the anchor to soluble component was usually employed. The average number of graft segments on each backbone was claimed to be between 0·5 and 5 and preferably between 1 and 2. In any random reaction of this type, it is inevitable that there will always be some ungrafted anchor and soluble component molecules, particularly when the extent of the grafting reaction is limited by the need to achieve the correct balance between the size of each graft segment and the total proportion of graft segments which can be tolerated in the composition. One feature of the particular mixture of species obtained from the above preparation, was the ease with which the polymer dispersion could be dissolved and then reprecipitated without loss of stability. As a consequence of this feature, coating compositions based on this type of graft copolymer dispersion were stated to be stable to the addition of both strong and weak solvents.

Although strictly outside the scope of this section, it is convenient to mention here a method used for the synthesis of graft dispersants by attaching a limited number of peroxide or azo groups to the preformed soluble or anchor polymer molecule[73]. On heating the modified polymer with the monomer of the second component, the desired graft copolymer was formed.

### 3.7.3. Grafting through randomly distributed copolymerizable groups

In the transfer grafting processes described above, the graft formed starts or terminates at the junction of the grafting component with the polymer backbone. An alternative method of graft formation is to attach a group, capable of copolymerizing with the grafting monomer, to a polymeric component. This component then becomes, in effect, a 'macro-monomer' which can participate in a conventional copolymerization and form graft by 'growing through' the polymeric backbone (Figure 3.9). Since copolymerization is inherently a more

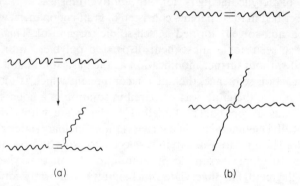

Figure 3.9. Schematic representation of radical grafting by transfer (a) and by copolymerization (b)

predictable process than transfer grafting, it is therefore possible to produce a wider range of graft copolymers of controlled composition by this method. Grafting by copolymerization is not, of course, limited to vinyl polymerization but can be readily extended to other types of addition polymerization processes[74].

Copolymerizable groups have been introduced into a polymeric component by two general methods. Firstly, by the copolymerization of a monomer containing two polymerizable groups, such as glycidyl methacrylate or allyl methacrylate, in the preparation of the first polymeric component of the graft copolymer. Secondly, by incorporating into the first component a reactive site to which the copolymerizable group can subsequently be attached. For example, the condensation reaction between an unsaturated acid chloride and a hydroxyl group on the polymeric chain can be used. The two methods are illustrated schematically in Figure 3.10.

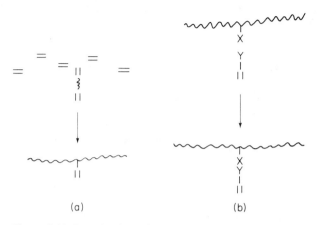

Figure 3.10. Introduction of copolymerizable groups into a polymer by copolymerization with a difunctional monomer (a) or by incorporating a reactive site (b). (=~~= difunctional monomer; Y—= co-reactive monomer)

A number of examples of the first method have been reported. Copolymers of lauryl methacrylate with minor amounts of glycidyl methacrylate were prepared and used as precursors for graft dispersants in subsequent dispersion polymerizations in hydrocarbon media[74]. A variety of monomers were utilized, including ethylene oxide, phenyl glycidyl ether, trioxane, epichlorhydrin, β-propiolactone and bis-chloromethyl oxacyclobutane. The glycidyl groups attached to the soluble copolymer participated in the subsequent ring-opening polymerization of the main monomer to yield graft copolymers which stabilized the resulting polymer dispersions. Unmodified poly(lauryl methacrylate) was shown to have a negligible stabilizing effect in these systems. This method of stabilizing dispersions of poly(formaldehyde), derived from the polymerization of trioxane[74], has been used as a variant of the commercial process for its manufacture[75]. It was known that the copolymerization of small proportions of epoxy-containing compounds conferred the desired thermal stability on the products. The formation of graft copolymer was visualized as shown by the following sequence of reactions, using boron trifluoride as catalyst:

*Initiation*

$$F_3B \leftarrow OR_2 + O\underset{CH_2-O}{\overset{CH_2-O}{\diagup\!\!\diagdown}}CH_2 \rightleftarrows$$

$$F_3B \leftarrow O\underset{CH_2-O}{\overset{CH_2-O}{\diagup\!\!\diagdown}}CH_2 + OR_2$$

$$\downarrow$$

$$F_3B \leftarrow \overset{-}{O}-CH_2-OCH_2-O\overset{+}{C}H_2$$

*Propagation*

$$-\overset{+}{O}CH_2 + O\begin{subarray}{c}CH_2-O\\ \diagup\quad\diagdown\\ \\ \diagdown\quad\diagup\\ CH_2-O\end{subarray}CH_2 \rightleftharpoons -O-CH_2-\overset{+}{O}\begin{subarray}{c}CH_2-O\\ \diagup\quad\diagdown\\ \\ \diagdown\quad\diagup\\ CH_2-O\end{subarray}CH_2$$

$$-O-CH_2-\overset{+}{O}\begin{subarray}{c}CH_2-O\\ \diagup\quad\diagdown\\ \\ \diagdown\quad\diagup\\ CH_2-O\end{subarray}CH_2 \rightarrow -O-CH_2-OCH_2-OCH_2\overset{+}{O}CH_2$$

*Grafting*

$$-\overset{+}{O}CH_2 + O\begin{subarray}{c}CH_2\\ \diagup\\ |\\ \diagdown\\ CH\\ \sim\end{subarray} \rightarrow O-CH_2OCH\overset{+}{C}H_2 \quad \text{etc.}$$
$$\qquad\qquad\qquad\qquad\qquad |$$
$$\qquad\qquad\qquad\qquad\qquad \sim$$

where $\sim$ = long polymer chain, e.g. poly(lauryl methacrylate).

An alternative process for the preparation of poly(epoxide) dispersions in hydrocarbons has been described in which a prepolymer was prepared either in solution or in microgel form in the hydrocarbon medium by the copolymerization of a linear 1-olefin oxide with a minor proportion of a diepoxide[76]. The poly(epoxide) forming the disperse phase was then obtained by the polymerization of a monoepoxide, such as epichlorhydrin or ethylene oxide, with an alkyl aluminium catalyst in the presence of the prepolymer. In the absence of any diepoxide in the prepolymer, a dispersion was not obtained. Instead, the polymer precipitated in large masses and coated the walls of the reactor.

It is clearly an advantage when difunctional monomers are used, that the mode of polymerization involved in the grafting stage should be different from that employed in the preparation of the first polymeric component. When the mode of polymerization is of the same type in both stages, the choice of a suitable monomer is very limited since it is generally difficult to polymerize a monomer containing two similar polymerizable groups and to leave one group unreacted. However, allyl methacrylate and allyl acrylate are examples of monomers in which the reactivity of the two polymerizable groups is sufficiently different to enable copolymers to be prepared and in which the allyl groups remain essentially intact. The copolymerization behaviour of an allylic monomer, allyl acetate, with some of the more important monomers, is shown in Table 3.20[77].

The data given in Table 3.20 would indicate that in some of the examples of transfer grafting involving allylic unsaturation described above (Section 3.7.2), it is likely that copolymerization reactions with the ethylhexyl acrylate monomer were also involved. Copolymers of ethylhexyl acrylate containing minor amounts of allyl acrylate have been used as dispersant precursors in the preparation of dispersions of vinylidene chloride copolymer of uniform particle size

Table 3.20. Reactivity ratios of allyl acetate. (Data from Mark, et al.[77] *Polymer Handbook*, 1966, pp. II–141 et seq., with permission of Interscience Publishers, New York.)

| $M_1$ | $r_1$ | $M_2$ | $r_2$ | $T$ (°C) |
|---|---|---|---|---|
| Allyl acetate | 0 | Methyl acrylate | 5 | 60 |
| | 0 | Methyl methacrylate | 23 | 60 |
| | 0 | Styrene | 90 ± 10 | 60 |
| | 0.45 ± 0.15 | Vinyl acetate | 0.60 ± 0.15 | 60 |
| | 0 | Vinyl chloride | 1.16 | 40 |
| | 0 | Vinylidene chloride | 6.6 | 60 |

in heptane[78]. The compositions were intended as moisture barriers and heat-seal coatings for cellulose films. The particle sizes of the polymer dispersion obtained showed some, albeit non-systematic, dependence on the amount of initiator (isobutyryl peroxide) employed. This suggests that transfer grafting may also have been playing a part in the process. However, a much more direct dependence of particle size on the allyl methacrylate content of the dispersant precursor was indicated, as shown in Table 3.21 below. The proportion of dispersant precursor on disperse polymer used was approximately 2%.

Table 3.21. Dependence of particle size of vinylidene chloride/methyl acrylate/acrylic acid copolymer on allyl methacrylate content of dispersant precursor[78]. [From Haskell, *U.S. Patent* 3,723,571 (1973).]

| Allyl methacrylate content (%) | Particle diameter (μm) |
|---|---|
| 0.25 | 2.0–2.6 |
| 0.50 | 2.2–2.6 |
| 1.00 | 1.7–2.0 |
| 2.00 | 1.5–1.8 |
| 4.00 | 1.0–1.5 |

As we might expect, the increased probability of graft formation arising from the higher proportion of pendant allyl groups on the soluble polymer, allows the stabilization of a larger particle surface area and consequently the particle size is reduced.

The second method of introducing a copolymerizable group into one component of the dispersant, by a post-polymerization reaction between complementary reactive groups in the polymer and the molecule carrying the polymerizable group [Figure 3.10(b), above] has been widely used and has resulted ultimately in the development of processes for making a wide variety of reasonably pure graft copolymers by commercially feasible routes in relatively unsophisticated chemical plant[24]. As originally described[79], the soluble group of the dispersant was modified to carry a small number of grafting sites and the resulting dispersant precursor was then added to the dispersion polymerization medium; alternatively, the precursor could be converted into a graft copolymer in a separate polymerization process in solution. Subsequently, a process was described in which the anchor group of the dispersant was modified to carry the reactive groups; in the grafting stage, soluble group was then polymerized in a solution of this anchor precursor[80].

The method most frequently cited for attaching the copolymerizable group to the polymer chain is by the reaction between an unsaturated carboxylic acid, such as methacrylic acid, with the glycidyl groups in a random copolymer containing a small proportion of glycidyl methacrylate[7,24]. A typical example was a 97:3 (w/w) lauryl methacrylate/glycidyl methacrylate copolymer. This reaction goes smoothly and controllably in the presence of tertiary amine catalysts at temperatures between 100 and 150 °C. Provided polymerization inhibitors are added, there is a negligible loss of unsaturation by the polymerization of the acidic monomer. Other suitable pairs of reactants which have been used include maleic anhydride–hydroxy (meth)acrylate, methacrylic acid–allyl glycidyl ether and methacrylic anhydride–hydroxy (meth)acrylate[79].

The performance of a soluble precursor of this type in a dispersion polymerization is very dependent on the number of copolymerizable groups attached to each polymer molecule. However, because of the succession of random processes which make up the whole preparation, this factor is difficult to control with any degree of precision.

If the dispersant is formulated with the aim of having no more than one copolymerizable grafting site per molecule, then inevitably there will be a large proportion of molecules of soluble polymer with no grafting sites at all. Calculations have shown (Section 3.7.4, below) that the ungrafted soluble species formed could constitute approximately 30% of the total composition in the case of the copolymerization of two methacrylate monomers of equal reactivity. On the other hand, if the dispersant is formulated in order to avoid the presence of ungrafted soluble component, then statistically it must contain a substantial fraction of polymer with at least four grafting sites per molecule. This number of sites would lead to overgrafting, which could result in gelation during the preparation of the latex or graft copolymer (if prepared separately) and ultimately to flocculation of the latex by the polyfunctional graft copolymer. When this method has been employed for the manufacture of dispersants, the calculated number-average functionality of copolymerizable double bonds per soluble molecule used has been near to unity. In practice, the presence of a large proportion of ungrafted soluble component is tolerated rather than risk the flocculation of the latex.

Analysis of poly(lauryl methacrylate-g-methyl methacrylate) copolymers prepared by this method[81] has enabled the theoretical calculations to be tested experimentally. A 97:3 (w/w) copolymer of lauryl methacrylate and glycidyl methacrylate, of molecular weight, $\overline{M}_n$, approximately 15,000, was prepared in a mixture of ester solvents. The glycidyl groups were reacted with methacrylic acid to give an average functionality of 1·5 polymerizable methacrylate groups on each polymer molecule. Methyl methacrylate and an azo initiator were added and polymerization continued to yield a polymer in which the grafted poly(methyl methacrylate) component was equal in molecular weight to the initial poly(lauryl methacrylate) copolymer. The weights of the two components in the final composition were the same. The copolymer formed was then fractionated by means of alternate precipitation in acetone and petroleum ether in order to remove the ungrafted components. A typical analysis showed that the material contained 21% acetone-insoluble polymer, 52% petrol-insoluble material and 27% graft copolymer, figures in good agreement with the calculated values.

Within the limits of the types of formulation which have already been discussed above, it has been found that by decreasing the functionality of the dispersant precursor in a latex preparation, a coarser latex is produced; correspondingly an increase in functionality gives a finer latex[82]. The magnitude of the effect is illustrated in Table 3.22 where the particle size distributions are

Table 3.22. Particle size distribution of poly(methyl methacrylate) latexes prepared with poly(lauryl methacrylate) dispersant precursor[82].

| Average number of copolymerizable groups per soluble molecule | Percentage of particles of given diameter (μm) | | | | | | | | |
|---|---|---|---|---|---|---|---|---|---|
| | 0.50 | 0.45 | 0.40 | 0.35 | 0.30 | 0.25 | 0.20 | 0.15 | 0.10 | 0.05 |
| 0.45 | 6.1 | 6.7 | 8.7 | 12.1 | 20.8 | 15.4 | 13.4 | 8.1 | 8.7 | — |
| 0.68 | 0.7 | 2.1 | 4.1 | 9.6 | 19.3 | 20.6 | 16.4 | 14.4 | 12.3 | 2.1 |
| 0.91 | — | — | 2.5 | 2.5 | 10.8 | 22.8 | 22.8 | 15.8 | 17.3 | 5.1 |
| 1.14 | — | — | 3.1 | 3.9 | 7.8 | 14.7 | 30.2 | 18.6 | 16.3 | 5.4 |
| 1.70 | Very fine seed; subsequently flocculated | | | | | | | | |

given for a series of poly(methyl methacrylate) latex preparations in aliphatic hydrocarbon. The graft dispersant was formed *in situ* from a lauryl methacrylate/glycidyl methacrylate copolymer which had been reacted with methacrylic acid. The same copolymer was used for each latex, only the extent of the reaction with methacrylic acid being varied. It can be seen that there was a systematic reduction in the peak particle size as the number of grafting sites on the dispersant precursor was increased.

The difficulties encountered in making pure graft copolymer by this route, when the grafting occurs on the preformed soluble component of the dispersant, restricted its use mainly to *in situ* grafting in dispersion polymerization. However, this method of latex preparation has much to commend it since the dispersant precursor can be 'tailored' precisely to suit a particular polymerizing system. If relatively pure graft copolymer is required from this type of synthesis, the preferred route is by grafting on to the modified anchor component rather than on to the soluble polymer[80].

It is perhaps not generally realized that, in non-aqueous dispersions, polyfunctional graft copolymers of the form shown in Figure 3.11 may act under some conditions as weak flocculants for the disperse phase. In small amounts, the polyfunctional graft causes a gel structure to be formed which may stabilize

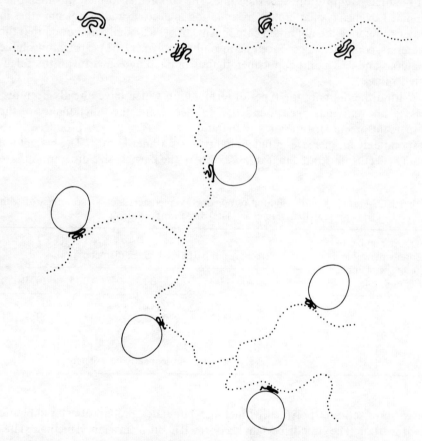

Figure 3.11. Polyfunctional graft dispersant and flocculation behaviour (schematic): (·····) soluble backbone; (———) insoluble anchor

the system against gravitational settlement. In larger amounts, or more especially when present during the dispersion polymerization, gross flocculation is observed, leading eventually to the complete loss of dispersion stability. It follows, therefore, that in the synthesis of graft copolymers for non-aqueous dispersion polymerization, especial care should be taken to avoid the formation of graft copolymers which contain a multiplicity of anchor groups. It may, however, sometimes be advantageous to add such a graft to a polymer or other particulate suspension, or to an emulsion, to prevent settlement on storage. A thorough study of the relationship between the graft copolymer structure and its effectiveness as a flocculant was carried out by Backhouse[83]. The graft

copolymers used were prepared by the method of grafting through randomly-distributed copolymerizable groups previously described. A series of copolymers, based on lauryl methacrylate and glycidyl methacrylate of varying molecular weights, was first prepared. The glycidyl groups were then reacted with methacrylic acid to give a range of precursors of different functionality with respect to the grafting sites. Finally, these precursors were copolymerized with methyl methacrylate, or with mixtures of methyl methacrylate and methacrylic acid to form grafted anchor moieties. The molecular weight of these graft chains was also varied. The resulting comprehensive series of graft copolymers enabled the following effects to be studied:
  (i) The number of anchor groups per soluble backbone,
 (ii) The molecular weight of the soluble backbone and the spacing of the anchor groups along it,
(iii) The molecular weight of the anchor groups,
(iv) The chemical composition of the anchor group.

Each polyfunctional graft copolymer was evaluated by assessing the nature of the gel structure formed when it was added (at a level of 1% by weight) to a 50% dispersion of poly(methyl methacrylate), of average particle diameter 0·2 µm, in a blend of aliphatic hydrocarbon, nonanol and dialkyl phthalate. The polymer dispersion itself was stabilized by the *in situ* formation of dispersant using a precursor based on poly(lauryl methacrylate) with pendant methacrylate groups. The results, using anchor chains with the composition 98:2 (w/w) methyl methacrylate/methacrylic acid, are summarized in Table 3.23.

Table 3.23. Effect of variations in backbone M.Wt., side chain M.Wt. and number of side chains per backbone on graft thickening properties[83].

| $\overline{M}_v$ side chains | | | | | | Gelling characteristics |
|---|---|---|---|---|---|---|
| Anchor chains per backbone | | | Anchor chains per backbone | | | |
| 3 | 4 | 5 | 3 | 4 | 5 | |
| $\overline{M}_v$ backbone = 60,000 | | | $\overline{M}_v$ backbone = 95,000 | | | |
| 10,000[a] | 10,500 | | 11,100 | 9,500 | 9,500 | No effect |
| 12,000 | | | 12,700 | | | |
| | | | 14,300 | | | |
| | | | 15,800 | | | |
| 13,000 | | | 19,000 | 11,900 | | Weak |
| 14,000 | | | | 13,100 | | |
| 16,000 | | | | | | Strong |
| 18,000 | | | | | | |
| | 11,200 | 10,400 | | 14,200 | 10,500 | |
| | 12,000 | 11,000 | | | 11,400 | |
| | 13,500 | 11,700 | | | 13,300 | Very strong |
| | 15,000 | 12,400 | | | | |
| | | 13,000 | | | | |

[a] Each column entry represents a separate preparation.

Although in an exercise of this kind it was virtually impossible to study one variable in complete isolation, a number of distinct trends could be distinguished in the results. The strength of the gel formed rose steadily with an increase in the number of anchor groups per soluble backbone. Increasing the molecular weight of the anchor group had the same effect. Increasing the molecular weight of the soluble backbone, did not, as one might have expected, increase the gelling tendency. This may perhaps be accounted for by the fact that for a given number of grafting sites per backbone, the molar concentration of copolymerizable groups is greater in the lower molecular weight backbone which in turn should lead to a higher molar concentration of polyfunctional graft in the final product. In a separate series of experiments, it was also shown that both the rate and strength of gel formation increased as the methacrylic acid content of the anchor groups was raised from zero up to 4% by weight.

For such a graft flocculant to be effective, it was preferable that it formed a clear solution in the solvent medium of the dispersion. Under these conditions, irreversible aggregation of the anchor moieties is presumed not to occur and there is an equilibrium between graft copolymer micelles, free graft molecules and graft flocculant adsorbed on the particle surfaces. In the absence of particles, no gel structure was observed at low concentrations of graft flocculants. The structure is therefore most probably associated with the bridging of particles by the polyfunctional graft and closely parallels the flocculation of aqueous dispersions by high molecular weight polymers[84].

It is puzzling at first sight to understand how an apparently stable non-aqueous dispersion can be so readily flocculated in this manner. The explanation probably lies in the wide range of surface coverage of stabilizing groups which can be tolerated without any apparent change in colloidal stability (see Section 3.4). At the polymerization temperature used (80 °C), the equilibrium concentration of dispersant on the particle surface is likely to be lower than the equilibrium concentration at room temperature. Space is therefore available for further adsorption of graft copolymer after the preparation cools down. An alternative explanation may be that at least some of the dispersant adsorbed is in equilibrium with dispersant or flocculant in solution and that the extent of bridging observed will depend on the actual position of this equilibrium.

### 3.7.4. Formation of graft copolymer from soluble polymers carrying terminal polymerizable groups

An important development of the method of grafting through randomly distributed copolymerizable groups attached to the soluble component of the dispersant (Section 3.7.3) was the synthesis of a polymer in which every molecule carried a single terminal unsaturated group. In this way, the possibility of either ungrafted soluble component or graft dispersant with a multiplicity of anchor groups being formed was eliminated. Waite and Thompson[24,85] devised an elegant method which made use of the glycidyl/carboxylic acid reaction route for the attachment of the terminal unsaturated group but which was also developed from an existing process employing 4,4'-azobis-(4-cyanopentanoic acid) as a free radical initiator for the production of block co-polymers[86]. When it decomposes, this initiator yields two radicals each carrying a carboxyl group. The problem was to produce an addition polymer which contained just one terminal carboxylic acid group on each polymer molecule. When the initiator is used in a polymerization in which the termination step is not controlled, a mixture of species is obtained. Termination by combination produces polymer chains with carboxyl groups at both ends of the molecule. Termination by disproportionation, on the other hand, produces two polymeric species—polymer with a carboxyl group at one end and a crotonate double bond at the other is formed at the same time as the preferred structure consisting of a polymer molecule with a carboxyl group at one end and a saturated structure at the other. In order to increase the yield of the species required at the expense of the other two, the concept of a 'matched' transfer agent was introduced, i.e. an efficient transfer agent which itself contains a carboxyl group. Thioglycollic acid proved to be particularly effective for producing the desired species in high yield. The main steps of the reaction sequence are set out below:

$$CH_3-\underset{\underset{\underset{COOH}{|}}{\underset{CH_2}{|}}}{\overset{\overset{CN}{|}}{C}}-N=N-\underset{\underset{\underset{COOH}{|}}{\underset{CH_2}{|}}}{\overset{\overset{CN}{|}}{C}}-CH_3 \rightarrow CH_3-\underset{\underset{\underset{COOH}{|}}{\underset{CH_2}{|}}}{\overset{\overset{CN}{|}}{C}}\cdot + N_2$$

$$CH_3-\underset{\underset{\underset{COOH}{|}}{\underset{CH_2}{|}}}{\overset{\overset{CN}{|}}{C}}\cdot + n(M) \rightarrow CH_3-\underset{\underset{\underset{COOH}{|}}{\underset{CH_2}{|}}}{\overset{\overset{CN}{|}}{C}}-(M)_n^{\cdot}$$

$$CH_3-\underset{\underset{\underset{COOH}{|}}{\underset{CH_2}{|}}}{\overset{\overset{CN}{|}}{C}}-(M)_n^{\cdot} + HSCH_2COOH \rightarrow CH_3-\underset{\underset{\underset{COOH}{|}}{\underset{CH_2}{|}}}{\overset{\overset{CN}{|}}{C}}-(M)_n-H + \dot{S}CH_2COOH$$

$$HOOCCH_2\dot{S} + m(M) \rightarrow HOOCH_2S-(M)_m^{\cdot}$$

Although undesirable difunctional molecules can still be formed by radical combination, this was minimized by a correct choice of reaction conditions and by passing the final product down an alumina column to remove low molecular weight, acidic by-products. Subsequent reaction of the carboxyl-terminated polymers with glycidyl methacrylate produced a monofunctional dispersant precursor substantially free from unwanted material. Precursors of this type, based on poly(lauryl methacrylate), when used in the dispersion polymerization of methyl methacrylate or vinyl acetate, were approximately three times more efficient than the corresponding precursors with randomly distributed co-polymerizable groups, as indicated by the amount required to produce latexes of similar particle size[24].

A further method for the production of a soluble component of a dispersant carrying a terminal unsaturated group also utilized a functionally active chain transfer agent[87]. In this process, the chain transfer agent used carried a hydroxyl or an amine group, e.g. mercapto-ethanol or 2-aminoethyl mercaptan. The chain transfer agent was used in amounts which substantially precluded kinetic chain termination, thereby ensuring, as far as possible, the presence of a terminal hydroxyl or amine group on each polymer molecule. Following the addition polymerization step, a di-isocyanate, such as 2,4-tolylene di-isocyanate, was reacted with the terminal group so as to produce a terminal isocyanate group, which in turn was reacted with hydroxymethyl methacrylate. The resulting terminally unsaturated polymer was used as the soluble component of a graft copolymer in the preparation of a 'self-stable organosol' of the type referred to in Section 3.7.2.

An alternative route to the preparation of a polymer with a single terminal carboxyl group is by the use of the condensation polymerization of hydroxy acids. The example most frequently described has been the polymerization of 12-hydroxystearic acid[24]. The pure acid yields a polymer with a hydroxyl group at one end of the chain and a carboxyl group at the other. The more commonly used commercial grade of 12-hydroxystearic acid contains between 8% and 15% of stearic and palmitic acids, which act as chain terminators and limit the molecular weight achievable to the range 1500–1800. Approximately one in five molecules produced under these conditions has a terminal hydroxyl group. This polymer has proved an ideal soluble component in the preparation of graft copolymers for non-aqueous dispersion polymerization in a wide range of aliphatic, aromatic and even ester solvents. By co-condensing other hydroxy acids or amino acids, it is possible to alter the solubility of the polymer to make it suitable for use with more polar diluents whilst still retaining the single terminal carboxyl group[88]. The poly(hydroxy acids) are readily converted to graft copolymers by esterifying the terminal carboxyl group with glycidyl methacrylate, followed by copolymerization of the resulting macromonomer with acrylic or vinyl monomers to form the anchor group. This type of macromonomer has been used successfully as a dispersant precursor in the preparation of poly(methyl methacrylate) dispersions in aliphatic hydrocarbon. However, since the macromonomer can readily be converted to almost pure graft copolymer of controlled composition by solution polymerization, this route has been preferred almost exclusively for the preparation of dispersants. The polyester macromonomer formed is of a relatively low molecular weight when compared with the poly(lauryl methacrylate) precursor made by conventional solution polymerization techniques in which each soluble molecule was approximately half the molecular weight of the final graft copolymer. As a result, graft copolymer molecules incorporating the polyester, unless of very low overall molecular weight, carry several polyester moieties on each molecule, so forming a 'comb-like' structure[24,89] illustrated in Figure 3.12.

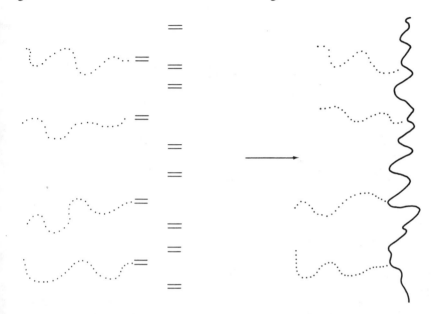

Figure 3.12. Formation of a 'comb' graft copolymer (schematic): (····) soluble component; (——) insoluble component; (══) anchor monomer

It is to some degree self-evident, that the larger the number of polyester side-chains in the average molecule on which the graft copolymer formulation is based, the smaller will be the content of anchor polymer carrying no soluble polyester groups in the final product. A mathematical treatment by Cornish and Lilley[90] of this type of copolymerization, following the basic method of Stockmayer[91], has confirmed this supposition and given some quantitative basis to the formulation of comb graft copolymers.

They calculated the weight fraction of anchor homopolymer produced during the formation of a methacrylate graft copolymer of defined number average degree of polymerization ($\bar{P}_n$) and molecular weight distribution. In the calculation, the exclusive termination of radicals by disproportionation was assumed and, by analogy with the published values for the reactivity ratios of higher alkyl methacrylates, it was also assumed that in the copolymerization of methyl methacrylate (A) and the soluble polymer molecule bearing a terminal methacrylate group (B), the reactivity ratio, $r_1$, was 1. The average number of branch points (B units), $q$, in the graft copolymer was defined as follows:

$q$ = total number of B units/total number of polymer molecules

Thus $q$ was the average number of branches on an average molecule containing just $\bar{P}_n$ units. The results for the calculation of homopolymer A content are given in Table 3.24, for the two cases where the molecular weight of the B unit, $M_B$,

Table 3.24. Weight fraction of anchor homopolymer A formed in copolymerization of methacrylate monomers[90].

| | $M_B = 3M_A$ | | | $M_B = 12M_A$ | |
|---|---|---|---|---|---|
| $q$ | $\bar{P}_n = 30$ | $\bar{P}_n = 100$ | $q$ | $\bar{P}_n = 30$ | $\bar{P}_n = 100$ |
| 1 | 0.234 | 0.245 | 1 | 0.183 | 0.225 |
| 2 | 0.096 | 0.106 | 2 | 0.063 | 0.090 |
| 3 | 0.049 | 0.058 | 3 | 0.028 | 0.046 |
| 4 | 0.029 | 0.036 | 4 | 0.015 | 0.027 |
| 6 | 0.012 | 0.017 | 6 | 0.005 | 0.012 |
| 8 | 0.006 | 0.010 | 8 | 0.002 | 0.006 |
| 10 | 0.004 | 0.006 | 10 | 0.001 | 0.004 |

was $3M_A$ and $12M_A$ respectively. The latter case corresponds roughly to a tetramer of hydroxystearic acid. Values calculated for $\bar{P}_n$ of 30 and 100 are listed for a range of values of $q$ from 1 to 10. It is apparent that at low values of $q$, substantial quantities of ungrafted anchor polymer will be formed, but at values of $q$ greater than 4, the amounts of anchor polymer become very small indeed.

The weight fraction of branched (graft) polymer formed was also calculated for the case of a copolymer with a weight ratio of side-chain polymer (B unit) to backbone polymer (A unit) of 1:1 and a number average degree of polymerization ($\bar{P}_n$) of 50. Results calculated for the weight fraction of copolymer with 5, 10 and 20 branches for average values of $q$, in the range 1 to 10, are given in Table 3.25.

Table 3.25. Weight fraction of branched copolymers formed in copolymerization of methacrylate monomers[90].

| $q$ | 5 or more branches | 10 or more branches | 20 or more branches |
|---|---|---|---|
| 1 | 0.04 | 0.02 | — |
| 2 | 0.26 | 0.06 | — |
| 3 | 0.43 | 0.14 | 0.01 |
| 4 | 0.58 | 0.23 | 0.04 |
| 6 | 0.73 | 0.48 | 0.14 |
| 8 | 0.74 | 0.60 | 0.19 |
| 10 | 0.74 | 0.61 | 0.20 |

Since in most preparations of the side-chain macromonomer, there is always a finite chance of forming a small proportion of material which is difunctional with respect to unsaturated groups, it is advisable to restrict the amount of highly-branched polymer in the composition. There are additional disadvantages in the use of highly-branched, higher molecular weight graft copolymer of the comb-type. The actual molar concentration of dispersant is less and there is also a greater chance of forming monomolecular 'self-stable' micelles which may show little tendency to adsorb on the surface of polymer particles. The calculations indicate that for values of $q$ up to 4, the amount of such highly-branched graft copolymer formed remains low. In general therefore, although no two preparative methods are likely to give the same results, values of $q$ in the range 3 to 5 should produce the most satisfactory graft copolymers for dispersion polymerization.

The graft copolymer dispersant is most conveniently prepared by feeding a mixture of the soluble component macromonomer, the anchor monomers and the initiator into refluxing ester solvents over a period of between 1 and 3 hours and maintaining the temperature of the reaction mixture until polymerization is complete. Making the additions in this way assists the formation of a copolymer with a uniform rather than a skewed composition so that the yield of effective dispersant is optimized. This feed method of polymerization also gives products of relatively low molecular weight. Molecular weights of the comb dispersants which have been prepared are typically in the range $\bar{M}_w = 20,000-35,000$; $\bar{M}_n = 6000-15,000$ (by gel permeation chromatography[92]). Comb graft copolymers based on poly(hydroxystearic acid) and prepared in this way contain a high proportion of active dispersant, usually of the order of at least 80%, as indicated by the amount adsorbed from solution in a dispersion polymerization[10,18]. Gel permeation chromatography both of the comb dispersant and of the supernatant liquor obtained from centrifuging the latex, has suggested that it is the higher molecular weight fraction of the dispersant which is preferentially adsorbed by the polymer particles[18]. The fraction of polymer not adsorbed, when it is used as a replacement for the original dispersant in a latex preparation, produces a much coarser particle size latex than the original material, demonstrating that it is a less efficient dispersant[93].

The relative proportions of the anchor and the soluble components in a comb graft copolymer of this type are not critical and good results have been reported with ratios between 30:70 and 70:30[94]. For most purposes, a ratio of 1:1 is satisfactory, particularly with components which have little specific adsorbing potential or ability to form hydrogen bonds, e.g. anchor groups based on methyl methacrylate or vinyl acetate. If acidic, amido or nitrile-containing monomers form a proportion of the anchor monomer mixture, it may be necessary to increase the proportion of the soluble component to maintain the equilibrium between the micellar aggregates formed and the single molecules of graft copolymer in solution (see Section 3.6).

The comb graft type of copolymer is capable of extensive variation in its fine structure to suit the requirements of a particular dispersion polymerization. Reference has already been made to the possible variations in the composition of the soluble component. The anchor group is capable of even more variation by an appropriate selection of the monomers from which it is to be formed in order to suit the medium or the polymer particle with which it is to be used. A further possibility is the incorporation of groups which are capable of co-polymerizing or otherwise reacting with the latex monomer, with the object of promoting stability in the latex formed, against the subsequent addition of strong solvents used in the formulation of surface coatings[32]. This may be achieved for example, by copolymerizing a minor proportion of glycidyl methacrylate with the anchor monomers, followed by the familiar reaction of a proportion of these glycidyl groups with methacrylic acid. Alternatively the methacrylic acid may be incorporated in the latex particle and the reaction between the acid and the glycidyl groups may be followed by the rate of disappearance of the acid (see Section 3.7.3). Another method of achieving stability to active solvents, while simultaneously improving adhesion to metallic and organic substrates, utilizes the reaction between glycidyl groups in the comb stabilizer and a combination of an unsaturated carboxylic acid and an alkyleneimine in the latex monomers[95].

Comb stabilizers can also be modified by the incorporation of polar groups in order to improve their performance as dispersants for pigments and other fine particles. Glycidyl groups in the anchor component have been reacted with aromatic or aliphatic acids[96] and strongly polar tertiary amino groups, introduced into the anchor group by the copolymerization of t-amino-methacrylates, have been reacted with propane-sultone to yield sulpho-betaines, or with benzyl halides, to generate quaternary ammonium pendant groups[93].

Comb dispersants are also excellent emulsifiers provided that the anchor component is soluble in the disperse phase (see Section 3.5.2). For the emulsification of water or hydroxy-containing compounds, 'double comb' dispersants are particularly advantageous[8,97]. The poly(hydroxystearic acid)/glycidyl methacrylate adduct is copolymerized with the acrylate of methoxy poly-(ethylene oxide) and other acrylic monomers to give dispersants, which may be schematically represented as in Figure 3.13. The double comb molecule can be visualized as a series of AB block copolymer molecules, joined through their mid-points. The cumulative anchoring effect appears to give a more stable emulsion than the individual block copolymer units. It is also claimed that the powders which result from the drying of particulate dispersions stabilized by double comb dispersants are re-dispersible in either the original medium or water[98].

It has already been mentioned that the solubility characteristics of a soluble component based on polyester may be varied by appropriate compositional changes. It is also possible to introduce reactive groups into the soluble component in the same fashion. Co-condensates of dimethylol propionic acid with hydroxystearic acid contain free hydroxyl groups which may be reacted with melamine formaldehyde resins in coating compositions or may be further esterified with unsaturated fatty acids, after the formation of the graft copolymer, to give air-drying and cross-linking resins[99].

The preparation has also been described[46] of microparticle dispersions which are derived from comb graft copolymers by controlled precipitation procedures.

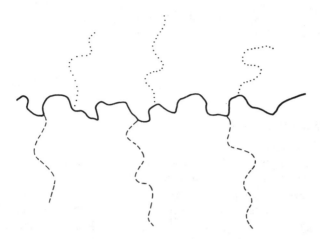

Figure 3.13. 'Double-comb' graft copolymer: (· · · ·) oil-soluble component; (– – – –) water-soluble component; (———) insoluble component

These may be considered as spherical micellar aggregates of comb dispersant molecules which are formed well above the environmental glass transition point (or melt temperature) of the graft copolymer (see Section 3.6) but which are below the transition point when cooled to room temperature. In the latter conditions, free-flowing dispersions of low viscosity and up to 50% weight content of graft copolymer may be obtained. Particles with a core diameter between 2·5 and 25 nm were claimed, the core being composed of the insoluble components of the graft copolymers. At these particle diameters, the volume of the solvated sheath of stabilizing polymer is of the same order as the core volume and has a major effect on the rheology of the dispersion[20]. The related 'self-stable' organosols, which are also formed by the addition of a non-solvent to a solution of an amphipathic copolymer, have been described elsewhere (Sections 3.7.2 and 5.1).

### 3.8. THE SYNTHESIS OF BLOCK COPOLYMERS AS DISPERSANTS

Since the preparation and properties of block copolymers have been the subject of extensive recent reviews[100,101], the scope of this section is limited to their use as dispersants in dispersion polymerization (see Table 3.1 above, for examples). Some of the possible dispositions of the anchor and soluble groups of amphipathic block copolymers are illustrated in Figure 3.7, types f–i (Section 3.7).

Although methods for the preparation of block copolymers by free radical polymerization are well-established[55,100], these have not been much used in preparing dispersants. An early patent[102] described the preparation of styrene/methacrylic acid and lauryl methacrylate/methacrylic acid block copolymers for this purpose by the mastication of polymer in the presence of another monomer. More recently, the use of polymeric initiators for the formation of graft copolymer dispersant has been disclosed[73] (Section 3.7.2). However, although both free radical and condensation reactions (Section 3.8.1) have been used to some extent for preparing block copolymers as dispersants, in practice more use has been made of ionic polymerization methods (Section 3.8.2) for this purpose.

### 3.8.1. Formation by coupling with condensation reactions

Relatively few examples of the use of this type of block copolymer in dispersion polymerization have been reported. The majority of commercial nonionic surfactants, such as condensates of fatty alcohols with poly(ethylene oxide), have anchor groups which are too weakly associated with polymer surfaces for their direct use as dispersants in dispersion polymerization. However, by increasing the molecular weight and hence reducing the solubility of the anchor group, while at the same time making commensurate changes in the soluble group, it is possible to produce satisfactory dispersants of this general type.

Examples have included the reaction products of poly(12-hydroxystearic acid) with commercial epoxy resins, which are the products of the reaction between diphenylolpropane and epichlorhydrin[103]. The two components were linked by the base-catalysed reaction of the terminal carboxyl group of the polyester with the terminal glycidyl ether group of the epoxy resin. The molecular weight of each polymeric component was varied over the range 500 to 1500 and the resulting block copolymers were shown to act as effective dispersants in the dispersion polymerization of acrylic and vinyl monomers. As a variant of this method of synthesis, addition polymers with terminal carboxyl groups were reacted in a similar manner with epoxy resins[85].

### 3.8.2. Formation by ionically-initiated polymerization

The anionic polymerization of monomers in hydrocarbon media is an ideal method for the preparation of block copolymers. By utilizing the 'living' polymer technique[100], it is possible to prepare the soluble component of the block copolymer at the molecular weight desired and by the subsequent growth of the anchor component, to produce a copolymer of the required anchor:soluble group balance.

An early publication[102] described the preparation of a poly(vinyltoluene-b-methyl methacrylate) dispersant using a sodium naphthalene complex which would probably have produced an A–B–A block copolymer. More recently, polymers of isoprene, butadiene and t-butyl styrene prepared with lithium butyl as initiator, have been employed as soluble components linked with styrene or methyl methacrylate as anchor groups[104]. AB blocks of this type proved effective dispersants in both anionic and in free radical dispersion polymerization. Soluble groups of poly(t-butyl styrene) or poly(isoprene) with molecular weights in the range 5000 to 10,000, combined with anchor groups of poly(styrene) (molecular weight 10,000–20,000) were found to give a suitable anchor/soluble group balance for dispersion polymerization in aliphatic hydrocarbon at room temperature.

Other workers have synthesized a wide range of t-butyl styrene/styrene block copolymers, again using lithium butyl initiation[105]. In this case, the ratio of t-butyl styrene to styrene was varied between 1:3 and 9:1 at an overall molecular weight of the block copolymer in the range 80,000 to 100,000. When styrene was polymerized using lithium butyl in the presence of a block copolymer in hexane, a stable dispersion of poly(styrene) was produced. Preparations of polymer dispersions were also described in which a second monomer, such as vinyl pyridine or ethylene oxide, was added to the polymer dispersion after the completion of the styrene polymerization. In this way, particles were formed which were themselves composed of block copolymers.

A further method of carrying out the dispersion polymerization process is to prepare the 'living' soluble polymer block first, to which the disperse phase monomer is added to continue growth from the 'living' end[104]. When the insoluble portion reaches the size at which the critical micelle concentration is exceeded, aggregates of block copolymer are formed. Dispersion polymerization then proceeds within the aggregate until eventually a particulate dispersion is formed. Additional block copolymer dispersant may be required if the surface area of the polymer particle formed exceeds the surface area which can be stabilized by the original soluble component. Using this method, the polymer particles formed are entirely composed of block copolymer. If desired a third monomer may be added so that the final product is a dispersion of an A–B–C block copolymer. Clearly a wide variety of compositions can be synthesized by this general method.

The synthesis of block copolymers with Ziegler–Natta catalysts has recently been reviewed[100] but there are only a few examples of their deliberate use to stabilize polymer dispersions. One example is in the preparation of colloidal dispersions of poly(propylene) using a titanium/aluminium catalyst[106] (see Section 5.4.1). Titanium tetrachloride was first reduced with an aluminium alkyl to produce an active catalyst containing titanium trichloride. A colloidal dispersion of this catalyst was then prepared by first suspending it in hydrocarbon diluent, adding a dialkyl aluminium halide and following this with the addition of an $\alpha$-olefin containing at least six carbon atoms, typically octene-1 or hexadecene-1. The very fine dispersion of catalyst particles produced was presumably stabilized by chains of poly($\alpha$-olefin) attached to the particles. This catalyst dispersion was then used to polymerize propylene to give a sub-micron dispersion of predominantly crystalline polymer.

Under these conditions, it is probable that the poly($\alpha$-olefin) soluble component formed a block copolymer with the poly(propylene). However, the possibility also exists that the poly(propylene) could grow quite independently from other activated sites on the catalyst surface eventually engulfing the poly($\alpha$-olefin) present. The polymer content of the dispersion appears to have been limited in practice to a value of 30%, a concentration at which it was to some extent flocculated.

Ethylene/propylene copolymers prepared using homogeneous Ziegler–Natta catalysts, e.g. vanadium chloride–aluminium alkyl complexes, yield block copolymers in which blocks of poly(propylene) and poly(ethylene) are randomly distributed along the polymer molecule[107]. Such polymers have been found to act as dispersion stabilizers for poly(ethylene) in aliphatic hydrocarbon[108]. An ethylene/propylene copolymer (21 mol.-% propylene) was prepared in pentane at $-47\,°C$ with a vanadium chloride/aluminium alkyl catalyst in the presence of hydrogen as transfer agent to limit the molecular weight. The resulting polymer was isolated and dissolved in hexane at 37 °C. A vanadium chloride/diethyl aluminium chloride catalyst was added and ethylene gas passed through. A sub-micron dispersion of poly(ethylene), free from flocculation, was obtained in this way.

Of the ionic methods described here, the anionic mode of polymerization is a particularly useful method for the preparation of block copolymer dispersants with well-defined structures. It is clearly capable of application to a wider range of monomers for dispersant preparation than has been reported so far.

## 3.9. SELECTED RECIPES FOR THE PREPARATION OF GRAFT COPOLYMER DISPERSANTS

Detailed formulations for the preparation of graft copolymer dispersants have up to now appeared only in the patent literature. For this reason, a selection of well-tried recipes is given here to enable the preparation of non-aqueous dispersants to be carried out on the laboratory scale (see Section 5.7 for dispersion polymerization recipes).

A few general notes on the use of these dispersants may be helpful at this point. These are aimed at overcoming the type of problem which is most frequently encountered in practice.

(1) The monomers employed in a dispersion polymerization should be free from dissolved polymer; even small traces of polymer can result in unwanted precipitation on dilution with the continuous phase, either when making up the initial charge or more probably during the monomer-feed stage of a dispersion polymerization.

(2) The particle size of a latex can be controlled by a suitable adjustment of the precipitation conditions at the start of a dispersion polymerization. The onset of precipitation is influenced by:

(i) The concentration of dispersant.
(ii) The concentration of monomer. Monomer usually behaves as a strong solvent for the polymer. Increasing the monomer concentration coarsens the latex.
(iii) The solubility of the precipitating polymer. Inclusion of a small quantity of a comonomer which lowers the solubility of the polymer, reduces the particle

size. This approach can lead to too fine a seed which then flocculates to a stable, coarser dispersion or to coarse agglomerates.
(iv) The anchor/soluble balance of the dispersant (Section 3.6).
(v) The solvency of the dispersion medium for the disperse polymer.

(3) Sometimes, the first attempts at a dispersion polymerization result in partial flocculation associated with sudden coarsening of particle size or even in actual solidification together with the evolution of heat. These effects are usually observed beyond 50% conversion of monomer when the latex formed has already reached a polymer concentration of 30–40%. The flocculation or coarsening effect often occurs when the disperse polymer is in a soft and semi-liquid state under the conditions of the dispersion polymerization and hence readily coalesces. Conversely, solidification is normally observed with hard polymers. Both phenomena are the result of a shortage of dispersant (or inadequate surface coverage by the dispersant used). The remedy can often be sought in the addition of further amounts of dispersant. However, this may not necessarily be a complete solution to the problem. It often happens that shortly before the sudden onset of stability failure, a fresh nucleation of particles has taken place due to either a temporary excess of dispersant or to a drop in the solvency of the medium (perhaps caused by monomer consumption). Growth of the new crop of particles quickly leads to the rapid consumption of all the available dispersant. In such circumstances, the cure may often be sought in an actual reduction in the amount of dispersant used.

### 3.9.1. Preparation of graft dispersant precursor based on poly(lauryl methacrylate)

(See Section 5.7.1 for example of its use; see also Section 5.7.3 for the preparation of a polyester stabilizer precursor.)

*3.9.1.1. Reactants*

*Initial charge* (wt-%): Petroleum (boiling range 100–120 °C), 36·54; petroleum (boiling range 70–90 °C), 24·25.

*Monomer feed* (wt-%): Lauryl methacrylate, 32·18; glycidyl methacrylate, 1·00; azo-di-isobutyronitrile, 0·43; ethyl acetate, 5·00.

*Ring-opening stage* (wt-%): Methacrylic acid, 0·48; hydroquinone, 0·02; 'Armeen' DMCD, ($N,N$-dimethyl laurylamine) 0·10.

*3.9.1.2. Procedure*

A copolymer of lauryl methacrylate and glycidyl methacrylate (97:3, w/w) is prepared in the hydrocarbon mixture. When conversion is complete, the lower boiling hydrocarbon is distilled off and replaced by the higher boiling fraction to raise the reflux temperature for the epoxide–acid reaction of the next stage. A proportion of the glycidyl groups is then esterified with methacrylic acid using a tertiary amine ('Armeen' DMCD) catalyst.

A reaction vessel of either glass or stainless steel can be used, fitted with a stirrer, reflux condenser and means for distillation. The polymerization is carried out at 90 °C and the temperature raised to 105–110 °C in the ring-opening stage. An accurate monomer metering pump is required together with a premix vessel for the monomer feed.

(a) *Preparation of copolymer*: The stirred hydrocarbon mixture is raised to reflux (about 90 °C) and the monomer feed metered in over 5 h and reaction continued for a further 2 h until the solids content reaches 30–32%. The reduced viscosity at this stage should be between 0·07 and 0·08 (0·5 g polymer/100 ml butyl acetate at 25 °C in an Ostwald 'A' viscometer). The low boiling petrol is removed by distillation and replaced with an equal weight of the high-boiling petrol.

(b) *Ring-opening stage*: Hydroquinone, methacrylic acid and 'Armeen' DMCD are added and the reaction mixture raised to reflux (105–110 °C) and maintained at this temperature (6–15 h) until the initial acid value (5·6–6·0 ml of 0·1 M alcoholic KOH/10 g sample) has fallen to a value of 2·0 ± 0·1 ml.

### 3.9.2. Preparation of preformed 'comb' graft copolymer dispersant from the copolymerization of the poly(hydroxystearic acid)/glycidyl methacrylate adduct with methyl methacrylate/methacrylic acid

(See Sections 5.7.2 and 5.7.4 for examples of its use.)

*3.9.2.1. Reactants*

*Preparation of poly(12-hydroxystearic acid)* (wt-%): 12-hydroxystearic acid (commercial grade), 90; xylene, 10.

*Reaction of poly(hydroxystearic acid) with glycidyl methacrylate* (wt-%): Poly(12-hydroxystearic acid), 47·02, and xylene 5·23 (solution); xylene, 41·76; hydroquinone, 0·05; 'Armeen' DMCD (*N,N*-dimethyl laurylamine), 0·09; glycidyl methacrylate, 5·85.

*Preparation of 'comb' graft copolymer dispersant* (wt-%): *Initial*—ethyl acetate, 17·64; butyl acetate, 8·82; *Feed*—methyl methacrylate, 24·02; Methacrylic acid, 0·49; poly(hydroxystearic acid)/GM adduct solution, 48·03; azo-di-isobutyronitrile, 1·00.

*3.9.2.2. Procedure*

(a) *Preparation of poly(12-hydroxystearic acid)*: The preparation can be carried out in the laboratory in a glass apparatus fitted with a Dean & Stark apparatus to remove the water of esterification. Xylene is added and heated to 60–70 °C. Hydroxystearic acid is added and the reaction mixture raised to reflux (195–198 °C) under nitrogen and the water of reaction removed by azeotropic distillation until the volume collected corresponds approximately to the theoretical amount and the acid value of the polyester is 33–35 mg KOH/g resin solids.

(b) *Reaction of poly(hydroxystearic acid) with glycidyl methacrylate*: The terminal carboxyl groups of the polyester are converted to methacrylate residues by reaction with a 1·5 molar excess of glycidyl methacrylate in the presence of a tertiary aliphatic amine. A glass reactor fitted with stirrer, reflux condenser and sample line is used. The reaction mixture is heated under reflux (c. 140 °C) until the acid content of a 10 g sample reaches 0·2 ml 0·1 M-KOH. The final non-volatile content is 51 %.

(c) *Preparation of 'comb' graft copolymer dispersant*: This is prepared by the copolymerization in solution of equal weights of the poly(12-hydroxystearic acid)/glycidyl methacrylate adduct and a 98:2 (w/w) mixture of methyl methacrylate and methacrylic acid. The initial charge is raised to reflux (82–84 °C) in a reaction vessel fitted with a stirrer, thermometer, reflux condenser and accurate monomer metering pump. The monomers are fed in at a constant rate over 3 h. After the addition is complete, the total charge is maintained at reflux (97–101 °C) for a further 2 h.

The final polymer solids content is 50 %. For use, this is diluted with petroleum (boiling range 100–120 °C) to a solids content of 33 %. In any event, it is preferable to dilute the 50 % solution obtained to a maximum of 40 % solids before it is cooled in order to prevent 'gelation' by micellar aggregation effects. The reduced viscosity of the polymer in butyl acetate (0·5 % by wt, 25 °C) is 0·10–0·12.

The product should be a clear solution of graft copolymer. Preferably, further dilution with low- or medium-boiling aliphatic hydrocarbons should not alter its clarity but at the most only a blue haze should develop. The diluted solution should be entirely free of actual precipitated insoluble polymer.

### 3.9.3. Preparation of dispersant precursor based on poly(lauryl methacrylate-co-methacrylic acid) reacted with allyl glycidyl ether

(See Sections 5.7.5 for example of its use.)

#### 3.9.3.1. Reactants

*Initial charge* (wt-%): white spirit, 33·15; hexane, 15·77.

*Monomer feed* (wt-%): Lauryl methacrylate, 33·15; methacrylic acid, 0·68; hexane, 16·55; azo-di-isobutyronitrile, 0·09.

*Reaction with allyl glycidyl ether* (wt-%): Allyl glycidyl ether, 0·60; 'Armeen' DMCD ($N,N$-dimethyl laurylamine), 0·01.

#### 3.9.3.2. Procedure

A glass reaction vessel, fitted with a stirrer, reflux condenser and means for distillation under water vacuum is used. A monomer metering pump and monomer premix vessel is also required.

The initial charge is heated to reflux (95–98 °C) and the monomer feed metered in over 2 h. Reflux is continued (c. 6 h) until the solids content reaches at least 30 %. Solvent is removed by distillation to raise the reflux temperature to 130 °C. The allyl glycidyl ether and 'Armeen' DMCD are added and reflux continued until the acid content is 80 % of the original value.

### 3.9.4. Preparation of block copolymer dispersant by anionic polymerization of t-butyl styrene and styrene[104]

#### 3.9.4.1. Reactants

All wt-%: Heptane, 61·60; tetrahydrofuran, 0·34; s-butyl lithium, 0·15 and hexane, 2·93 in 5% (w/w) solution; t-butyl styrene, 11·66; styrene, 23·32.

#### 3.9.4.2. Procedure

The reactor (Figure 3.14) is fitted with stirrer, inlet ports, thermometer and sampling line and connected to an inert gas supply. An external water-bath enables the reactor contents to be cooled or warmed as required.

Before use, the reactor is cleaned out with hot toluene and finally rinsed with hexane. The reactor is evacuated and refilled with argon or nitrogen—this procedure is repeated three times. For use, a slight excess pressure of inert gas is maintained in the reactor.

The heptane is passed through successive columns containing calcium alumino-silicate molecular sieve, type 5A, and BASF catalyst, R3-11, to remove water and oxygen respectively and then charged to the reactor. t-Butyl styrene is distilled from calcium hydride and copper wire under a reduced pressure of 3 mmHg. Inert gas is drawn through the boiling monomer. The distilled monomer is collected above the reactor in another vessel in which its volume can be measured.

s-Butyl lithium solution is added to the reactor by syringe. The external water-bath is filled with ice. The t-butyl styrene is charged to the reactor and the cooling controlled such that the exotherm of polymerization raises the internal temperature to 50 °C.

After 1 h, a sample is withdrawn for molecular weight characterization by gel permeation chromatography. Styrene, which has been distilled as the first monomer, but at a pressure of 30 mmHg, is then slowly charged to the reactor with external cooling, such that the internal temperature can be maintained at 50 °C. This addition is completed within 20 min. Polymerization is complete within a further 30 min.

Ethyl alcohol is added slowly by syringe until the red colour of the 'living' poly(styrene) is discharged.

### 3.10. REFERENCES

1. Napper, D. H. and Hunter, R. J. in *Surface Chemistry and Colloids*, Vol. 7 (Ed. Kerker, M.), Butterworths, London, 1972.
2. Koelmans, H. and Overbeek, J. Th. G., *Discuss. Faraday Soc.*, **18**, 52 (1954).
3. Osmond, D. W. J., ICI Paints Division, unpublished results, 1964.
4. Parfitt, G. D. (Ed.), *Dispersions of Powders in Liquids, with Special Reference to Pigments*, 2nd Edition, Applied Science Publishers, London, 1973.
5. Napper, D. H., *Trans. Faraday Soc.*, **64**, 1701 (1968).
6. Alexander, A. E. and Napper, D. H. in *Progress in Polymer Science*, Vol. 3 (Ed. Jenkins, A. D.), Pergamon Press, Oxford, 1971.
7. Osmond, D. W. J. and Walbridge, D. J., *J. Polym. Sci. C*, No. 30, 381 (1970).

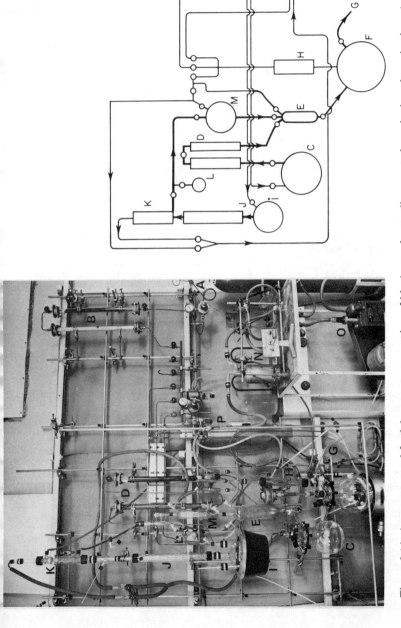

Figure 3.14. Apparatus used for laboratory preparation of block copolymer dispersants by anionic polymerization: A, inert gas supply; B, columns to remove oxygen and moisture; C, diluent reservoir; D, columns to remove oxygen and moisture; E, measuring vessel—diluent; F, reactor; G, sample line; H, condenser—reactor; I, monomer boiler; J, fractionating column; K, condenser—monomer; L, fore-cut collection; M, measuring vessel—monomer; N, cold-trap; O, vacuum pump; P, mercury lutes

8. Thompson, M. W., *I.U.P.A.C. Internat. Symp. on Macromols.*, *Aberdeen*, Abstracts p. 366, 1973.
9. Imperial Chemical Industries, *British Patent*, 1,108,261 (1968).
10. Walbridge, D. J. and Waters, J. A., *Discuss. Faraday Soc.*, **42**, 294 (1966).
11. Napper, D. H., *Proc. Roy. Aust. Chem. Inst.*, **36**, 137 (1969); **38**, 327 (1971).
12. Evans, R. and Napper, D. H., *Kolloid-Z.u.Z. Polymere*, **251**, 409, 329 (1973).
13. Napper, D. H., *J. Colloid Interface Sci.*, **32**, 106 (1970).
14. Napper, D. H., *Polymer Letters*, **10**, 449 (1972).
15. Elias, H. G., Adank, G., Dietschy, H., Etter, O., Gruber, U. and Ibrahim, F. W. in *Polymer Handbook* (Eds. Brandrup, J. and Immergut, E. H.), Interscience Publishers, New York, 1966.
16. Doroszkowski, A. and Lambourne, R., *J. Colloid Interface Sci.*, **26**, 214 (1968).
17. King, J. M., ICI Paints Division, unpublished results, 1966.
18. Fitch, R. M. and Kamath, Y. K., *J. Indian Chem. Soc.*, **49**, 1209 (1972).
19. Blundell, D. J. and Waters, J. A., ICI Corporate Laboratory, unpublished results, 1971.
20. Barsted, S. J., Nowakowska, L. J., Wagstaff, I. and Walbridge, D. J., *Trans. Faraday Soc.*, **67**, 3598 (1971).
21. Doroszkowski, A. and Lambourne, R. in *Chimie, Physique et Applications Pratiques des Agents de Surface*, Vol. II, Ediciones, S.A., Barcelona, 1969, p. 73.
22. Doroszkowski, A., ICI Paints Division, unpublished results, 1970.
23. Doroszkowski, A. and Lambourne, R., *J. Polym. Sci. C.*, 253 (1971).
24. Waite, F. A., *J. Oil Col. Chem. Assoc.*, **54**, 342 (1971).
25. Imperial Chemical Industries, *British Patent* 1,123,611 (1968).
26. Smith, H. M., ICI Paints Division, unpublished results, 1966.
27. Waite, F. A. and Waters, J. A., ICI Paints Division, unpublished results, 1966.
28. Molau, G. E. in *Block Polymers* (Ed. Aggarwal, S. L.), Plenum Press, New York, 1970.
29. Riess, G., Periard, J. and Banderet, A. in *Colloidal and Morphological Behaviour of Block and Graft Copolymers* (Ed. Molau, G. E.), Plenum Press, New York, 1971.
30. Periard, J., Riess, G. and Neyer-Gomez, M. J., *Eur. Polym. J.*, **9**, 687 (1973).
31. Imperial Chemical Industries, *British Patent* 1,211,532 (1970).
32. BALM Paints, *British Patent* 1,231,614 (1971).
33. BALM Paints, *British Patent* 1,269,964 (1972).
34. PPG Industries, *United States Patent* 3,666,710 (1972).
35. Imperial Chemical Industries, *British Patent* 1,198,052 (1970).
36. Doroszkowski, A. and Monk, C. J. H., *J. Sci. Instrum.*, **2**, 536 (1969).
37. Davies, J. T. and Rideal, E. K., *Interfacial Phenomena*, Academic Press, New York, 1961, p. 235.
38. Ottewill, R. H. and Tiffany, J. M., *J. Oil Colour Chem. Assoc.*, **50**, 844 (1967).
39. Pimentel, G. C. and McClellan, A. L., *The Hydrogen Bond*, W. H. Freeman and Company, San Francisco, 1960.
40. Reilly, J. and Rae, W. N., *Physicochemical Methods*, Vol. I, Methuen, London, 1954.
41. Sadron, C., *Pure Appl. Chem.*, **4**, 347 (1962).
42. Inoue, T., Soen, T., Hashimoto, T. and Kawai, H., *J. Polym. Sci. A-2*, **7**, 1283 (1969).
43. Gallot, B. and Sadron, C., *Macromol.*, 514 (1971).
44. Sadron, C. and Gallot, B., *Makromol. Chem.*, **164**, 301 (1973).
45. Elias, H.-G., *J. Macromol. Sci.-Chem.*, **A7**, 601 (1973).
46. Imperial Chemical Industries, *British Patent* 1,317,249; 1,319,448; 1,319,781 (1973).
47. Becher, P., *Emulsions, Theory and Practice*, 2nd Edn., Reinhold, New York, 1965.
48. Tankey, H. W., ICI Paints Division, unpublished results, 1967.
49. Wagstaff, I., ICI Paints Division, unpublished results, 1966.
50. Imperial Chemical Industries, *British Patent* 1,206,398 (1970).
51. I. G. Farbenindustrie, *British Patent* 434,783 (1934).

52. British Rubber Producers' Research Association, *British Patent* 797,346 (1958).
53. Imperial Chemical Industries, *British Patent* 893,429 (1962).
54. Rohm and Haas, *British Patent* 934,638 (1963).
55. Ceresa, R. J., *Block and Graft Copolymers*, Butterworths, London, 1962.
56. Osmond, D. W. J. and Rhind-Tutt, A. J., ICI Paints Division, unpublished results, 1962.
57. du Pont, *British Patent* 1,101,983; 1,109,984 (1968).
58. Rohm and Haas, *British Patent* 1,002,493; 1,007,723 (1965).
59. Cook Paint and Varnish, *British Patent* 1,134,997 (1968).
60. Ford Werke, *German Patent* 2,151,782 (1972).
61. Badische Anilin- und Soda-Fabrik, *United States Patent* 3,551,525 (1970).
62. Celanese Coatings, *United States Patent* 3,654,201 (1972).
63. Meredith, C. L., *Rubber Chem. Technol.*, **44**, 1130 (1971).
64. Copolymer Rubber and Chemical, *United States Patent* 3,538,190 (1970).
65. Copolymer Rubber and Chemical, *United States Patent* 3,671,608 (1972).
66. Firestone Tyre and Rubber, *British Patent* 1,008,001; 1,008,188 (1965).
67. Stampa, G. B., *J. Appl. Polym. Sci.*, **14**, 1227 (1970).
68. Dow Chemical, *German Patent* 2,142,598 (1972).
69. Bueche, F., *J. Colloid Interface Sci.*, **41**, 374 (1972).
70. du Pont, *British Patent* 1,223,343 (1971).
71. du Pont, *South African Patent* 71/4337 (1971).
72. du Pont, *United States Patent* 3,660,537 (1972).
73. Rohm and Haas, *German Patent* 2,008,991 (1971).
74. Imperial Chemical Industries, *British Patent* 1,095,931 (1967).
75. Celanese Corporation, *United States Patent* 3,027,352 (1962).
76. Hercules, *United States Patent* 3,634,303 (1972).
77. Mark, H., Immergut, B., Immergut, E. H., Young, L. J. and Beynon, K. I. in *Polymer Handbook* (Eds. Brandrup, J. and Immergut, E. H.), Interscience Publishers, New York, 1966.
78. du Pont, *United States Patent* 3,723,571 (1973).
79. Imperial Chemical Industries, *British Patent* 1,052,241 (1966).
80. Imperial Chemical Industries, *British Patent* 1,174,391 (1969).
81. Moody, A. G., ICI Paints Division, unpublished results, 1964.
82. Duell, E. G. and Walbridge, D. J., ICI Paints Division, unpublished results, 1964.
83. Backhouse, M. P., ICI Paints Division, unpublished results, 1963.
84. Kitchener, J. A., *Br. Polym. J.*, **4**, 217 (1972).
85. Imperial Chemical Industries, *British Patent* 1,096,912 (1967).
86. Bamford, C. H. and Jenkins, A. D., *Nature*, **176**, 78 (1955).
87. du Pont, *United States Patent* 3,689,593 (1972).
88. Imperial Chemical Industries, *British Patent* 1,305,715 (1972).
89. Imperial Chemical Industries, *British Patent* 1,174,391 (1969).
90. Cornish, G. R. and Lilley, H. S., ICI Paints Division, unpublished results, 1966.
91. Stockmayer, W. H., *J. Chem. Phys.*, **13**, 199 (1945).
92. Walbridge, D. J., ICI Paints Division, unpublished results, 1968.
93. Waite, F. A., ICI Paints Division, unpublished results, 1966.
94. Imperial Chemical Industries, *British Patent* 1,122,397 (1968).
95. Dowbenko, R. and Hart, D. P., *Ind. Eng. Chem. Prod. Res. Develop.*, **12**, 14 (1973).
96. Imperial Chemical Industries, *British Patent* 1,159,252 (1969).
97. Imperial Chemical Industries, *British Patent* 1,211,532 (1970).
98. BALM Paints, *British Patent* 1,211,344 (1970).
99. Imperial Chemical Industries, *British Patent* 1,325,927 (1973).
100. Allport, D. C. and Janes, W. H. (Eds.), *Block Copolymers*, Applied Science Publishers, London, 1973.

101. Ceresa, R. J. (Ed.), *Block and Graft Copolymerization*, Vol. I, John Wiley and Sons, London, 1973.
102. Imperial Chemical Industries, *British Patent* 941,305 (1963).
103. Imperial Chemical Industries, *British Patent* 1,123,611 (1968).
104. Imperial Chemical Industries, *South African Patent* 72/7635 (1973).
105. Mobil Oil, *Netherlands Patent* 72/06366 (1972).
106. Hercules, *British Patent* 1,165,840 (1969).
107. Suminoe, T., Yamazaki, N. and Kambara, S., *Chem. High Polymers* (*Japan*), **20**, 461 (1963).
108. Downing, S. B., ICI Paints Division, unpublished results, 1970.

# CHAPTER 4
# *Kinetics and Mechanism of Dispersion Polymerization*

K. E. J. BARRETT AND H. R. THOMAS

| | |
|---|---|
| 4.1. GENERAL FEATURES OF DISPERSION POLYMERIZATION | 115 |
| 4.2. POLYMER PRECIPITATION AND POLYMER SOLUBILITY | 119 |
| 4.2.1. Solubility parameters as criteria of miscibility | 120 |
| 4.2.2. Solubility parameters of polymers | 121 |
| 4.2.3. Flory–Huggins theory of polymer solutions | 124 |
| 4.2.4. Phase separation in polymer solutions: solubility and molecular weight | 126 |
| 4.2.5. Phase separation in the presence of solvent or monomer | 128 |
| 4.2.6. Effect of particle size of monomer partition and polymer solubility | 129 |
| 4.3. PARTICLE FORMATION IN DISPERSION POLYMERIZATION | 130 |
| 4.3.1. Methods of investigation | 131 |
| 4.3.2. Factors controlling particle number and size | 134 |
| 4.3.3. Theories of particle formation | 143 |
| 4.3.4. The theory of homogeneous nucleation | 151 |
| 4.3.5. Nucleation controlled by capture of oligomers | 159 |
| 4.3.6. Dispersant-limited nucleation | 170 |
| 4.3.7. Dispersant-limited agglomeration | 171 |
| 4.3.8. Present status of theories of particle formation | 173 |
| 4.4. PARTICLE GROWTH IN RADICAL-INITIATED DISPERSION POLYMERIZATION | 177 |
| 4.4.1. The mode of polymerization | 177 |
| 4.4.2. Polymerization within a polymer matrix | 179 |
| 4.4.3. A kinetic model for dispersion polymerization | 184 |
| 4.4.4. Results of kinetic studies | 187 |
| 4.4.5. Copolymerization parameters | 192 |
| 4.4.6. Control of molecular weight | 194 |
| 4.4.7. Continuous processes | 195 |
| 4.5. DISPERSION POLYMERIZATION BY OTHER THAN FREE RADICAL MECHANISMS | 196 |
| 4.6. REFERENCES | 197 |

## 4.1. GENERAL FEATURES OF DISPERSION POLYMERIZATION

In principle, the conditions for successful dispersion polymerization are quite straightforward. The main requirements are the presence of an inert diluent which dissolves the monomer but precipitates the polymer and a polymeric dispersant to stabilize the polymer particles as they are formed by the attachment of a dissolved protective layer around them. Provided these conditions are fulfilled, polymer dispersions may be prepared using any type of polymerization mechanism–free radical and ionic addition, condensation, ring-opening and so on. Since the main field of practical application has so far been with free radical dispersion polymerization, detailed study of the kinetics and mechanism has largely been confined to this area, although many of the principles which have been established are of wider application. We shall, therefore, be concerned primarily with the free radical dispersion polymerization of vinyl and acrylic monomers, such as vinyl acetate, vinyl chloride, methyl methacrylate and acrylonitrile, in mainly aliphatic hydrocarbon diluents. However, other modes of dispersion polymerization which have not been studied in such detail are also discussed briefly.

In the simplest form of dispersion polymerization, a monomer such as methyl methacrylate, together with a small amount of the polymeric dispersion stabilizer and an azo or peroxide initiator, is dissolved in a hydrocarbon diluent and heated to reflux with stirring. Initially the mixture is clear and transparent, but after a short induction period (often of only a few seconds duration) a faint opalescence appears. At this stage, the colloidal dispersion of initially-formed polymer particles often has a bluish tinge by reflected light and appears red by transmitted light (Tyndall effect). At first slowly, and then faster, the opalescent liquid whitens and eventually an opaque white latex is formed. Polymerization then continues without much further visible change and its progress can be monitored by a measure of total solids, density, heat output or unused monomer. A typical dispersion polymerization begins slowly and gradually accelerates to a maximum rate somewhere between 20% and 80% conversion followed by a diminution in rate as monomer is consumed. Under normal conditions, conversions above 99% are attained over a period of an hour or so[1].

A number of features therefore, can be regarded as characteristic of free radical dispersion polymerizations:

(i) The insoluble polymer precipitates from an initially homogeneous reaction mixture.
(ii) Polymer particles are formed at a very early stage of the polymerization. Usually little or no polymerization can be detected before the first appearance of opalescence.
(iii) The rate of polymerization steadily increases to a maximum value. Figure 4.1 shows the typical form of the sigmoid curve of conversion against time which is obtained. The auto-acceleration of polymerization rate is quite distinct from the delayed onset of polymerization during the induction period, which results from inhibition due to impurities. Thorough purification and deoxygenation eliminates the induction period but leaves the auto-acceleration unchanged.
(iv) In many cases, the rate of dispersion polymerization is much faster than the corresponding polymerization in solution using the same quantities of reactants. However, the degree of enhancement in rate depends very much on the nature of the monomer. The effect is particularly marked with methyl methacrylate, as shown in Figure 4.2, below.

It is worthwhile at this point to contrast dispersion polymerization with other well-known types of heterogeneous polymerization. In precipitation

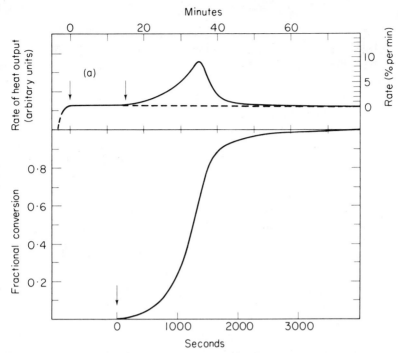

Figure 4.1. Rate of dispersion polymerization of methyl methacrylate[1]: (a) Rate of output of heat of polymerization against time; (b), Corresponding time/conversion curve obtained by integration of (a). Conditions: methyl methacrylate 50%, azodiisobutyronitrile 0.2%, graft dispersant 2.5% in n-dodecane at 80 °C. Arrows mark establishment of reaction temperature and start of polymerization. (From Barrett and Thomas, *J. Polym. Sci.* A1, **7**, 2626 (1969), with permission)

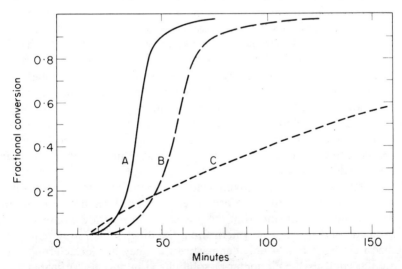

Figure 4.2. Comparison of rates of dispersion polymerization (A), precipitation polymerization (B) and solution polymerization (C) of methyl methacrylate at 80 °C[1]. (From Barrett and Thomas, *J. Polym. Sci.* A1, **7**, 2627 (1969), with permission)

117

polymerization, monomer and initiator are dissolved in a liquid which is a non-solvent for the polymer. Polymer precipitates out as it is formed. In certain cases, such as vinyl chloride or acrylonitrile, the monomer itself is a precipitant for its own polymer so that precipitation polymerization takes place in the absence of any added diluent. In emulsion polymerization, monomer is emulsified in a non-solvent, usually water, in the presence of a surfactant. A water-soluble initiator is used and particles of polymer form in the aqueous diluent as the reservoir of monomer in the emulsified droplets is gradually used up. According to Harkins and others[2,3], surfactant micelles are the site of particle formation, although Roe has suggested that they merely serve as a source of surfactant and monomer[4]. In suspension polymerization, monomer is emulsified in water using a surfactant or suspending agent, but the initiator is dissolved within the monomer droplets rather than in the aqueous diluent. The monomer droplets themselves are gradually converted into insoluble polymer particles but no new particles are formed in the aqueous phase.

Dispersion polymerization therefore, may be regarded as a form of precipitation polymerization modified by the presence of a polymeric dispersant to prevent flocculation and aggregation of the precipitated particles. When the process is carried out in the absence of an agent for stabilizing the dispersion, although an opalescent dispersion of polymer particles is formed initially, this rapidly flocculates (within seconds), aggregation continues and polymerization accelerates, sometimes with explosive violence. Precipitation polymerization was studied many years ago by Norrish, Smith and others[5-7] who demonstrated that the enhancement in rate could be attributed to the hindered termination of growing polymer radicals within the precipitated polymer. In addition, the uncontrolled aggregation restricts the access of monomer to the polymer radicals and also prevents the even dissipation of the heat of polymerization, leading to runaway reactions and generally erratic behaviour. When the problems of heat transfer are minimized, for example, by reducing the bulk of the reaction mixture to a few milligrams in a metal vessel, the polymerization follows a similar course to that of the corresponding stabilized dispersion polymerization, albeit somewhat retarded, as shown in Figure 4.2. This suggests that the basic mode of polymerization is the same in both cases. Since aggregation is prevented in dispersion polymerization however, both heat and mass transfer can take place without restriction resulting in a highly reproducible and controllable process.

In this respect, dispersion polymerization is essentially similar to emulsion polymerization although there are, of course, important differences. For example, all the complications arising from the presence of a separate monomer phase are avoided; in dispersion polymerization, the dispersant is only required to stabilize the dispersion of polymer particles as they are formed and not in addition to emulsify the monomer or solubilize it in micelles. A further major difference concerns the relation between the number of polymer particles and the rate of polymerization. In both emulsion and dispersion polymerization, the number of polymer particles formed in a given volume of latex depends directly on the concentration of surfactant or dispersant used. In emulsion polymerization however, the rate is usually strongly dependent on the number of polymer particles per unit volume and often directly proportional to it. According to the classical theory of emulsion polymerization due to Harkins[2] and to Smith and Ewart[3], this is because two or more free radicals can only coexist in a single, very small polymer particle for a very short time, so that the number of polymerizing radicals at any given moment of time averages out at just over half the total number of polymer particles. In dispersion polymerization however, the rate is virtually independent of both particle size and number over a very wide range, for reasons which will be discussed later. This is not only of theoretical significance in its own right but has an important practical outcome in that the rate of dispersion polymerization is much more reproducible and readily controlled for this reason. It has certainly facilitated experimental kinetic studies in this field since the factors controlling polymerization rate and the factors affecting particle size and number can be treated quite independently. A similar situation often exists in suspension polymerization since the polymer particles formed by polymerization within the emulsified monomer droplets containing oil-soluble initiator are usually large enough to contain many radicals simultaneously. This is clearly different from the situation in emulsion polymerization where the coarse emulsified monomer droplets merely serve as a reservoir of monomer while the much finer polymer particles are formed in the aqueous phase in which the initiator is dissolved. The similarities and differences between dispersion polymerization and the other types of heterogeneous polymerization described are summarized in Table 4.1.

Table 4.1. Comparison of properties of heterogeneous polymerization systems.

| Property | Heterogeneous polymerization process | | | |
| --- | --- | --- | --- | --- |
|  | Dispersion | Precipitation | Suspension | Emulsion |
| 1. Separate monomer phase | − | − | + | + |
| 2. Initiator dissolved in diluent | + | + | − | + |
| 3. Particles formed in diluent phase | + | + | − | + |
| 4. Particles stabilized | + | − | + | + |
| 5. Particle number dependent on stabilizer concentration | + | − | + | + |
| 6. Polymerization rate dependent on particle number | − | − | − | + |

## 4.2. POLYMER PRECIPITATION AND POLYMER SOLUBILITY

Clearly, the processes involved in the formation and growth of polymer particles during dispersion polymerization are strongly dependent on the solubility characteristics of the polymer. While a semi-empirical description in terms of the polar or non-polar character of the polymer and solvent can serve as a rough qualitative guide for experimental purposes, it is desirable to characterize the solubility properties of polymers in a more precise and quantitative manner. For this purpose, the Solubility Parameter has proved most useful in relating solubility to the chemical structure of polymer and solvent[8]. This in turn can be linked with the interaction parameters employed in the theory of polymer solutions developed by Flory and Huggins, which explains the dependence of solubility on molecular weight as well as many other aspects of polymer solutions.

### 4.2.1. Solubility parameters as criteria of miscibility

The concept of the solubility parameter arises out of the theory of regular solutions developed by Hildebrand and Scatchard[9-12].

For two liquids to mix, the free energy change $\Delta G_m$ for the process of mixing must be negative; this arises from the enthalpy or heat of mixing $\Delta H_m$ with a contribution from the entropy of mixing $\Delta S_m$ at absolute temperature $T$:

$$\Delta G_m = \Delta H_m - T \Delta S_m \tag{4.1}$$

The entropy of mixing is always positive, since mixing involves increased disorder, and therefore the criterion for miscibility is that the heat of mixing $\Delta H_m$ must be smaller than the entropy factor $T \Delta S_m$. The theory of Hildebrand and Scatchard provides a means of calculating $\Delta H_m$ in terms of parameters for pure liquids.

The heat of mixing is attributed to interactions between unlike molecules of exactly the same nature as the interactions between like molecules which hold a pure liquid together. Each pure liquid is therefore characterized by its cohesive energy density $E$, i.e. the total energy required to disperse the molecules contained in unit volume of liquid to a distance infinitely far from each other. This can be determined experimentally from the heat of vaporization per mole $\Delta H_v$ and the molar volume of the liquid $V_m$:

$$E = (\Delta H_v - RT)/V_m \tag{4.2}$$

The corresponding parameter defining the energy of interaction between unlike molecules is taken to be the geometric mean of the cohesive energy densities of the two components, and for this reason it is convenient to define the solubility parameter $\delta$ as the square root of the cohesive energy density, since then:

$$E_1 = \delta_1^2; \quad E_2 = \delta_2^2; \quad E_{12} = \delta_1 \delta_2 \tag{4.3}$$

Assuming that contributions to the total cohesive energy density of a mixture are proportional to the volume fractions $\phi_1$, $\phi_2$ of each pair of components taken separately, the heat of mixing per mole can be obtained by subtracting this total from the cohesive energies of the separate unmixed components:

$$\Delta H_m/V_m = (E_1\phi_1 + E_2\phi_2) - (E_1\phi_1^2 + E_2\phi_2^2 + 2E_{12}\phi_1\phi_2) \tag{4.4}$$

where $V_m$ is the average molar volume of the mixture, and the factor 2 arises because interactions between like molecules are in effect counted twice. Bearing in mind that $\phi_1 + \phi_2 = 1$, the heat of mixing can be expressed in terms of the solubility parameters as:

$$\Delta H_m = V_m \phi_1 \phi_2 (\delta_1 - \delta_2)^2 \qquad (4.5)$$

Obviously, according to this theory $\Delta H_m$ can never be negative but is minimized when $\delta_1 - \delta_2$ is made as small as possible. Solubility and miscibility are most favoured when the solubility parameters of the two components are most closely matched.

### 4.2.2. Solubility parameters of polymers

Although amorphous polymers may be regarded as liquids for the purpose of this theory, the entropy of mixing a given volume of polymer with solvents is in general much less than that for mixing a similar volume of a low molecular weight solute, so that the requirements for polymer solubility are much more stringent: the solubility parameters must match much more closely.

In fact, using the expression developed by Flory and Huggins[13-15] (Figure 4.3; see also Section 4.2.3), it is easy to show that the maximum value of the

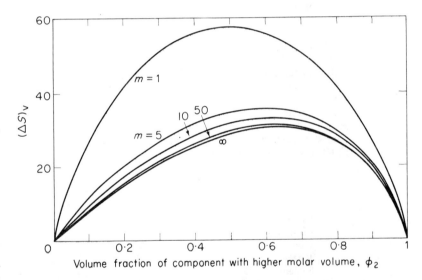

Figure 4.3. Configurational entropy of mixing per unit volume of mixture, $(\Delta S)_v$, calculated by Flory–Huggins theory for various values of $m$, the ratio of the molar volumes of the two components [$(\Delta S)_v$ in J/litre deg. C]

entropy of mixing for a pair of liquids of typical molar volume around 100–150 ml corresponds to a difference of at least 4 solubility parameter units, so that most organic liquids are mutually miscible. However, if one of the components is a polymer the theoretical maximum difference for full miscibility is around 2–3 units. The practical limits are often considerably less, for the reasons discussed below.

Solubility parameters of polymers cannot of course, be determined from their heats of vaporization. However, Small and others[16,17] have shown that the product of solubility parameter and molar volume is an additive function of 'molar attraction constants' for the various chemical groups in a compound, which have been tabulated from extensive data on small molecules. Values calculated from composition, in this way, agree well with empirical estimates obtained by finding the solvent which gives maximum swelling of the corresponding cross-linked polymers[18].

Values of solubility parameters for a number of polymers, monomers, diluents and solvents are listed in Table 4.2. In aliphatic hydrocarbons, with

Table 4.2. Solubility parameters of monomers, polymers, solvents and diluents[a]. (From *Polymer Handbook*, 1966, pp. IV–347 *et seq.*, with permission.)

| Monomer | Monomer $\delta(cal/cc)^{\frac{1}{2}}$ | Polymer $\delta(cal/cc)^{\frac{1}{2}}$ |
|---|---|---|
| Acrylonitrile | 10.5 | 12.7 |
| Methacrylonitrile | 9.9 | 10.7 |
| Methyl acrylate | 8.9 | 10.1 |
| Vinyl chloride | 7.8 | 9.7 |
| Vinyl acetate | 9.0 | 9.4 |
| Methyl methacrylate | 8.8 | 9.3 |
| Ethyl acrylate | 8.6 | 9.4 |
| Styrene | 9.3 | 9.1 |
| Butyl acrylate | 8.8 | 8.9 |
| Butyl methacrylate | 8.3 | 8.7 |
| Lauryl methacrylate | 8.3 | 8.2 |
| 12-Hydroxystearic acid | 9.3 | ~8.0 |

| Solvents and diluents | $\delta(cal/cc)^{\frac{1}{2}}$ |
|---|---|
| Methanol | 14.5 |
| Ethanol | 12.7 |
| n-Butanol | 11.4 |
| Acetone | 9.9 |
| Benzene | 9.2 |
| Toluene | 8.9 |
| n-Butyl acetate | 8.5 |
| Cyclohexane | 8.2 |
| n-Octane | 7.6 |
| n-Pentane | 7.0 |

[a] From Reference 19 or using method of Reference 16. Revised values for many liquids are listed in Reference 17.

$\delta$ around 7–8, polymers with $\delta$ close to 8 are soluble and can function as the soluble portion of polymeric dispersants; the polymers with $\delta > 9$ are insoluble and in most cases readily form dispersions, while those with $\delta$ around 8.5–9 have marginal solubility and can usually only participate in polymer dispersions as comonomers. On the other hand, polymers can also be prepared as dispersions in diluents of higher solubility parameter, if the difference is large enough: examples are poly(styrene) in methanol, or poly(lauryl methacrylate) in methanol or acetone.

It is also obvious that most monomers are good solvents for their own polymers because their similar chemical structure results in similar solubility parameters. The major exceptions are acrylonitrile and vinyl chloride, both small molecules which therefore show an exceptionally large difference between the densities of monomer and polymer and a correspondingly large difference in solubility parameter. Most monomers therefore, raise the solvency of aliphatic hydrocarbon diluents in which they are dissolved and their concentration has a marked effect on the size of the polymer particles produced in dispersion polymerization. The exception is vinyl chloride, which has a solubility parameter very similar to that of aliphatic hydrocarbons and can therefore be present in large amounts with little or no effect on particle size.

In using tabulated values, it should be borne in mind that the solubility parameter falls with rising temperature roughly in proportion to the density[11,18]. Values are normally calculated for 25 °C, and since solvents and diluents have a higher coefficient of expansion than polymers, the difference in solubility parameter between a polymer and hydrocarbon diluent will be appreciably larger at typical polymerization temperatures (70–100 °C); on the other hand, poly(styrene) becomes more soluble in alcohols as the temperature rises.

Even after making such allowances, Table 4.2 reveals a few anomalies: for example, poly(acrylonitrile) will not dissolve in ethanol, cyclohexane is a poorer solvent for moderately polar polymers and poly(styrene) is somewhat more soluble than their respective places in the Table would suggest. These exceptions are usually the result of specific interactions such as dipole attraction or hydrogen bonding which do not conform to the assumption in the theory that forces between unlike molecules can be taken as the geometric mean of the forces between like molecules. The presence of such specific interactions normally results in a lowering of the heat of mixing and therefore makes the requirements for miscibility more stringent. For this reason, most acrylic polymers are much less soluble in hydrocarbons than their solubility parameters would suggest, while poly(styrene) behaves more in accordance with the simple theory and shows marginal solubility though its solubility parameter is very close to that of poly(methyl methacrylate).

The solubility parameter concept has been extended to take account of these anomalies by introducing additional qualitative factors, such as tendency to hydrogen bonding[17,18], quantitative parameters such as fractional polarity[20,21], or by splitting the solubility parameter into several components attributable to dispersion, hydrogen bonding, induction and dipole–dipole interactions and so on ('three-dimensional solubility parameters')[22,23]. In each case, miscibility is maximal when all the parameters of both solvent and solute are as alike as possible. These approaches have resolved anomalies in solubility characteristics with considerable success, but will not be described here in further detail.

### 4.2.3. Flory–Huggins theory of polymer solutions

The theory developed by Flory and Huggins[13-15] explains many of the special features of polymer solutions by showing that the entropy component of the free energy of mixing is strongly dependent on the ratio of the molar volumes $V_2/V_1$ of a long chain molecule and its solvent respectively. This ratio, which we shall call $m$, is clearly proportional to the molecular weight of the polymer.

The calculation starts from a statistical mechanical approach and postulates that the entropy of a system depends on the number of distinguishable states in which it can exist. The main contribution to the entropy of mixing is taken to be the configurational entropy $(\Delta S_m)_{\text{config.}}$: using a three-dimensional lattice model to represent a mixture containing volume fractions $\phi_1$ and $\phi_2$ of solvent and polymer respectively, this value is calculated from the number of distinguishable ways $W$ in which flexible chains occupying $m$ adjacent lattice sites can be placed on the lattice so as to fill a fraction $\phi_2$ of the total number of sites. A corresponding value $W_0$ is found for the pure amorphous polymer, placing the same number of chains on a lattice with just enough sites to be filled completely. The configurational entropy of mixing derived in this way is found to be:

$$(\Delta S_m)_{\text{config.}} = k \ln(W/W_0) = -R(n_1 \ln \phi_1 + n_2 \ln \phi_2) \qquad (4.6)$$

where $k$ is Boltzmann's constant, $R$ the gas constant per mole, and $n_1$, $n_2$ the number of moles of the two components.

The expression for free energy of mixing is completed by an interaction term:

$$\Delta G_m = RT(n_1 \ln \phi_1 + n_2 \ln \phi_2 + \chi n_1 \phi_2) \qquad (4.7)$$

or for the free energy of mixing per unit volume:

$$(\Delta G_m)_v = (RT/V_1)(\phi_1 \ln \phi_1 + (\phi_2/m) \ln \phi_2 + \chi \phi_1 \phi_2) \qquad (4.8)$$

Compared with the corresponding expression for low molecular weight solutes, in which $m \simeq 1$, this yields a much larger entropy of mixing per mole of solute at the same molar concentration, but a much smaller entropy of mixing per unit volume at the same volume concentration (Figure 4.3).

$\chi$ is the polymer–solvent interaction parameter, which includes the heat of mixing together with an entropic component $\mu_s$. It can therefore be related to the solubility parameters of the components by[24]:

$$\chi = \mu_s + (\delta_1 - \delta_2)^2 V_1/RT \qquad (4.9)$$

$\mu_s$ generally has a value in the range 0·2–0·4 where the liquid is a good solvent for the polymer, but may be considerably higher for non-solvents since it reflects the extent of non-randomness in the distribution of nearest neighbours around each molecule or segment.

It may be readily shown that $\chi$ must be less than 0·5 to ensure miscibility in all proportions at all molecular weights, whereas for low molecular weight solutes ($m = 1$) miscibility results if $\chi$ is less than 2. For a typical solvent of molar volume around 100 ml, this corresponds to a maximum difference in solubility parameters of 3·5 for low molecular weight solutes, but less than 1·7 for polymers (the actual value depending on $\mu_s$).

The interaction parameter $\chi$ depends on the temperature; for any given solvent–polymer pair, an important constant is the theta ($\theta$) temperature[15], at which $\chi = \frac{1}{2}$; under these conditions the system behaves as if the molecules and their configurations were randomly distributed, and theoretical treatment is simplified. It is also close to separation into two phases, and one method of determining $\theta$ depends on extrapolation from cloud points at a series of concentrations[25]. Values of $\chi$ may be obtained from measurements of osmotic pressure, viscosity or vapour pressure of polymer solutions[15,26], and some typical values are listed in Table 4.3. Values for non-solvents may be estimated from the

Table 4.3. Typical Flory–Huggins interaction parameters ($\chi$) and $\theta$-temperatures.

| | $\chi$ | $T(°C)$ | | $\chi$ | $T(°C)$ |
|---|---|---|---|---|---|
| Poly(styrene) in[29,30]: | | | Poly(methyl methacrylate) in[28,35–37]: | | |
| Styrene | 0·42 | 23 | Methyl methacrylate | 0·55$^a$ | 25 |
| Chloroform | 0·43 | 23 | Methyl methacrylate | 0·47 | 25 |
| Toluene | 0·44 | 23 | n-Butyl acetate | 0·50 | 25 |
| Cyclohexane | 0·60 | 23 | Ethyl acetate | 0·48 | 25 |
| Cyclohexane | 0·48 | 60 | Toluene | 0·45 | 25 |
| Acetone | 0·74 | 23 | Acetone | 0·48 | 25 |
| n-Heptane | 1·19 | 23 | Chloroform | 0·34 | 25 |
| n-Hexane | 2·67 | 23 | Poly(vinyl chloride) in [35,38,39]: | | |
| Poly(vinyl acetate) in[28,31–33]: | | | Vinyl chloride | 0·88$^a$ | 50 |
| Vinyl acetate | 0·33$^a$ | 30 | Tetrahydrofuran | 0·36 | 25 |
| Vinyl acetate | 0·28$^a$ | 60 | Butyl acetate | 0·40 | 53 |
| Benzene | 0·46 | 50 | Acetone | 0·60 | 53 |
| Ethanol | 0·47 | 55 | Benzene | 0·77 | 76 |
| Acetone | 0·38 | 50 | n-Butanol | 1·74 | 53 |
| n-Propanol | 1·04 | 50 | Dioctyl ether | 2·6 | 53 |
| Water | 2·0 | 40 | Poly(butyl acrylate) in[28]: | | |
| Natural rubber in[34]: | | | Butyl acrylate | 0·57$^a$ | 25 |
| n-Octane | 0·48 | 25 | Poly(isobutylene) in[35]: | | |
| Benzene | 0·44 | 25 | Cyclohexane | 0·42 | 30 |
| Cyclohexane | 0·40 | 25 | Benzene | 0·50 | 25 |
| Ethyl acetate | 0·77 | 25 | Toluene | 0·49 | 30 |
| Acetone | 1·36 | 25 | n-Heptane | 0·47 | 25 |

$\theta$-Temperatures (°C) of poly(methyl methacrylate) solutions[37]: Benzene $-223 \pm 50$; methyl methacrylate $-163 \pm 50$; ethyl acetate $-98$; toluene $-65 \pm 10$; acetone $-65 \pm 10$; butyl acetate $-20$; carbon tetrachloride $+27$; amyl acetate $+41$.
$\theta$-Temperature of poly(styrene) in cyclohexane[40]: $+34$ °C.

$^a$ Data from aqueous emulsion systems.

extent of the swelling of polymers in their presence. The concept provides a useful insight into the behaviour of polymers even in the absence of quantitative parameters.

While the Flory–Huggins theory successfully explains many properties of polymer solutions, it can only be regarded as a first approximation to a quantitative treatment. In particular, experimental values of $\chi$ are not entirely independent of concentration. Considerable progress has been made in developing improved theoretical treatments[27], but these will not be described further here since they introduce considerable complexity, and do not in any way affect the main conclusions of the Flory–Huggins treatment.

### 4.2.4. Phase separation in polymer solutions: solubility and molecular weight

Precipitation and phase separation in polymer solutions can, in principle, be fully described in terms of an equilibrium process by the Flory–Huggins theory[15,41]. For this purpose, it is convenient to derive the partial molal free energy $\Delta \bar{G}$ of each component (also known as the chemical potential $\mu$), which is directly related to its thermodynamic activity $a$ ($\Delta \bar{G} = RT \ln a$). This quantity is defined as the free energy change per mole of the component added, when an infinitesimal quantity of it is mixed into the system with the amounts of all the other components kept constant. It is obtained by partial differentiation of the expression for free energy of mixing (putting $\phi_1 = n_1 V_1/(n_1 V_1 + n_2 V_2)$ etc.) and for the Flory–Huggins model is given by:

Solvent:

$$\Delta \bar{G}_1 = (\partial \Delta G/\partial n_1)_{n_2} = RT(\ln \phi_1 + \phi_2(1 - 1/m) + \chi \phi_2^2)$$

Polymer:

$$\Delta \bar{G}_2 = (\partial \Delta G/\partial n_2)_{n_1} = RT(\ln \phi_2 + \phi_1(1 - m) + m\chi \phi_1^2) \tag{4.10}$$

At equilibrium, the total free energy of the system must be a minimum, and the transfer of an infinitesimal quantity of any component from one phase to another must produce zero change in free energy, so the primary condition of equilibrium is that the chemical potential of each component must be equal in the two phases.

For the case we are most concerned with in dispersion polymerization, $\chi$ is large and the two phases consist of nearly pure diluent and slightly-swollen polymer. Good approximations for the volume fraction of polymer in the diluent and diluent in the polymer are therefore obtained by putting the corresponding expression for $\Delta \bar{G}$ equal to that of the pure component, which is zero. In this way we find:

Volume fraction of diluent in the polymer phase:

$$\ln \phi_1 = -\phi_2(1 - 1/m) - \chi \phi_2^2 \simeq 1/m - \chi - 1$$

and since $m$ is large, $\quad \phi_1 \simeq e^{-(\chi + 1)} \tag{4.11}$

Volume fraction of polymer in the diluent phase

$$\ln \phi_2 = -\phi_1(1 - m) - m\chi\phi_1^2 \simeq m(1 - \chi - 1/m)$$

and since $m$ is large, $\quad \phi_2 \simeq e^{-m(\chi-1)} \quad (4.12)$

Thus the solubility of a polymer in a liquid diluent falls off very rapidly as its molecular weight is raised, and increasingly steeply with larger values of the interaction parameter $\chi$ (i.e. larger difference between solubility parameters). The amount of diluent in the swollen polymer also falls with rising $\chi$, but quite slowly, and is almost unaffected by its molecular weight.

The extreme sensitivity of polymer solubility to molecular weight and the value of $\chi$ is illustrated in Figure 4.4. In practice, of course, the polymer formed

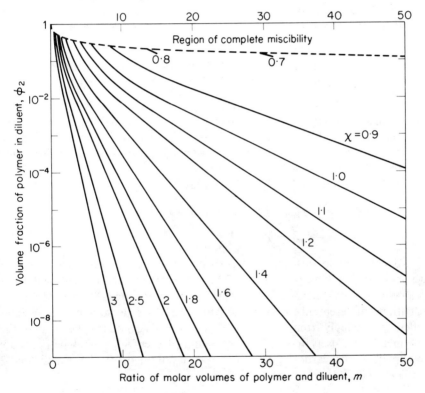

Figure 4.4. Dependence of polymer solubility on molecular weight for various values of the interaction parameter $\chi$ (calculated from Flory–Huggins equations)

in any free radical process is not uniform but comprises a wide distribution of molecular weights. The diluent phase will therefore have a higher polymer content than the vanishingly low solubilities implied by its average molecular weight; this is contributed almost entirely by very low molecular weight fractions, and may reach quite appreciable levels especially with monomers which readily undergo chain transfer during polymerization. On the other hand, the fraction of highest molecular weight will obviously precipitate out first and be preferentially partitioned into the precipitated phase. This behaviour of course, forms the basis of one well-known method of polymer fractionation[42,43].

### 4.2.5. Phase separation in the presence of solvent or monomer

Dispersion polymerization usually involves the precipitation of a polymer from the diluent in the presence of a third component, namely monomer. As we have seen, the monomer is usually a good solvent for its own polymer. Scott[44], and Krigbaum and Carpenter[45], have set out the appropriate expressions for partial molal free energy in systems containing a polymer and two solvents. For this purpose, it is necessary to use three interaction parameters:

$\chi_{dp}$ for interaction between diluent and polymer,
$\chi_{mp}$ for interaction between monomer and polymer and
$\chi_{md}$ for interaction between monomer and diluent

If the following simplifying assumptions are made that $m$ is large, that the diluent phase contains negligible polymer and *vice versa*, that the molar volumes of monomer and diluent are about equal, and that the volume fractions of monomer in the polymer and diluent phase $\phi_p$, $\phi_d$ are fairly small, the following approximation is obtained:

Partition of monomer between the phases:

$$\ln(\phi_p/\phi_d) \simeq \chi_{md} - \chi_{mp} - 1 + 1/m + \phi_p(1 + 2\chi_{mp}) - 2\phi_d\chi_{md}$$
$$\simeq \chi_{md} - \chi_{mp} - 1 \qquad (4.13)$$

The partition ratio is therefore essentially independent of both the molecular weight of the polymer and the polymer–diluent interaction parameter, and can be treated as roughly constant at low monomer concentrations. Monomer partitions preferentially into the polymer phase as its solubility parameter differs increasingly from that of the diluent (high $\chi_{md}$) or matches that of the polymer more closely (low $\chi_{mp}$). Experimental studies with methyl methacrylate in aliphatic hydrocarbons give values of $\phi_p/\phi_d$ of the order of 1, and around 2 for vinyl acetate (see Section 4.4.1).

#### 4.2.5.1. Effect on polymer solubility

With the same assumptions, an approximate expression can be derived for the effects of added monomer or strong solvent on the solubility of the polymer:

$$\ln \phi_2 = \ln(\phi_2)_0 + m[\phi_d(\chi_{dp} + \chi_{md} - \chi_{mp}) - \phi_p] \qquad (4.14)$$

where $(\phi_2)_0$ is the solubility in the absence of monomer. This represents a somewhat greater increase in solubility than would be obtained by replacing diluent and monomer by a single diluent, with $\chi$ corresponding to the average value for the monomer/diluent mixture: for that situation the same expression is obtained omitting only $\chi_{md}$. In both cases the overall result is a marked increase in solubility as the concentration of monomer or strong solvent is raised, since $\chi_{mp}$ is small compared with both the other interaction parameters.

*4.2.5.2. Two monomers or monomer and solvent*

To a first approximation the partition of each component between the phases can be treated separately. For a pair of monomers, this leads to an expression for the differential partition factor, which markedly affects the apparent relative reactivity (see Section 4.4.5).

$$\ln[(\phi_p/\phi_d)_1/(\phi_p/\phi_d)_2] = \chi_{m_1d} - \chi_{m_2d} + \chi_{m_2p} - \chi_{m_1p} \qquad (4.15)$$

This can be taken as constant for a given copolymer composition, but is not necessarily constant over a range of copolymer compositions, since $\chi_{m_1p}$ and $\chi_{m_2p}$ are the interaction parameters of the two monomers with the corresponding copolymer and not with their own homopolymers. It will tend to be nearly constant if the two homopolymers have very similar solubility parameters and polarity. The overall effect is to favour swelling of the polymer by the monomer with the solubility parameter nearest to the average for the copolymer and furthest from the diluent.

## 4.2.6. Effect of particle size on monomer partition and polymer solubility

In principle, all equilibria between very small particles and a bulk phase are dependent on the particle size, because additional work must be done whenever any component is added to the particles, in order to increase their surface area against the interfacial tension $\gamma$[46]. For spherical particles the extra work per mole of component of molar volume $V_m$ is inversely proportional to the radius $r$:

$$dE/dn = 2\gamma V_m/r \qquad (4.16)$$

This quantity represents the difference between the chemical potentials of the component in the bulk and particle phase:

$$\Delta \bar{G}_{\text{particle}} = \Delta \bar{G}_{\text{bulk}} - 2\gamma V_m/r \qquad (4.17)$$

so that the equilibrium of all components is displaced towards the bulk phase, that is the diluent.

In aqueous emulsion polymerization, this effect is very important because the monomer absorbed in the polymer particles is in equilibrium with bulk monomer in the form of droplets, and in fact the particles would swell until all bulk monomer had disappeared but for the opposing force provided by the interfacial tension (except of course where the monomer is not a solvent for its polymer, e.g. vinyl chloride). Morton, Kaizerman and Altier[47], and Gardon[28], have shown that the typical near-constant values of the order of 50–70% monomer concentrations found in the particles during emulsion polymerization can be explained very satisfactorily on the basis of the observed particle sizes and interfacial tensions using the Flory–Huggins expression for $\Delta \bar{G}$.

In dispersion polymerization in organic media, monomer in the particles is in equilibrium with a dilute solution of monomer in the diluent, and the effect of particle size on the partition ratio is given by:

$$\ln(\phi_p/\phi_d) = \ln(\phi_p/\phi_d)_0 - 2\gamma V_m/rRT \qquad (4.18)$$

where the subscript '0' refers to partition between bulk polymer and diluent. Since $RT$ is about $3 \times 10^{10}$ ergs/mol at typical polymerization temperatures, and the interfacial tension between swollen polymer and hydrocarbon diluent is unlikely to be more than about 10 dyn/cm, it follows that the particle diameter would have to be less than about 0.01 μm to produce a 10% change in the partition ratio of a typical monomer with a molar volume of 100 ml. The polymer particles formed in a typical dispersion polymerization are usually at least an order of magnitude larger, except in the very short period immediately after a new particle nucleus appears, and the interfacial tension is reduced to still lower levels in the presence of amphipathic dispersants, so for all practical purposes the effect of particle size on the monomer partition ratio may be taken as negligible. An exactly similar argument applies to the swelling of polymer by diluent, which is diminished by the same negligible factor as the monomer partition ratio.

On the other hand the effect on polymer solubility is much greater, since the molar volume to be used in the calculations is now that of the polymer. The approximation found earlier (equation 4.12) is therefore modified to:

$$\ln \phi_2 \simeq m(1 - \chi - 1/m + 2\gamma V_1/rRT) \qquad (4.19)$$

so the effect is equivalent to lowering the value of the Flory–Huggins interaction parameter $\chi$ by $2\gamma V_1/rRT$. The conditions which would lower the absorption of monomer in the particles by 10% thus have an effect on solubility equivalent to reducing $\chi$ by 0.1. The effective overall solubility of the polymer in a dispersion is defined by the least soluble particles present, i.e. the largest ones, so for the reasons given above any modification due to particle size can, in practice, be neglected. However, the increased solubility of very small particles may well be a contributory factor in hindering the precipitation of new nuclei in the presence of existing particles, since growing oligomer molecules formed in the diluent phase would tend to precipitate preferentially on to the existing particles (see Section 4.3).

## 4.3. PARTICLE FORMATION IN DISPERSION POLYMERIZATION

From a practical point of view, it is obviously of the utmost importance to know how to control the particle size and particle size distribution in a polymer dispersion. This will affect the rheology of the latex and its tendency to settlement on storage, as well as other properties.

The particle size distribution is essentially determined by the numbers of new particles which form at each moment during the course of a polymerization, together with their subsequent growth and any aggregation which may occur.

Our primary aim, therefore, is to identify the main factors which affect particle formation, the numbers of particles formed and the time of their appearance.

The most important factors have proved to be the solvency of the dispersion medium for the polymer, the amount of dispersant and its nature, particularly the size and solubility of its anchoring and stabilizing portions. These will be discussed in more detail in Section 4.3.2, after a preliminary discussion of the experimental methods available.

In broad outline, the mechanisms of particle formation are well enough understood to explain the facts and trends which have been established. However, the position has not yet been reached where a precise model can be formulated which predicts quantitatively the numbers and sizes of particles formed in any given conditions, although several semi-quantitative approaches will be discussed in Sections 4.3.3–7 which could eventually perhaps form the basis for a consistent theoretical treatment.

### 4.3.1. Methods of investigation

The direct study of polymer particles as they are being formed presents great difficulty, since the decisive processes take place over a very short period of time on a submicroscopic scale. Some investigators have attempted to follow the formation and early growth of polymer particles in aqueous systems by measurement of light scattering but the interpretation of these measurements is not always clear-cut and estimates of particle size distribution are at best indirect[48].

Consequently, nearly all our knowledge of particle formation is based upon measurements of particle size and its distribution in samples of dispersions taken after the initial period of their formation, and the way in which these change during the subsequent course of the polymerization. Any of the methods well known in the study of aqueous emulsion polymers (with the exception of soap titration) can be readily used for dispersions in organic media: these include light scattering, sedimentation, centrifugation, etc.[49–54]. However, a much greater wealth of useful information can usually be obtained by electron microscopy, and most of the findings reported in this Chapter result from the use of this method.

The experimental techniques using the electron microscope are well known and will not be described in detail here[55,56]. Two basic methods are suitable for examining particles of polymer dispersion. The dispersion is normally first diluted with volatile diluent to ensure a reasonable separation of the particles:

(i) A small sample is evaporated on a very thin supporting film of carbon, 'Formvar' etc. (fairly transparent to electrons) mounted on a metal grid, then shadowed at a fixed angle with heavy metal, usually a gold/palladium or platinum/carbon alloy, and viewed without further treatment in the electron microscope.

(ii) A small sample is evaporated on a glass surface and shadowed at a fixed angle with heavy metal. An even layer of electron-transparent material (carbon or aluminium) is evaporated on to the preparation to form a coherent film, which is usually backed by a thicker film of methyl cellulose or other soluble polymer laid down from solution. The whole preparation can then be floated off from the glass surface on water or other solvent and transferred to a metal grid for viewing in the electron microscope. During the course of this transfer process, the film of methyl cellulose (and even the original particles if desired) is dissolved away, leaving a shadowed replica supported solely by the carbon or aluminium.

With either technique, photographic enlargement is usually the final stage. The shadowing process is sometimes omitted, but is essential for precise measurements, especially on small particles, because organic polymers absorb electrons only weakly, providing poor contrast, and are eroded or even depolymerized by electron bombardment. Shadowing thus preserves a record of the original dimensions. For some softer polymer particles, both evaporation of the sample and shadowing must be carried out at very low temperatures to avoid coalescence and distortion.

What is actually measured, of course, is the diameter of the particles. If these are uniform in size, the number per unit volume, $N$, is given by

$$N = 3v/4\pi r^3 \tag{4.20}$$

where $v$ is the volume fraction of polymer particles in the dispersion and $r$ is their radius. These refer, of course, to non-volatile polymer; the particles may be appreciably larger in the liquid dispersion due to absorption of monomer and components of the diluent. Clearly, any errors in the measurement of $r$ will be multiplied roughly three-fold in calculating $N$; such errors are likely to become proportionately larger as the particle size falls, since some of the sources of error depend on their absolute size.

If the particles vary in size, $N$ can be calculated from equation (4.20), using the radius corresponding to the average volume of the particles:

$$N = 3vn/4\pi \Sigma r^3 \tag{4.21}$$

where $n$ is the number of particles counted. With a wide distribution of particle sizes, very large numbers of particles must be measured to obtain a precise result, since a very small fraction of large particles contributes a large fraction of the total volume. Computerized automatic scanning systems have been developed to carry out this laborious operation[57].

However, much important information can be obtained even by a qualitative examination of electron micrographs:

(i) A highly uniform particle size normally indicates that the particles have been formed within a very short period of time and subsequent growth has taken place without the formation of further particles or involving agglomeration processes [Figure 4.5(a)].

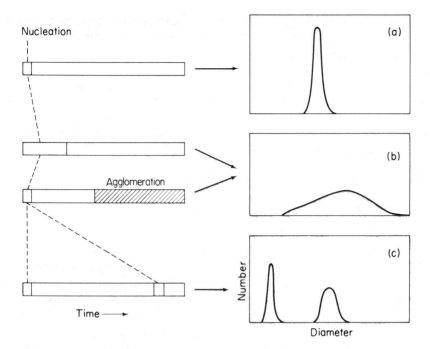

Figure 4.5. Schematic illustration of typical particle size distributions in dispersion polymerization (right) and characteristic course of the processes which give rise to them (left)

(ii) A wide and fairly random distribution of particle sizes usually indicates that particle formation has continued over a long period of time, corresponding to a wide range of periods of growth [Figure 4.5(b)]. Occasionally, however, this distribution can also result from a process of agglomeration.

(iii) The presence of many very small particles among a fairly uniform population of larger ones usually indicates that renucleation has occurred, i.e. a fresh crop of new particles has been formed long after the first period of particle formation [Figure 4.5(c)].

(iv) Comparison of particle sizes at various stages during the course of polymerization, in conjunction with independent measurements of the amount of polymer formed, can distinguish between processes of agglomeration and extended periods of particle formation, depending on whether the total number of particles decreases or increases. The time of renucleation and the conditions for its occurrence can be assigned with some precision, and the rates of growth of large and small particles compared.

(v) In some cases, inferences can also be drawn from the appearance and micro-structure of the particles, which provide clues to the manner of their formation. Spherical particles usually result from uniform growth of polymer, especially if it is fairly plastic under the conditions of polymerization. Irregular or protuberant particles may arise from processes of aggregation, or by growth from isolated points in a very rigid polymer matrix (Figure 4.15, Section 4.4.1).

Overall therefore, electron microscopy provides a tool which can be of great diagnostic value in studying the processes of particle formation and growth, especially if used in conjunction with other approaches, such as kinetic studies. Nevertheless, it must be emphasized that conclusions about particle formation are essentially inferences from the end results of processes which we do not yet have the means to study directly: the facts are often open to a variety of interpretations, and could fit more than one theoretical model.

### 4.3.2. Factors controlling particle number and size

*4.3.2.1. Period of particle formation: constancy of numbers*

In a typical dispersion polymerization producing polymer which is very insoluble in the diluent—say methyl methacrylate in an aliphatic hydrocarbon—the stage of particle formation is normally completed very quickly, within a few seconds or tens of seconds after the start of reaction. All the evidence from electron micrographs indicates that very few particles form after this stage unless the solvency of the medium is drastically altered or substantial amounts of additional dispersant are added. The number of particles therefore, remains virtually unchanged throughout the remaining course of the polymerization, unless some form of aggregation intervenes, as may happen if the total particle surface formed outruns the amount of dispersant available.

The first particles observable in the electron microscope are usually of the order of 10–20 nm in diameter, but growth is very rapid at this stage and the processes are difficult to follow experimentally.

Those conditions which extend the particle-formation stage over a longer period of time, usually also slow down the rate of particle formation to an even greater extent, so that relatively fewer and larger particles are obtained. This is the normal result of increasing the solvency of the diluent for the polymer. An unusual situation is found with vinyl chloride in aliphatic hydrocarbons, where new particles may continue to form throughout the course of polymerization, provided the supply of dispersant is adequate. Such behaviour however, is exceptional and will be discussed later together with other peculiarities of vinyl chloride (Section 4.4.4).

*4.3.2.2. Solvency of the medium and solubility of the polymer*

The processes of particle formation are strongly influenced by increasing the solvency of the medium for the polymer which is being produced:

(i) The onset of particle formation is often retarded;
(ii) The duration of particle formation is prolonged;
(iii) Fewer particles are produced;
(iv) They are correspondingly larger and usually of wider particle size distribution.

For a given polymer, solvency may be varied by an appropriate choice of diluent, addition of strong solvents, or by altering the concentration of the monomer. In most polymer–diluent systems, the effects of changes in temperature are small compared with the results of even minor changes in composition, but solvency generally increases with rising temperature.

Since in most cases, the monomer is a good solvent for its own polymer, the effect of changing the composition of the diluent is most clearly demonstrated at low monomer concentrations. In an extreme form, this can be achieved by slowly feeding the monomer into hot (refluxing) diluent in the presence of dispersant and initiator, as in the example shown in Table 4.4[58].

Table 4.4. Diameter of polymer particles obtained by feeding monomer into refluxing diluent[a,58].

| Diluent composition | Aromatic content | Particle diameter (μm) |
|---|---|---|
| Petroleum distillate, b.p. 70–90 °C (100%) | 6% | 0·2 (uniform) |
| Petroleum distillate, b.p. 70–90 °C (80%) <br> Toluene (20%) | 25% | 5 (range 2–14) |

[a] Feed monomer contained 3% stabilizer, 0·4% ADIB, 1% octyl mercaptan in methyl methacrylate, fed in over 1 h; monomer:diluent = 1:1.

The monomer itself behaves as a strong solvent, and so relatively large particles are obtained at high monomer concentrations, and fairly small variations can have a marked effect, as shown in Table 4.5[58,59]. The effect of a small increase in monomer content in coarsening and widening the range of particle sizes is particularly marked in the presence of fairly high concentrations of dispersant and near to the limiting concentration at which the polymer is

Table 4.5. Diameter of particles obtained with varying monomer concentrations[a,58,59].

| Stabilizer and content | Monomer content (%) | Particle diameter (μm) Mean value | Range |
|---|---|---|---|
| Comb stabilizer | 5[b] | 0·16 | uniform |
| 5% on monomer | 50 | 1·0 | uniform |
|  | 60 | ~7 | 0·1–35 |
| Stabilizer precursor | 40 | 2·2 | uniform |
| 0·3% on monomer | 45 | 4 | uniform |
|  | 50 | 7 | nearly uniform |
|  | 55 | ~8 | 4–20 |
|  | 60 | ~9 | 1–30 |

[a] Methyl methacrylate in petroleum distillate, 70–90 °C. BP, containing 0·4% ADIB and 0·5% octyl mercaptan on monomer; refluxed 2 h.
[b] Conditions for 5% monomer as in Table 4.6, line 2.

soluble. At even higher monomer concentrations, of course, polymerization takes place in solution until the monomer content falls to the threshold value (around 55% for methyl methacrylate in aliphatic hydrocarbons) below which polymer precipitates as a very coarse dispersion of highly swollen particles with a wide range of sizes.

Addition of extra solvent or monomer after the first crop of particle nuclei has been formed does not usually reduce the number of particles, unless so much is added as to desorb stabilizer causing flocculation. On the other hand, a drastic fall in solvency, in an initially coarse latex may produce a fresh crop of fine particles; this may happen when the initial monomer concentration is very high and then falls as polymerization continues. This phenomenon has been utilized in a process for making polymer dispersions with a very wide range of particle sizes (see Section 5.6.2).

Of course, the solvency of the medium depends as much on the nature of the polymer as on the nature of the diluent. The insolubility of polymers in aliphatic hydrocarbon diluents runs closely parallel to their polar character and the value of their solubility parameter (see Section 4.2). We can arrange the corresponding monomers in a hierarchy on this basis which also indicates their tendency to form fine particles in dispersion polymerization, thus:

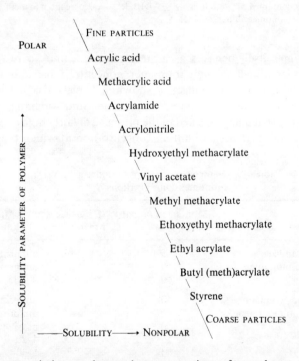

These characteristics apply to the preparation of copolymers as well as homopolymers in dispersion, and, for example, the incorporation of 1–5% of methacrylic acid in methyl methacrylate has a very marked effect in reducing the particle size of the latex formed under given conditions. Copolymerized butyl acrylate has the opposite effect.

In fact, it is very difficult to prepare stable dispersions of polymers with high proportions of monomers at either of the extreme ends of the hierarchy, by dispersion polymerization in aliphatic hydrocarbons. On the one hand, excessive amounts of dispersant are needed to stabilize the large total surface area produced when ultra-fine particles are formed. On the other hand, the less polar polymers are too near complete solubility and their large particles readily aggregate because the dispersant is only weakly anchored to their surfaces. With more polar monomers, these difficulties can be overcome to some extent by using a correspondingly polar diluent such as an aromatic hydrocarbon or an ester.

The effects of solvency on particle size have been demonstrated in precipitation polymerization in the absence of any dispersant, although this is often masked by subsequent aggregation processes. For example, Slavnitskaya and co-workers[60] polymerized methyl methacrylate in cyclohexane at 55 °C and measured the size of particles formed at very low conversion (3%). At monomer concentrations around 10%, particles of 0.1 μm diameter were formed, but at a concentration of 25–35%, the particle size was about 1 μm, apparently consisting of coalesced primary particles of 0.2–0.3 μm diameter.

However, in dispersion polymerization there is an important additional effect of solvency in modifying the operation of the dispersant and its influence on the number of particles formed. Higher solvency for the polymer moiety which anchors the dispersant to the polymer particles probably reduces the tendency of the stabilizer to associate with growing polymer chains during the process of particle formation, as well as impairing the efficiency of anchoring to the particles which have been formed (see Section 3.5). Whatever the molecular mechanism, the practical outcome is fewer and larger particles. This is clearly shown by the effect of changing the solubility of the anchor polymer of the dispersant without altering monomer or medium, as in Table 4.6, below. When a dispersant precursor is used in place of a pre-formed dispersant, the anchor polymer produced *in situ* has of course, the same composition as the polymer particles being formed, so that both effects of the solvency of the medium reinforce each other (but cannot easily be distinguished).

*4.3.2.3. Concentration and nature of the dispersant*

The concentration of the dispersant is one of the most important factors controlling nucleation: the number of particles increases and their size decreases as the stabilizer content is raised. In any given system, the exact relationship must be established empirically.

Table 4.6[59] demonstrates the effect on particle size of using varying amounts of four poly(hydroxystearic acid) 'comb'-type dispersants, differing only in the composition of the backbone anchor polymer chain (see Section 3.6). The dependence on dispersant content is very marked up to a certain level, beyond which further additions have much less effect.

Table 4.6. Diameter (μm) of polymer particles obtained with varying concentration and composition of dispersant[59].

| Backbone composition[a] | Dispersant concentration (% on diluent) | | | | | |
|---|---|---|---|---|---|---|
| | 0.25 | 0.5 | 1.0 | 1.5 | 2.0 | 2.5 | 3.0 |
| MM/MA (98:2) | 0.12 | 0.07 | 0.051 | 0.037 | 0.035 | — | — |
| MM (100) | 0.16 | 0.14 | 0.057 | 0.055 | 0.048 | — | — |
| MM/EA (50:50) | — | 0.48 | — | 0.27 | 0.22 | — | 0.23 |
| EA (100) | — | 1.44 | — | 1.30 | — | 1.45 | — |

[a] MM = methyl methacrylate, MA = methacrylic acid, EA = ethyl acrylate. Ratio of soluble PHSA chains to anchor polymer = 1:1 (w/w). Reaction mixture contained 5% methyl methacrylate, 0.23% ADIB, 2% butyl acetate in 4:1 mixture of cyclohexane and high-boiling petroleum fraction; refluxed 1–5 h.

The dramatic effect of increasing the non-polar character and solubility of the dispersant anchor, as discussed in the previous section, is also clearly shown. At very high levels of anchor chain solvency, the efficiency of stabilization is seriously impaired and it is possible that secondary aggregation of primary particles occurs under these conditions, so that the final particle size is not very sensitive to dispersant concentration.

The dispersions produced in Table 4.6 have very low total solids content. The additional examples in Table 4.7[61] illustrate the way in which such dispersions may be converted to high solids dispersions of uniform particle size by feeding in additional monomer and initiator over several hours (cf. Section 5.6.2). The original 'seed' particles grow, and in the absence of fresh nucleation

Table 4.7. Diameter of polymer particles produced in seed and final dispersion with varying concentrations of dispersant in the seed stage[a,61].

| | | | | | |
|---|---|---|---|---|---|
| % dispersant (non-vol) in seed stage | 0.190 | 0.255 | 0.378 | 0.613 | 0.867 |
| % monomer in seed stage | 5 | 5 | 5 | 4 | 4 |
| Diameter of seed particles (μm) | 0.106 | 0.070 | 0.053 | 0.035 | 0.018 |
| Diameter of final particles (μm) | 0.30 | 0.20 | 0.15 | 0.10 | 0.05 |
| Final % solids | 51 | 51 | 51 | 45 | 43 |

[a] Dispersant as in Table 4.6 (MM/MA 98:2 backbone), 50% solution in mixed esters. Seed stage contained 0.2% ADIB and methyl methacrylate monomer in 5:1 mixture of 70–90 °C b.p. and 230–250 °C b.p. petroleum distillates, refluxed 45 min. Monomer ratio (feed:seed) = 22:1.

the ratio of the diameters of seed and final particles is directly proportional to the cube root of the ratio of amounts of polymer formed in the seed stage and in the total dispersion. A carefully controlled addition of dispersant is normally needed during the early part of the feed stage to prevent flocculation as the total surface area increases without at the same time producing renucleation.

When the initial concentration of monomer in a dispersion polymerization is very high, the particle size (of course, much larger) is somewhat less sensitive to changes in dispersant concentration. In the examples shown in Table 4.8[58], there is a range of dispersant concentrations over which the particles are uniform and almost constant in size; at higher levels, the particles are smaller and some very fine particles are also present. At lower levels of dispersant, the particles have a much wider size distribution with a higher average size. Table 4.8 also shows that variations in initiator concentration have a negligible effect on particle size, at least in this type of formulation.

Table 4.8. Diameter of polymer particles obtained with varying dispersant and initiator content, using a high monomer concentration[a,58].

| Dispersant precursor on monomer (%) | ADIB[b] on monomer (%) | Particle diameter (μm) Mean | Range |
|---|---|---|---|
| 0·2 | 0·4 | 7 | 2–28 |
| 0·3 | 0·4 | 7 | 4·5–10 |
| 0·5 | 0·4 | 4·5–5 | almost uniform |
| 0·8 | 0·4 | 4·5 | uniform |
| 1·0 | 0·4 | 4·5 | uniform |
| 1·5 | 0·4 | 4 | uniform |
| 2·0 | 0·4 | 2·5 + 0·1 | bimodal |
| 3·0 | 0·4 | 2 + 0·1 | bimodal |
| 0·8 | 0·4 | 4 | uniform |
| 0·8 | 0·2 | 4 | uniform |
| 0·8 | 0·1 | 4 | uniform |

[a] 49% methyl methacrylate in petroleum distillate (b.p. 70–90 °C) containing 0·5% octyl mercaptan on monomer. Refluxed 2 h.
[b] Azo di-isobutyronitrile.

It is tempting to look for a simple reciprocal relationship between dispersant content and particle size. The data of Table 4.7 in fact fit extremely well an expression of the form

$$\text{Diameter} = Kd^{-1} \tag{4.22}$$

where $d$ is the dispersant content in the seed stage, but this must be regarded as to some extent fortuitous, since there are variations in solvency due to different monomer contents and different amounts of ester solvent added with the dispersant.

To a fair approximation, the first three dispersants used in Table 4.6 give a variation of particle size with dispersant content which fits the form of equation (4.22) but with a coefficient $d^{-0.6}$ and with a different value of $K$ for each dispersant. The data of Table 4.8 correspond very roughly to a coefficient of $d^{-0.5}$.

These relationships are found by plotting the data on a double logarithmic graph as in Figure 4.6, which also reveals the considerable experimental scatter. The extent of fit of the corresponding curves to the experimental data plotted directly is also shown.

### 4.3.2.4. Reactivity of dispersant precursors

Whatever the type of dispersant used, a broadly similar relationship is found between its concentration and the particle size of the dispersion formed: examples have been drawn from both preformed 'comb'-type grafts and precursor-type dispersants. However, in the latter case the true dispersant is formed only gradually from the precursor during polymerization, so that its effective concentration at the critical period when particles are forming depends on the relative rates of this grafting reaction and the polymerization of the monomer. For this reason, particle size falls as the proportion of reactive groups in the precursor is raised, or if the type of reactive group is replaced by one with reactivity ratios more favourable to copolymerization.

For example, in the dispersion polymerization of a mixture of vinylidene chloride, methyl acrylate and acrylic acid (weight ratio: 76/19/5) in heptane, using a copolymer of allyl methacrylate and 2-ethylhexyl acrylate as dispersant precursor[62], the particle size varied smoothly from 2·0–2·6 to 1·0–1·5 µm as the allyl methacrylate content of the dispersant was raised from 0·25% to 4%. Similarly, larger particles were obtained by polymerizing methyl methacrylate in heptane in the presence of nominally pure poly(isobutene) than with a commercial sample containing 1·5–2·0% unsaturation[63]; as expected, smaller particles were produced by raising dispersant content or lowering monomer content.

### 4.3.2.5. Renucleation, agglomeration and flocculation

So far we have been mainly concerned with the factors controlling the numbers of particles formed at the beginning of a dispersion polymerization, i.e. the nucleation stage, which in turn defines the size and numbers of particles finally present if other processes do not intervene.

Renucleation is the formation of fresh particles at a later stage in the polymerization. It is most likely to occur in feed processes, in which the original seed particles grow by further addition of monomer together with more dispersant, since the latter may at times exceed the amount required to protect the increasing surface area of the growing particles.

Renucleation is inhibited by the presence of existing particles, and in this respect the total number of particles seems to be more important than the total polymer concentration, so that coarse dispersions renucleate much more readily than fine ones of equivalent concentration.

Renucleation is also promoted by all the factors which tend to produce fine particles, so that we may summarize the situation as follows:

| *Renucleation favoured by:* | *Renucleation inhibited by:* |
| --- | --- |
| Coarse particle size | Fine particle size |
| Low total solids | High total solids |
| Low solvency medium | High solvency medium |
| Low free monomer content | High free monomer content |

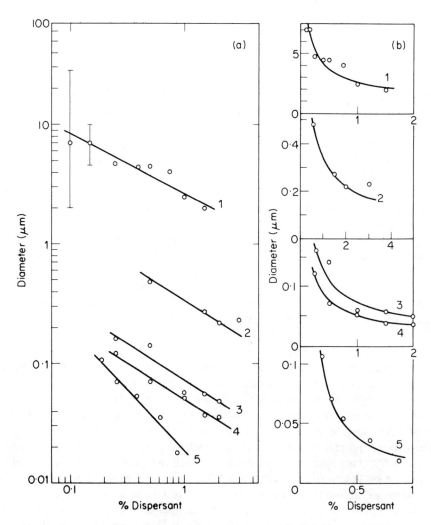

Figure 4.6. Dependence of particle size on dispersant concentration: (a), Double logarithmic plot; (b), Direct plot, with curves corresponding to straight lines in (a). 1, Data of Table 4.8; 2, 3, 4, Data of Table 4.6; 5, Data of Table 4.7

| More polar monomers | Less polar monomers |
| High dispersant content | Low dispersant content |
| Strongly anchored dispersant | Weakly anchored dispersant |
| (more polar anchor chain) | (less polar anchor chain) |

The effect of renucleation, of course, is to produce a bimodal or polymodal particle size distribution, whilst at the same time preventing the original particles from growing to as large a size as if renucleation had not taken place.

The aggregation of polymer particles occurs in the absence of sufficient effective dispersant to protect the total particle surface, and can take the form of flocculation or agglomeration.

Agglomeration is the formation of new larger particles by the aggregation of limited numbers of primary particles, which may be followed by partial or complete fusion.

Flocculation is the formation of loose aggregates or flocs over extended regions, sometimes throughout the whole of a dispersion, which may then appear to gel or coagulate.

Agglomeration of primary particles can in principle afford a means of increasing the size of particles in a dispersion and reducing their number. Such controlled secondary aggregation has rarely been demonstrated conclusively in practice. It would not be expected to be revealed in the shape of the particles formed since agglomeration would usually be followed by complete coalescence in the conditions normally used in dispersion polymerization. It is therefore necessary to demonstrate that the particle size after aggregation has increased by an amount more than could be accounted for from the particle size before aggregation and the extra amount of monomer polymerized, or in other words, that the number of particles has decreased.

In principle, we should expect agglomeration to be favoured by the use of amounts of dispersant which are inadequate to provide complete stabilization and this has for example been demonstrated in the surfactant-stabilized dispersion polymerization of methyl methacrylate in very dilute aqueous solution[56]. However, in the conditions which are normally used for the dispersion polymerization of methyl methacrylate in hydrocarbons, the use of inadequate amounts of dispersant usually results in gross flocculation and coagulation of the dispersion, rather than in agglomeration. In fact, one of the most important aspects in developing practical formulations, is the establishment of practical limits to the amount and rate of addition of dispersant sufficient to prevent flocculation, but small enough to avoid too much renucleation. Paradoxically, excessive renucleation can also lead to subsequent flocculation because the additional dispersant cannot keep pace with the extra surface formed by growth of the new particles.

Other polymers which form fairly rigid particles behave similarly, but polymer particles which are semi-liquid under polymerization conditions [e.g. poly(vinyl acetate) or poly(ethyl acrylate)] more readily undergo agglomeration, and flocculation, if it occurs, is a much more gradual process.

However, controlled agglomeration has been shown to occur occasionally with methyl methacrylate when using a type of dispersant which can anchor irreversibly to the surface of the particles through copolymerizable groups (see Section 3.5.3). Under some conditions, the ultra-fine polymer particles produced appear to become partially destabilized at a certain stage in their growth, but agglomeration rather than gross flocculation occurs, with complete coalescence to form larger spherical particles[64]. Under these circumstances, the initial dispersant concentration may have relatively little effect on particle size.

A similar phenomenon with only partial fusion occurs in the dispersion polymerization of vinyl chloride under some conditions, producing particles with characteristic protuberances[65] (Figure 4.15, Section 4.4.1).

A possible explanation for the behaviour may be related to the fact that such irreversibly anchored dispersant molecules are held fast to fixed points on the particle surface and cannot move about, so that particle growth causes a uniform attenuation of the stabilizer barrier, whereas dispersant merely adsorbed at the surface, even if fairly strongly anchored, can be displaced on the particle surface, so that completely bare patches can appear when the surface area increases sufficiently in size. Even in the dilute aqueous polymerization system referred to above, part of the stabilizing action may arise from comparable fixed ionic groups, in the form of terminal carboxyl and sulphate groups introduced into the polymer from the breakdown of the initiator and subsequent oxidation (initiators: persulphate/bisulphite redox or hydrogen peroxide with ferrous catalyst).

It is conceivable that agglomeration processes may occur after primary particle formation in the more typical conditions of dispersion polymerization in aliphatic hydrocarbons. However, the weight of evidence from electron micrographs points to constancy of particle numbers from a very early stage and if any such agglomeration takes place it must be almost immediately after the primary nucleation stage, or even concurrent with it.

### 4.3.3. Theories of particle formation

There is as yet no single definitive theory able to predict quantitatively the number of particles formed in a dispersion polymerization under defined conditions. It is worthwhile bearing in mind that this is still also the situation in the field of aqueous emulsion polymerization: whilst the theory put forward by Smith and Ewart in 1948[3] appears to explain the kinetics of polymerization satisfactorily in a wide range of experimental systems, the numbers of particles found conform to their predicted relationships only for a very few monomers under very limited conditions[66–68]. Here we will examine several approaches which can contribute to a theoretical treatment of particle formation.

*4.3.3.1. Equilibrium systems and irreversible processes*

Some systems are known in which a phase separates out in the form of particles or domains of well-defined size and number according to a true thermodynamic equilibrium: the outstanding example is the formation of micelles in solutions of amphipathic substances, including both low molecular weight surfactants[69] and amphipathic polymers[69,70]. Microphase separation in block copolymers is a similar phenomenon in which the equilibrium may sometimes be 'frozen' as the material solidifies from a molten state or from evaporating solutions[70,71]; the spontaneous emulsification of some acidic copolymers in aqueous alkali is a further example[72]. In all of these systems, growth beyond a certain size is limited because it would force parts of the added molecules into energetically unfavourable situations, and equilibration is possible only because the mobility or solubility of the separating phase is high enough to permit it.

In contrast to such true equilibrium systems, the formation of particles in dispersion polymerization must clearly be regarded as an irreversible process: once formed, the particles are thermodynamically stable and continue to grow. Moreover, further polymerization within the particles lowers their solubility to a point where transfer of material from one particle to another is virtually impossible. The numbers and sizes of particles are therefore determined by the balance of several competing rate processes and the way they vary during the course of polymerization. The most important of these is the rate of nucleation, i.e. the rate of formation of new particle nuclei and the way in which this is related to the number and size of existing particles, the presence of a dispersant and the rate at which new polymer chains are initiated in the diluent phase.

*4.3.3.2. Qualitative models of the nucleation process*

The nucleation process starts in an essentially homogeneous solution containing monomer, initiator and usually an amphipathic dispersant. Exactly as in conventional solution polymerization, the first steps involve production of radicals by breakdown of the initiator, which in turn react with monomer to form growing oligomeric chains with a reactive free radical at the end.

Three different models have been proposed for the formation of particle nuclei from these growing oligomer chains (Figure 4.7):

*Self-nucleation*: Each individual oligomer chain as it grows, at first has an extended configuration in solution, but then collapses into a condensed state when it reaches a certain threshold molecular weight depending on its solubility in the medium. This condensed oligomeric chain therefore constitutes a new particle nucleus. According to this view, proposed by Fitch and Tsai[56], the behaviour of each oligomer chain is unaffected by the presence of other oligomer molecules, so every chain initiated forms a new particle unless it is captured by diffusion to an existing particle before it reaches the threshold molecular weight.

*Aggregative nucleation*: Growing oligomer chains associate with each other increasingly as their molecular weight and concentration rise, at first reversibly. Aggregates below a certain critical size are unstable, but above this critical

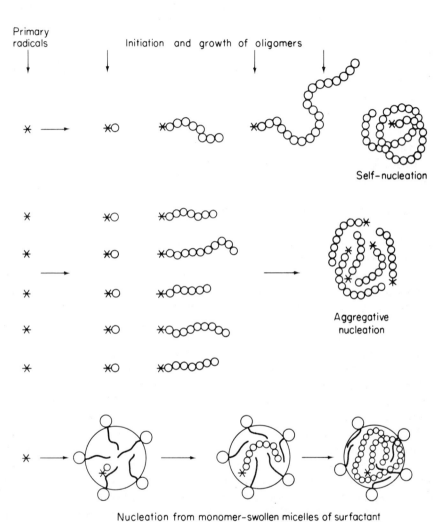

Figure 4.7. Three models for the formation of particle nuclei

size they are stable and tend to grow, constituting new particle nuclei. According to this view, which corresponds to the classical theory of homogeneous nucleation developed by Volmer[73], and by Becker and Döring[74], the rate of nucleation is dependent on the activation energy required to form a critical aggregate, which can in principle be calculated from the interfacial tension between the two phases and the relative supersaturation of the solute. Both rising concentration and rising molecular weight of the oligomer chains therefore result in a sharply increasing rate of nucleation, but as in the previous model, capture by existing particles competes with the process of aggregation to form nuclei.

*Nucleation from micelles*: Particle nuclei are formed by growth of oligomer chains initiated in monomer solubilized in micelles of surfactant or amphipathic dispersant. This, of course, is the model proposed by Harkins for emulsion polymerization[2].

In aqueous emulsion polymerization, the micellar model appeared plausible in view of the very low concentration of molecularly dissolved monomer in the aqueous phase. However, even in this case, it is now increasingly recognized[75,76] that the primary initiation step probably takes place in the aqueous phase between a water-soluble radical and the low concentration of dissolved monomer; growing oligomers then either enter a micelle or acquire a layer of surfactant around them, provided there is excess surfactant present. In fact, Roe has shown that quantitative conclusions based on the micellar nucleation model are equally well derived by treating the micelles merely as a reservoir of available surfactant and monomer[4].

Although the types of amphipathic copolymers used as dispersants in dispersion polymerization are capable of forming micelles (Sections 3.6 and 6.4), there is even less reason to believe that they play any significant part in the formation of particle nuclei in a situation where monomer is completely soluble in the diluent. We shall therefore, give no further consideration to this model, except in so far as it is formally equivalent to an extreme version of dispersant-limited nucleation. (See Section 4.3.6.)

The other two models are best regarded as complementary rather than mutually exclusive. A wide range of possible situations can be envisaged, depending on the solubility of the polymer, its dependence on molecular weight and the rate and mechanism of polymerization. At the 'nearly soluble' end of the range, a substantial concentration of polymer will build up in solution before critical supersaturation is reached, with formation of nuclei involving many polymer molecules; at the other extreme, self-nucleation of individual oligomer molecules may take place, especially if the initiation rate and polymer solubility are both very low: this model may well be valid for some aqueous systems such as the one studied by Fitch and his associates. In the intermediate situation which is probably more typical of dispersion polymerization in hydrocarbon diluents, oligomer chains may aggregate to form nuclei while still growing, but the rate of nucleation and numbers of molecules per nucleus will be strongly dependent on their concentration.

Before passing on to the processes which limit nucleation, mention should be made of the special type of condensation polymerization involving predispersed monomers (see Section 5.5.2). Here there is no true nucleation process, since the emulsified droplets of monomer are transformed directly into particles. The process is therefore essentially a form of suspension polymerization, and as in the familiar aqueous systems of this type, the size of droplets and therefore the size and number of polymer particles are determined by the amount and nature of the amphipathic emulsifier, the viscosities of the two phases and the degree of agitation used in the emulsification stage.

### 4.3.3.3. Suppression of nucleation: capture of oligomers

If no competing process came into action, the formation of particle nuclei might be expected to continue throughout the course of polymerization as long as free monomer remains. In the aggregation model, the build-up of oligomers above the supersaturation threshold would result in an initial burst of nucleation followed by a fall as supersaturation is lowered due to precipitation of particles, reaching a steady-state level in which the rate of initiation of the oligomers is balanced by their rate of loss to form nuclei. Alternatively, according to the self-nucleation model, the rate of nucleation would follow the rate of initiation throughout, falling gradually as monomer becomes depleted.

In practice, of course, the rate of formation of particles usually falls to a negligible level very early in the course of polymerization, unless conditions are substantially altered: once many particles are present, the formation of new particles is strongly inhibited. The most likely reason is that nearly all the oligomer molecules forming in the diluent phase are captured by existing particles before they can form new nuclei. Indeed they are probably captured while still very small and continue to grow within the particles, serving essentially as a source of radicals from the diluent.

The rate of oligomer capture clearly increases with the rising number and the size of particles and since it must largely determine the number of particles formed it is obviously important to try and quantify such relationships. Two approaches have been adopted, depending on whether capture is regarded as controlled by a diffusion or by an equilibrium process:

(i) *Diffusion capture*: According to the model proposed by Fitch and Tsai[56], any oligomer which reaches an existing particle by diffusion before it has attained the critical size for self-nucleation, is irreversibly captured. They derive an expression for the rate of capture which is proportional to the surface area of the particle and the number of particles; the difference between this and the rate of initiation defines the rate of nucleation. The result incorporates a parameter $L$ representing the average distance an oligomer can diffuse before attaining the critical size for self-nucleation, which therefore, characterizes the solubility of the polymer (see Figure 4.11, Section 4.3.5). With minor modifications the theory can be adapted to allow for nucleation by aggregation of oligomers.

However, the method of deriving the basic equation for rate of capture appears to be at variance with classical diffusion theory, which would indicate that a very much higher proportion of oligomers is captured, at a rate proportional to the number and the diameter, rather than the surface area, of the particles[77]. The diffusion equations assume that the diffusing molecules are removed from the system irreversibly when they reach a particle surface and if this were rigorously true, far fewer particles could be formed than are found in practice.

(*ii*) *Equilibrium capture*: In fact, there is every reason to believe that at least the lower oligomers of growing polymer chains are subject to a dynamic equilibrium between diluent and surface, perhaps moving from one to the other many times between the addition of each monomer unit. There is, in addition, the possibility of further interchange with temporary aggregations of oligomers on the one hand and the interior of particles on the other (Figure 4.8). Effectively irreversible removal of an oligomer from the diluent can then take place when (*a*), it becomes part of an aggregation above the critical size for nucleation (or grows large enough for self-nucleation), or when (*b*), it passes from the loosely adsorbed surface layer to the interior of a particle when it is sufficiently large and insoluble that its chance of escape before adding another monomer unit is negligible.

The proportions of oligomer molecules undergoing capture or nucleation are therefore decided by their equilibrium distribution immediately prior to this process of irreversible removal, rather than by diffusion. This approach is essentially analogous to the treatment of chemical reaction rates by transition state theory[78].

According to the equilibrium theory, the parameter which controls nucleation is the total surface area of the particles, as in the theory of Fitch and Tsai, although in a slightly different form. (A further nucleation parameter could be the total volume of the particles if they are sufficiently fluid to allow a rapid equilibration with their interiors.)

All of the approaches described above of course, require a knowledge of the rate of growth of particles before they can be applied to actual situations, and plausible assumptions can be made about this in the light of the mechanism of polymerization (Section 4.4). However, the theories postulated also involve parameters which cannot easily be estimated independently, although Fitch and Tsai attempted to measure the molecular weight of polymer just prior to precipitation[56]. Nevertheless, useful conclusions can be drawn about the general form of the relationships to be expected.

Except for the micellar nucleation model, all of the mechanisms discussed so far would yield unstabilized particle nuclei, which rapidly flocculate while continuing to grow, followed by coalescence or agglomeration to an extent dependent upon the physical properties of the polymer and its plasticization by monomer etc. The typical product of an unstabilized precipitation polymerization is therefore a coarse aggregate of particles which themselves consist of smaller primary particles largely fused together.

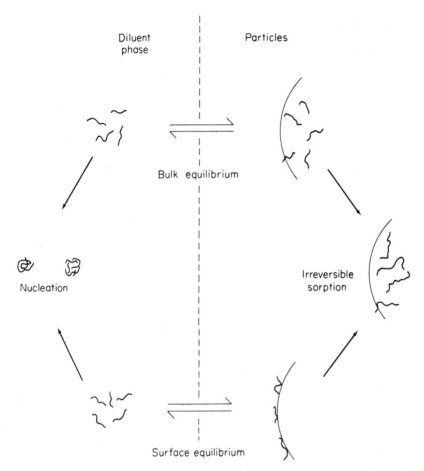

Figure 4.8. Equilibrium capture of oligomers. Models for bulk (above) and surface equilibration (below). Irreversible sorption or nucleation takes place when the oligomers grow above a threshold molecular weight

### 4.3.3.4. Particle formation in the presence of dispersants

The nature and the amount of amphipathic dispersant used in a dispersion polymerization has a profound effect on the occurrence of nucleation and the numbers of particles formed as well as preventing flocculation. It is possible to propose a plausible role for the dispersant molecules in modifying each of the processes postulated in the various theoretical models, and a great deal of further work is needed to establish exactly the complete role which a dispersant plays in any given polymerization system. In each case, enhanced nucleation is accounted for by some form of association between the dispersant and the growing oligomer which raises the probability of forming a nucleus and in turn, lowers the probability of capture by existing particles. The effects of increasing polarity and insolubility of both the anchoring portion of the dispersant and the polymer being formed therefore fall naturally into place since these will increase the tendency to associate and hence increase the rate of nucleation (Figure 4.9).

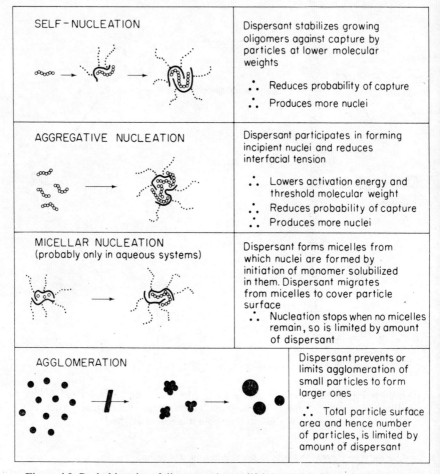

Figure 4.9. Probable roles of dispersant in modifying particle formation processes

So far, it has been tacitly assumed that nucleation is brought to a halt by prior capture of oligomers while there is still excess unused dispersant present, so that the concentration of dispersant can be taken as roughly constant throughout the nucleation period. However, there are two possible situations in which depletion of dispersant could become the limiting factor determining particle size and number:

(i) *Dispersant-limited nucleation*: The dispersant is so efficient that nucleation continues at a high rate until there is very little dispersant remaining unattached to particles; thereafter nearly all oligomers are captured by particles. This is essentially the model proposed by Roe for emulsion polymerization[4], and predicts relationships equivalent to the micellar model of Smith and Ewart[3]. Whilst this behaviour has been taken as the normal situation in emulsion polymerization, there is little evidence for its widespread occurrence in organic media, at any rate in such an extreme form.

(ii) *Dispersant-limited agglomeration*: Primary particles grow in size and in number until the dispersant available is insufficient to cover the surface effectively and prevent flocculation. In principle, this may occur either while nucleation is still under way, or well after it has been brought to a halt by prior capture of oligomers. The usual practical result of running out of dispersant in a dispersion polymerization is gross flocculation throughout the dispersion, but in certain circumstances controlled agglomeration occurs instead, to form fewer and larger particles with a smaller total surface. In its extreme form, this model assumes that the dispersant serves merely to prevent secondary aggregation and plays no part in the formation of primary particles[56].

Controlled agglomeration of this type has only been clearly demonstrated with very soft polymers or under very special conditions. However, its onset would be difficult to detect if it took place very early in the course of polymerization, when particles are extremely small and growing rapidly, and its occurrence may be more widespread than is apparent at present.

*4.3.3.5. The overall picture*

The various theoretical approaches, taken together, provide a satisfactory description of particle formation and an explanation of the effects of solubility, molecular weight and the amount and nature of the dispersant. In the sections which follow an outline is given of each of the main theories, together with the quantitative and qualitative conclusions which can be drawn from them.

**4.3.4. The theory of homogeneous nucleation**

Many features of particle formation in precipitation or dispersion polymerization run closely parallel to the formation of inorganic precipitates from aqueous solutions and the formation of aerosol dispersions from the vapour phase. These include the suppression of new nucleation when many particles have been formed and the effects of increasing solubility in coarsening particle size and delaying nucleation.

Von Weimarn[79] showed that the controlling factor is the relative supersaturation $S_r$ (actual concentration divided by equilibrium solubility). Nuclei are not formed rapidly until a threshold value of $S_r$ is exceeded, and above this value the rate of nucleation rises very steeply. More insoluble substances produce more particles because the threshold value of relative supersaturation is exceeded more readily and by a greater margin, before the supersaturation is relieved.

The theory of homogeneous nucleation was developed by Volmer[73], and by Becker and Döring[74], to account for these relationships by calculating the activation energy required to form an irreversible association of molecules, i.e. a nucleus. While mainly applied to vapours and aqueous solutions[80], the theory has also been invoked to provide a qualitative explanation of some features of turbidimetric titration of polymer solutions[81] and the morphology of particles formed in heterogeneous bulk polymerization of vinylidene chloride[82]. The same principles have been used with great success in quantitative studies of crystallization of supercooled polymer melts and solutions[83].

For purposes of calculation, the incipient nuclei are treated as spherical globules. The net free energy of formation of a globule of radius $r$ is the difference between the work done against interfacial tension $\gamma$ to form its surface, and the free energy which would result from dilution of the molecules making up its volume to the same concentration as the surrounding solution:

$$\Delta G = \gamma(4\pi r^2) - \Delta G_v(4\pi r^3/3)$$

where $\Delta G_v$ is the bulk free energy of dilution per unit volume of the precipitated phase.

The interfacial energy term is always positive and opposes growth. Below the saturation concentration, the free energy of dilution is negative and also opposes growth. In supersaturated solutions, the free energy of dilution is positive, but very small globules shrink and disappear because the interfacial energy outweighs it. Since the volume of a sphere rises faster than its area, there is a critical size above which the net free energy of formation falls and the globule therefore tends to grow, becoming a permanent nucleus (Figure 4.10). As the degree of supersaturation rises, $\Delta G_v$ rises with it, so the critical size and free energy of formation fall rapidly. Their values, which correspond to the maximum net free energy of formation at a given degree of supersaturation, are readily found by differentiating and equating to zero:

Critical radius

$$r^* = 2\gamma/\Delta G_v \tag{4.23}$$

Critical free energy of formation

$$\Delta G^* = 16\pi\gamma^3/3\,\Delta G_v^2 \tag{4.24}$$

The essential step in the theory is to equate this final value with the activation energy for nucleation, so that the rate of nucleation is given by:

$$R_n = dN/dt = A\exp(-\Delta G^*/kT) \tag{4.25}$$

where $k$ is Boltzmann's constant and $T$ the absolute temperature.

For low molecular weight solutes, and to a fair approximation also for very dilute solutions of polymers (cf. equation 4.10), the free energy of dilution per unit volume is given by:

$$\Delta G_v = (RT/V_m)(\ln c - \ln c_s)$$
$$= (RT/V_m)(\ln S_r)$$

where $V_m$ is the molar volume of the solute, $c$ its actual concentration and $c_s$ its equilibrium solubility.

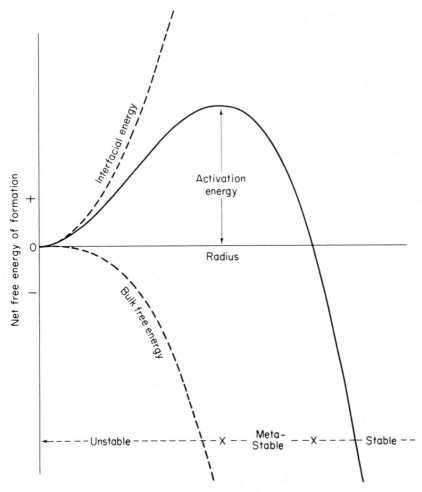

Figure 4.10. Activation energy for formation of a stable nucleus. Dependence of net free energy of formation of a globule on its radius

The rate of nucleation is therefore related to the relative supersaturation by an expression of the form

$$R_n = A \exp[-C\gamma^3(V_m/\ln S_r)^2] \tag{4.26}$$

where $C = 16\pi/3kT(RT)^2$ and has a value of $4 \times 10^{-7}$ at $80\,°C$.

The characteristic feature of this expression is that it increases from vanishingly small values to very high levels over quite a small range of supersaturation, so that for all practical purposes there is a threshold degree of supersaturation for nucleation, the actual value of which depends on the interfacial tension. The factor $A$ is a function of the collision frequency in solution and the concentration of the solute. Whilst it is difficult to estimate precisely, both theoretical considerations[83,84] and comparison with other systems, indicate that $A$ should be of the order of $10^{30}$, and fortunately the behaviour of the system is quite insensitive to variations in this parameter over many orders of magnitude.

Using these equations, we may draw certain conclusions about the threshold conditions for precipitation of polymers and the typical size of nucleus to be expected.

*4.3.4.1. Size of nucleus*

Combining equations (4.23–4.25), we obtain for the critical radius

$$r^* = [3kT\ln(A/R_n)/4\pi\gamma]^{\frac{1}{2}}$$

This implies that the size of a nucleus is dependent only on the interfacial tension at a given rate of nucleation (although this rate will itself depend on the concentration and molecular weight). Since the rate of nucleation increases very steeply with degree of supersaturation, we can take the threshold nucleation rate as some arbitrary value, say $10^6$ nuclei per ml per second, at which the formation of particles should begin to be apparent, so that $\ln(A/R_n)$ would be about 55. The interfacial tension between typical polymers formed by free radical polymerization in a mainly aliphatic hydrocarbon medium may be taken as in the range 5 to 15 dyn/cm[85], resulting in corresponding values of 3·6 nm and 2·1 nm for the critical radius.

These are values for the size of nucleus at the threshold of nucleation, but they vary only very slightly as the nucleation rate increases: even if the threshold supersaturation is exceeded sufficiently to give a nucleation rate of $10^{16}$ nuclei per ml per second (enough to complete the nucleation stage of a typical latex in a fraction of a second), $\ln(A/R_n)$ falls to 32 and the corresponding values for the critical radius are 2·7 nm and 1·6 nm. This illustrates the very small effect resulting from the choice of values taken for $A$ and $R_n$.

*4.3.4.2. Number of molecules per nucleus and the nucleation rate*

From the critical radius and molar volume, the number of molecules per nucleus is readily obtained as

$$Q = 4\pi r^{*3} N_A/3V_m = 2C\gamma^3 V_m^2/(\ln S_r)^3 \tag{4.27}$$

where $N_A$ is Avogadro's number.

It is readily shown from equation (4.26) that the number of molecules per nucleus defines the rate of change of nucleation rate with supersaturation:

$$Q = d(\ln R_n)/d(\ln S_r) \quad \text{(if } V_m \text{ remains constant)} \tag{4.28}$$

$$= d(\ln R_n)/d(\ln c) \quad \text{(if solubility remains constant)}$$

The very steep rise in nucleation rate with supersaturation therefore follows naturally from the fact that many molecules are needed to form a stable nucleus which will not break up again spontaneously. This relationship is not surprising since it corresponds to an equilibrium reaction of order $Q$ between single molecules and incipient nuclei, so that

$$R_n \propto \text{concentration of incipient nuclei} \propto c^Q$$

In the case of polymers, the number of molecules in a nucleus of a given size obviously depends on the molecular weight. Taking a typical molar volume of 100 ml per monomer unit, the number of monomer units which can be accommodated in nuclei of the size estimated above is 100–220 for an interfacial tension of 15 dyn/cm, or 500–1200 at 5 dyn/cm. Whilst the number of oligomers involved depends on their average threshold molecular weight for nucleation, it seems likely that in typical conditions of dispersion polymerization, some 5 to 50 oligomer chains may take part in forming a nucleus.

### 4.3.4.3. Threshold supersaturation conditions

Equation (4.26) defines the relative supersaturation corresponding to a given rate of nucleation by:

$$(\ln S_r)/V_m = [C\gamma^3/\ln(A/R_n)]^{\frac{1}{2}} = H \qquad (4.29)$$

For monomolecular solutes, the molar volume is constant and therefore a definite threshold supersaturation can be specified for a given interfacial tension. The treatment of polymer systems is much more complex since both the molar volume and the solubility depend on the molecular weight. If we introduce the approximation (4.12) for polymer solubility, the relative supersaturation is given by

$$\ln S_r = \ln \phi + m(\chi - 1) \qquad (4.30)$$

where $\phi$ is the volume fraction of polymer in the solution, $m$ its degree of polymerization and $\chi$ the polymer–solvent interaction parameter. It follows that the polymer concentration needed to produce a given rate of nucleation is given by

$$\ln \phi = -m(\chi - 1 - V_0 H) \qquad (4.31)$$

where $V_0$ is the molar volume per monomer unit of the polymer, typically about 100 ml. For the threshold nucleation rate assumed previously, the values of $V_0 H$ would be 0.1 or 0.5 for interfacial tensions of 5 and 15 dyn/cm respectively. Since for the insoluble polymers we are considering, $\chi$ is much greater than unity, it is clear that the threshold polymer concentration for nucleation will fall exponentially as its molecular weight increases, just like the polymer solubility but not quite as rapidly, even though the threshold relative supersaturation rises. The threshold concentration for nucleation is lowered closer to the equilibrium solubility as the interfacial tension falls.

### 4.3.4.4. Threshold molecular weight and nucleation rate

If polymer is precipitating during polymerization, not only its concentration but also the molecular weight and its distribution, are all varying continuously. The weight average degree of polymerization probably gives the fairest approximation to overall behaviour: obviously the highest molecular weight components will precipitate preferentially. At any overall concentration $\phi$, there will be a threshold degree of polymerization for nucleation approximated by

$$m = -\ln \phi/(\chi - 1 - V_0 H) \qquad (4.32)$$

According to circumstances either $m$, $\phi$, or even $\chi$, or any combination of them, may be the variable which reaches a threshold value for nucleation while the others remain more or less constant. However, the molecular weight is different from the other parameters in that a wide distribution of molecular weights is present at any moment, and each single oligomer molecule is constantly growing. While equations (4.30–4.32) strictly refer to the behaviour of a polymer of uniform molecular weight, similar relationships should apply to each species in the range of molecular weights present.

For the rate of increase in nucleation rate with molecular weight, an equation analogous to (4.28) is readily derived:

$$d(\ln R_n)/d(\ln m) = -Q \ln \phi \qquad \text{(if } \phi \text{ remains constant)} \qquad (4.33)$$

This implies that the nucleation rate rises extremely steeply with molecular weight, since at the low concentrations expected in the precipitation of an insoluble polymer, $\ln \phi$ is a large negative number.

In fact, at a molecular weight only a little higher than the threshold for nucleation, the rate of disappearance of oligomers in forming nuclei will overtake the rate at which oligomers are formed, i.e. the initiation rate $R_i$. Hardly any oligomers will survive much above this molecular weight.

An estimate of the range of molecular weight over which the oligomers take part in forming nuclei can be made from equation (4.32) since

$$m_1/m_2 = (\chi - 1 - V_0 H_2)/(\chi - 1 - V_0 H_1)$$

where subscripts 1 and 2 refer to a nucleation rate balancing initiation (typically of the order of $10^{16}$ per ml per second) and the threshold nucleation rate ($\sim 10^6$ per ml per second) respectively. Taking $\chi$ as 2, the ratio of molecular weights is 1·45 for an interfacial tension of 15 dyn/cm or only 1·035 for 5 dyn/cm. Obviously the range over which nucleation is completed may be very small, especially at low interfacial tensions and with polymers of very low solubility. This presents some theoretical justification for the conception used below (Section 4.3.5) of a threshold degree of polymerization $P$ above which all oligomers take part in forming nuclei unless they have been previously captured by existing particles.

### 4.3.4.5. Approach to a steady state

If the threshold degree of polymerization $P$ is so low that hardly any oligomer radicals are terminated before they precipitate, the average degree of polymerization will be $P/2$ and the time each takes to reach the threshold value is

$P/k_p[M]$, so that the volume fraction of polymer in solution is simply given by

$$\phi = V_0 R_i P^2 / 2k_p[M] N_A$$

where $R_i$ is the number of oligomers initiated per second per ml, $k_p$ the propagation rate constant for the monomer and $[M]$ its concentration. Steady-state conditions are then defined by substituting in (4.29) and (4.30), taking a value of $R_n$ corresponding to $R_i$ (strictly $R_n = R_i/Q$, but the difference is insignificant in view of the steep rise of nucleation rate with molecular weight):

$$(1/P)\ln(V_0 R_i P^2 / 2k_p[M] N_A) = -(\chi - 1 - V_0 H) \tag{4.34}$$

The logarithmic expression is very insensitive to changes in any of the parameters, and over a very wide range of conditions lies between $-10$ and $-20$, so that $P$ is inversely proportional to the right-hand side of the equation, which specifies the effect of increasing insolubility ($\chi$) or decreasing interfacial tension ($H$) in lowering the threshold molecular weight for nucleation.

For example, if $\chi = 2$, and values of the other parameters correspond roughly to the seed stage of a dispersion polymerization of methyl methacrylate at 80 °C ($V_0 = 83$ ml, $R_i = 10^{16}$ per ml per sec, $k_p = 800$ l/mol.sec, $[M] = 0.35$ mol/l), then $P = 36$ if the interfacial tension is 15 dyn/cm and $P = 16$ at 5 dyn/cm. As the interfacial tension falls further, $P$ approaches a limiting value of 14 for this particular value of $\chi$; corresponding limiting values of $P$ would be 60 for $\chi = 1.2$ or 10 for $\chi = 2.5$.

While this theoretical approach can be regarded as no more than a rough approximation, it does give some idea of the order of magnitude for the degree of polymerization reached before oligomers precipitate.

Equation (4.34), of course, only applies if the threshold degree of polymerization is well below the value which would be reached in solution polymerization at a similar initiation rate. At lower values of $\chi$ (less insoluble polymers) this is obviously no longer the case: higher molecular weight oligomers will be progressively depleted by prior termination, so that a correspondingly higher concentration must be reached before the rate of nucleation matches the rate of initiation. In the extreme case, practically all the oligomer radicals may be terminated in solution, and the steady-state concentration reaches a relatively high level defined by equation (4.31), $m$ now being the degree of polymerization reached in solution. For such low values of $\chi$, the approximation (4.12) is no longer valid, but a similar, nearly-exponential relation between solubility and molecular weight can be derived. An extreme case of this type occurs in the polymerization of vinyl chloride in dispersion or bulk. A value of 0.88 has been obtained for $\chi$ for this polymer/monomer system at 50 °C[38], and this corresponds to appreciable solubility at low molecular weights, falling only very slowly as the degree of polymerization rises. It follows that nearly all the polymer chains are terminated before they form nuclei, hence nearly all polymerization takes place in solution and nuclei grow only very slowly since hardly any active oligomer radicals are precipitated with them or absorbed by them later.

### 4.3.4.6. The effect of dispersants and solvency

The effects of varying dispersant concentration and the solubility of its anchor chain, which in turn affects the extent of its adsorption at an interface, may be formally incorporated into the theory through their effects on the interfacial tension between diluent and precipitating polymer. This can in principle be measured independently in macroscopic systems, although obviously it is a gross oversimplification to expect quantitative application to transient submicroscopic assemblages involving only a few molecules of oligomer and dispersant.

Qualitatively, the essential conclusion is that a higher dispersant concentration, or an anchor group adsorbing more strongly, results in a lower threshold for nucleation in terms of molecular weight or polymer concentration or both, but also in larger nuclei since the lower threshold implies a smaller energy of formation per unit volume. As indicated by equations (4.29), (4.31) and (4.34), the form of the relationships is comparatively complex and the threshold molecular weight $P$ is much more sensitive to changes in interfacial tension when $\chi$ is relatively low (less insoluble polymer) or when the interfacial tension is high. A law of diminishing returns therefore operates, increasing amounts of dispersant having less and less effect, with $P$ gradually approaching its limiting value. On the other hand, changes in solvency directly alter the value of $\chi$ and have a very strong effect on $P$ and its limiting value. As will be seen below (Section 4.3.5), the value of $P$ is the main parameter controlling the number of particles finally formed.

### 4.3.4.7. Delimitation of number of particles: the suppression of nucleation

The special feature of polymerizing systems, involving growth of individual molecules, permits rapid attainment of steady-state conditions of nucleation. However, as soon as particles have been formed they begin to capture some of the oligomer molecules before they can take part in forming new nuclei. As the number and size of the particles grow, this competing process becomes dominant and more or less rapidly reduces the rate of nucleation to negligible proportions. It is the way in which the rate of nucleation changes as particles form and grow which ultimately defines the total number of particles formed.

Several approaches have been adopted to try and quantify the relationships involved in suppression of nucleation, treating the capture of oligomers as either a diffusion process or an equilibrium adsorption. These will be discussed in the following section (4.3.5). Whatever the theoretical approach used, ultimately it is the threshold degree of polymerization $P$, taken in conjunction with the kinetics of particle growth, which determines the duration of the nucleation stage and hence the number of particles formed. The conclusions reached above about the factors which determine $P$ therefore lead directly to conclusions about particle numbers, lower values of $P$ corresponding to larger numbers of finer particles.

### 4.3.5. Nucleation controlled by capture of oligomers

#### 4.3.5.1. *The diffusion capture theory of Fitch and Tsai*

In the model of particle formation proposed by Fitch and Tsai[56], it is assumed that each oligomer molecule initiated in the diluent will form a fresh particle nucleus if it reaches some threshold degree of polymerization $P$ before being captured by an existing particle (Figure 4.11). The time taken to grow to this

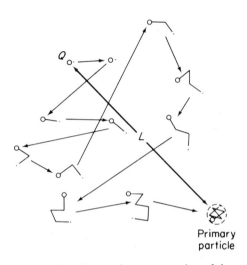

Figure 4.11. Schematic representation of the distance, $L$, through which a growing oligomeric radical diffuses before it precipitates out to form a new particle[56]. (From *Polymer Colloids* (Ed. R. M. Fitch), Plenum Press, 1971, p. 94, with permission)

size is taken as:

$$t_n = P/k_p[M] \qquad (4.35)$$

where $[M]$ is the monomer concentration in the diluent and $k_p$ is the propagation rate constant. According to the laws of diffusion, the average distance the oligomer can diffuse in this time is:

$$L = (2t_n D)^{\frac{1}{2}} = (2PD/k_p[M])^{\frac{1}{2}} \qquad (4.36)$$

where $D$ is its mean diffusion coefficient.

An equation is derived for the rate of capture of oligomers which have travelled less than the critical distance $L$, by a single particle of radius $r$, if oligomer molecules are generated at a rate $R_i$ per second per unit volume:

$$\text{Rate of capture} = \pi r^2 L R_i;$$

the total rate of capture by $N$ particles per unit volume is then,

$$R_c = \pi r^2 L R_i N$$

and the rate of nucleation is

$$R_n = dN/dt = R_i - R_c = R_i(1 - \pi r^2 LN) \tag{4.37}$$

The average radius $r$ can be expressed in terms of the number of particles $N$ and the volume of polymer $V$ formed in unit volume of diluent, giving

$$R_n = R_i[1 - L(\pi N)^{\frac{1}{3}}(3V/4)^{\frac{2}{3}}] \tag{4.38}$$

If an expression is known for the rate of polymerization $R_p$, $V$ can be found by integration, assuming that essentially all the polymer is present as particles and not appreciably swollen:

$$V = V_m \int_0^t R_p \cdot dt \tag{4.39}$$

where $V_m$ is the molar volume of the polymer repeating unit. The total number of particles is then obtained by integrating $R_n$ from the start, up to the moment when it falls to zero, i.e. $R_c = R_i$.

In their work, Fitch and Tsai used the equations for $R_p$ corresponding to the non-steady state initial period of free radical polymerization in solution, when radical concentration is building up to the steady state conditions, giving:

$$V = (k_p V_m[M]/k_t) \ln[\cosh(R_i k_t)^{\frac{1}{2}} t] \tag{4.40}$$

where $k_t$ is the termination rate constant (Section 4.4.2). The resulting expression for $R_n$ is mathematically intractable and can only be integrated numerically for each set of initial parameters. They claim good agreement with numbers of particles found in practice in the polymerization of methyl methacrylate in very dilute aqueous solution, but results are strongly dependent on an arbitrary selection of a low value for the initiator efficiency and a choice of a value for $L$ (estimated from the highest molecular weight found in solution just prior to visible precipitation).

### 4.3.5.2. A generalized form of diffusion capture equation

Much simpler equations result if the relation between the volume of polymer and time of formation can be expressed as a simple power law, since then the rate of nucleation takes the form

$$dN/dt = A(1 - Bt^x N^y) \tag{4.41}$$

and the final number of particles $N_f$ is obtained by integrating up to the limit at which $dN/dt$ is zero. If $n_f$ is the corresponding solution obtained when $A = B = 1$, it is easily shown that

$$N_f = n_f \cdot A^{x/(x+y)} B^{-1/(x+y)} \tag{4.42}$$

(This result is found by expressing the equation in terms of new variables $N' = pN$, $t' = qt$ and equating coefficients.)

It is therefore, only necessary to calculate $n_f$ once for any given exponents $x$ and $y$, in order to obtain a general equation for the final number of particles for any values of $A$ and $B$. For example,

$$\text{for } x = \tfrac{2}{3}, y = \tfrac{1}{3}, \quad N_f = 0.584 \ A^{\frac{2}{3}}B^{-1}$$
$$\text{for } x = \tfrac{4}{3}, y = \tfrac{1}{3}, \quad N_f = 0.662 \ A^{\frac{4}{3}}B^{-\frac{2}{3}}$$

We are now in a position to write down general relationships corresponding to a given mechanism of polymerization. Several which are worth consideration in the context of dispersion polymerization are listed in Table 4.9. The first

Table 4.9. Parameters of nucleation equation for various polymerization models.

|  | Initiation without termination | Steady state in solution | Dispersion polymerization (initial) | Constant radius |
|---|---|---|---|---|
| Polymerization rate, $R_p \propto$ | $[M]R_i t$ | $[M]R_i^{\frac{1}{2}}$ | $V^{\frac{1}{2}}[M]R_i^{\frac{1}{2}}$ | — |
| Polymer volume, $V \propto$ | $[M]R_i t^2$ | $[M]R_i^{\frac{1}{2}}t$ | $[M]^2 R_i t^2$ | $Nr^3$ |
| $t^x N^y =$ | $t^{\frac{1}{3}}N^{\frac{1}{3}}$ | $t^{\frac{1}{3}}N^{\frac{1}{3}}$ | $t^{\frac{1}{3}}N^{\frac{1}{3}}$ | $t^0 N$ |
| Particle number, $N_f \propto$ | $R_i^{\frac{2}{3}}$ $P^{-\frac{1}{10}}$ $[M]^{-\frac{1}{10}}$ | $R_i^{\frac{1}{3}}$ $P^{-\frac{1}{2}}$ $[M]^{-\frac{1}{6}}$ | $R_i^{\frac{2}{3}}$ $P^{-\frac{1}{10}}$ $[M]^{-\frac{1}{2}}$ | $R_i^0$ $P^{-\frac{1}{2}}$ $[M]^{\frac{1}{2}}$ |

two can be regarded as approximations to the non-steady state solution polymerization model chosen by Fitch and Tsai: if the period of particle formation is short compared to the average radical lifetime, termination of radicals may be neglected altogether, whilst if it is relatively long the conventional steady-state kinetic equations may be used. The third model assumes that particle growth is mainly due to polymerization within the particles rather than in solution, following equation (4.72) derived in Section 4.4.3. The time dependence has the same form as in the first model, since it is readily shown that the volume of polymer formed is proportional to $t^2$ in the initial period. The fourth model is not based on any particular polymerization mechanism, but corresponds to the situation which could arise if appreciable amounts of polymer accumulate in solution before nucleation starts, and the subsequent rate of formation of particles becomes much faster than their rate of growth, so that the radius can be taken as practically constant and only their numbers increase.

The first three models, as expected, yield very similar general relationships: in all cases the number of particles increases with the rate of initiation and decreases as the monomer concentration rises (quite apart from any further effect on the solvency of the medium and therefore on the threshold degree of polymerization for nucleation, $P$). The exponents are fractional, so that the number of particles is not very sensitive to variations in the parameters. Also as expected, the relationships estimated from the numerical integrations reported by Fitch and Tsai lie between those predicted for the first and second models, but generally nearer to the steady state approximation.

So far, very little experimental work has been reported which could be used to test the predicted relationships. Dependence on monomer concentration would be very hard to distinguish from indirect effects due to a change in solvency. The most revealing test would be provided by a study of the effect of the rate of initiation on particle number, which should be quite negligible in the constant radius model. Results obtained in the dispersion polymerization of methyl methacrylate in hydrocarbons suggest that the rate of initiation is relatively unimportant, but have not usually extended over a wide enough range to be conclusive. Furthermore, it must be remembered that by increasing the initiation rate, the average molecular weight is lowered and this might in turn reduce the proportion of oligomers which can reach the threshold molecular weight for nucleation.

The experimental work reported by Fitch and Tsai on very dilute aqueous systems also shows some evidence of a slight increase in particle number with monomer concentration, but to a much smaller extent than predicted by the constant radius model.

*4.3.5.3. Diffusion capture with aggregative nucleation*

As discussed in Section 4.3.3, the formation of particle nuclei in organic media may involve the aggregation of oligomers rather than self-nucleation. We can still envisage an effective threshold molecular weight of oligomer for nucleation and a corresponding value of $L$, though this will no longer be independent of the concentration of oligomers and will be at best an idealization of a statistical distribution of molecular weights of oligomers participating in nucleation. If the average number of oligomers forming a nucleus is $Q$, the basic equation (4.42) becomes

$$dN/dt = (R_i/Q)(1 - \pi r^2 LN) \tag{4.43}$$

Provided $Q$ and $L$ can be taken as roughly constant over the period of nucleation, this modification does not alter the general relationships shown in Table 4.9, but merely introduces a numerical factor to slow down the overall rate of nucleation and reduces the final number of particles formed by a factor ranging from $1/Q^{4/5}$ to 1.

*4.3.5.4. Diffusion capture modified by dispersant*

In its original form, the theory of Fitch and Tsai assumes that the threshold degree of polymerization for an oligomer to become the nucleus of a new particle depends solely on the nature of the polymer and diluent, while dispersant or surfactant serves only to hinder subsequent aggregation of the primary particles.

However, we may suppose that dispersant molecules increasingly tend to associate with oligomer molecules as they grow longer, so that the effective threshold degree of polymerization $P$ would be reached when the oligomer is sufficiently surrounded by dispersant molecules to become fully stabilized against capture by another particle. $P$ would then depend on the nature of the dispersant and its concentration, $S$. If, as might be expected, several dispersant molecules are necessary to stabilize the nucleus completely, this dependence may take the form

$$P = k/S^n \qquad (4.44)$$

although such a power law must only be regarded as a semi-empirical approximation to a more complex fundamental relationship and only valid over a limited range of dispersant concentration; $k$ of course is a measure of the weakness of the tendency of the dispersant to associate with oligomer.

According to the various polymerization models we have examined here, the number of particles is proportional to $P^{-\frac{1}{2}}$ or $P^{-\frac{3}{10}}$, so that we should expect a relationship of the form

$$N \propto S^{n/2} \quad \text{or} \quad S^{3n/10} \qquad (4.45)$$

As we have seen in Section 4.3.2, relationships have been found experimentally with exponents in the range 1·5 to 3, which would correspond to values of $n$ in the range 3 to 10, a not unreasonable order of magnitude. However, whilst the proposed modified model gives reasonable relationships, it must at best be regarded only as a semi-empirical framework, with the parameters $k$, $n$ and $Q$ depending on the nature of the polymer, diluent and dispersant but which cannot readily be estimated from independent data.

*4.3.5.5. Validity of the diffusion capture equations*

In deriving their basic equation for the rate of diffusion capture, Fitch and Tsai[56] argue that, since oligomers have an equal chance of diffusing in any direction, the fraction of the oligomers originating at a given distance from a particle, which ultimately collide with it, will be given by the 'cross-sectional' area presented by the particle, divided by the surface area of a sphere of radius equal to the distance of the cross-section from the point of origin: in other words, it is the fraction corresponding to the solid angle subtended by the particle at that point of origin.

By a complicated series of integrations, they arrive at an equation which can be more simply derived as follows. Oligomers reach a particle of radius $r$ uniformly from all directions, i.e. from directions contained within a solid angle $4\pi$, and also diverge uniformly in all directions from their points of origin. Those reaching the particle from an infinitesimal solid angle $\phi$ (effectively from a single direction) and originating within a distance $L$ from its surface are all generated within a cylindrical volume $\pi r^2 L$ (Figure 4.12). They constitute a fraction $\phi/4\pi$ of those generated within that volume, since they diverge in all directions, but also represent the same fraction $\phi/4\pi$ of the total number of oligomers reaching the particle from all directions, which is therefore equivalent to the total rate at which they are generated within the cylinder, i.e. $\pi r^2 L R_i$ (where $R_i$ is the rate of initiation per unit volume). The total rate of capture is then obtained by multiplying by $N$, the number of particles per unit volume: the final expression is of course proportional to the total surface area of the particles:

$$R_c = \pi r^2 L R_i N \qquad (4.46)$$

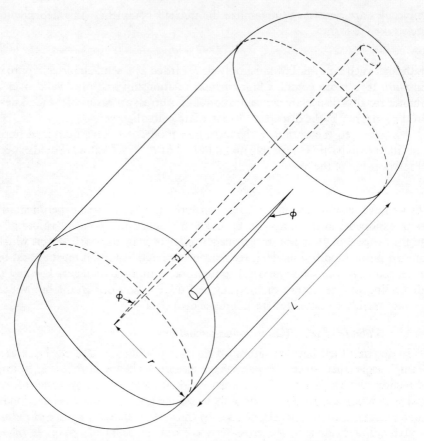

Figure 4.12. Rate of capture by a spherical particle, assuming oligomers behave as if they diffused in straight-line paths. The fraction $\phi/4\pi$ is both the proportion of oligomers reaching the particle which come from directions within the infinitesimal solid angle $\phi$, and also the proportion of oligomers generated at any point which diffuse within those same directions. The rate of capture of oligomers which have diffused less than a distance $L$ is therefore equivalent to their total rate of generation in a cylinder of length $L$, radius $r$

Fitch and Tsai apparently envisage the diffusion of growing oligomers as Brownian motion involving a 'random walk' path. However, the basic assumption used in the derivation, namely that the probability of an oligomer colliding with a particle is given by the fractional solid angle subtended at its point of origin, is really equivalent to assuming that the oligomers behave as if they travel in straight lines, like rays of light. This greatly under-estimates the probability of collision with a particle, because many of the oligomers which on this assumption are not captured because they move in other directions, will in fact have passed through the region occupied by the particle at some time or other. A similar assumption has often been made by others, for example

Gardon[86], who used it to deduce that the rate of capture of radicals by different-sized particles or micelles (in aqueous emulsion polymerization) should be proportional to their surface area. The classical theory of diffusion[77], based on Fick's laws, shows that the rate of flow of a substance with diffusion coefficient $D$ at a concentration $C$ to a particle of radius $r$ which irreversibly absorbs it, is $4\pi rDC$, and thus is proportional to the radius, and not the surface area of the particle. In effect, such a particle behaves as a 'sink' which can trap oligomers which otherwise would have diffused in other directions if it had not been there.

A modified diffusion capture equation can be derived by considering the fate of oligomers of any definite degree of polymerization $n$, which can be treated as a separate species from the point of view of diffusion. Each such population of $n$-mers will have an overall steady-state concentration $C_n$, and its rate of formation from $(n-1)$-mers will be equal to its rate of disappearance, which is the sum of two alternative fates:

$$\text{Rate of growth to } n+1\text{-mers} = k_p[M]C_n \tag{4.47}$$

$$\text{Rate of diffusion to particles} = 4\pi rD_nNC_n \tag{4.48}$$

where $N$ is the number of particles, of average radius $r$, and $D_n$ is the diffusion coefficient of $n$-mers.

The concentration and rate of formation of each successive $n+1$-mer therefore decreases by a factor

$$k_p[M]/(k_p[M] + 4\pi rD_nN) = 1/(1 + 4\pi rD_nN/k_p[M])$$

Since the rate of formation of 1-mers is equal to the rate of generation of oligomers $R_i$, the rate of formation of uncaptured $P$-mers, which according to this model is also the rate of nucleation, is given by:

$$R_n = R_i \prod_{n=1}^{n=P-1} 1/(1 + 4\pi rD_nN/k_p[M]) \tag{4.49}$$

The diffusion coefficient increases only very slowly with degree of polymerization (being roughly proportional to the cube root of the molar volume), so it is legitimate to take an average value $D$, when each term of the product becomes identical. Substituting the expression for average diffusion distance $L$ (equation 4.36), corresponding to $P$, the threshold degree of polymerization for self-nucleation, the rate of nucleation is:

$$R_n = R_i(1 + 2\pi rL^2N/P)^{1-P} \tag{4.50}$$

On the basis of this equation, the rate of nucleation never falls strictly to zero but becomes vanishingly small as $r$ and $N$ increase, and over the main period of rapid particle formation (when the expression in brackets is still not much greater than unity) can be approximated by:

$$R_n = R_i(1 - 2\pi rL^2N) \tag{4.51}$$

Here it can be seen that the rate of capture is proportional to the particle radius, as expected, and very much larger than the value obtained from the equation of Fitch and Tsai, since $L$ is typically several orders of magnitude larger than $r$.

In fact, the amended equation would reduce their calculated values for the number of particles per unit volume (which agree reasonably well with their experimental results) by a factor of the order of $10^3$. In order to explain the formation of appreciable numbers of new particles in an existing latex (renucleation)—as often happens in practice when dispersant as well as monomer is added—it would be necessary to assume that growing oligomers undergo self-nucleation when only two or three monomer units have been added to the initiating radical. This seems very unlikely in a hydrocarbon medium.

Flory[15] used a very similar argument in claiming that initiation in emulsion polymerization must take place in micelles, because the diffusion equation shows that nearly all radicals would reach micelles before they had a chance to add even one monomer unit.

However, the basic equation for rate of diffusion to particles is only valid if the molecules diffusing are taken up irreversibly by the particles. In the case of aqueous systems, there is no reason for the primary water-soluble radicals to be taken up by a surfactant micelle, and every likelihood that growing oligomers will not be permanently retained in a micelle until their hydrophobic polymer chains have grown at least large enough to displace a surfactant molecule. In the same way, there is every probability that growing oligomers in a hydrocarbon medium will show little tendency to adsorb irreversibly at a polymer surface until they have reached an appreciable size and, in fact, are in a state of dynamic equilibrium between diluent and particles.

One way of treating this situation theoretically would be to postulate a lower threshold degree of polymerization $P_a$, above which oligomers are irreversibly absorbed when they meet a particle. The corrected diffusion capture equation would then apply within the small range of further increase of molecular weight before the threshold for nucleation is reached, yielding an equation of the form

$$R_n = R_i(1 + 2\pi r L^2 N/P)^{P_a - P} \tag{4.52}$$

However, this would not be really valid, since the rate of forming oligomers of degree of polymerization $P_a$ would no longer be equivalent to the rate of generation of oligomers, but to the much smaller fraction of that rate, corresponding to the fraction remaining in solution when their molecular weight reaches the critical value $P_a$ at which absorption can be regarded as irreversible.

### 4.3.5.6. Equilibrium capture of oligomers

The simplest model using this concept treats the distribution of growing oligomers as a direct equilibrium between diluent and bulk polymer in the particles (Figure 4.8, Section 4.3.3) and defined by the relations between solubility and molecular weight as outlined in Section 4.2.4. For any particular degree of polymerization $n$, the concentrations in the diluent $c_d$ and in the particles $c_p$

are related by the solubility of the $n$-mer $S_n$:

$$c_d/c_p = S_n$$

If $v$, the volume fraction occupied by the particles, is small and the solubility is low, the overall concentration of $n$-mers in the dispersion is

$$c_n = c_d(1 + v/S_n)$$

Let us assume (as in the treatment of Fitch and Tsai) that every oligomer which reaches the threshold degree of polymerization $P$ while in the diluent phase becomes a fresh particle nucleus, whereas oligomers reaching this molecular weight while in a particle are irreversibly retained by it. The proportion of oligomers forming new nuclei is therefore the proportion of $P$-mers which are still in the diluent phase. Neglecting the small fraction of lower oligomers lost by cross-termination, an approximately steady state must exist in which the rate of formation of $n$-mers is equal to the overall rate of initiation of oligomers, $R_i$. The rate of formation of particle nuclei is therefore given by:

$$R_n = R_i/(1 + v/S_p) \tag{4.53}$$

For the initial period of dispersion polymerization, the kinetics can be represented (see Section 4.4.3) by

$$dv/dt \propto R_i^{\frac{1}{2}} v^{\frac{1}{2}}$$

or

$$v = kR_i t^2$$

so

$$R_n = R_i/(1 + t^2 kR_i/S_p) \tag{4.54}$$

On the basis of this model therefore, although the rate of nucleation does not fall to zero it is nevertheless rapidly reduced to very low levels because the very large values of $t^2/S_p$ dominate the expression. In fact, it is readily shown that the number of particles approaches a limiting value

$$N_f = \tfrac{1}{2}\pi(R_i S_p/k)^{\frac{1}{2}}$$

Typical experimental values for the parameters yield estimates for $S_p$ of the order of $10^{-6}$, in other words, the threshold molecular weight for an oligomer to form a nucleus would be that which has a bulk solubility around 1 mg/l. Whilst this may be regarded as a plausible value, a model in which the total volume of particles appears as the factor controlling nucleation is not fully consistent with the experimental evidence that fine particles inhibit nucleation more effectively than the same volume of coarse particles. Moreover, it is physically unsatisfactory, since a true equilibrium with bulk polymer would not permit the oligomers in the diluent phase to reach the degree of supersaturation necessary to produce new nuclei by aggregation.

It is probably therefore more realistic to envisage a rapid equilibrium between growing oligomer chains in the diluent and oligomers loosely adsorbed at the particle surfaces. A slower process of rearrangement would be involved in their passage into the interior of the particles, soon becoming effectively irreversible as the molecular weight reaches a value comparable with the threshold for nucleation (Figure 4.8, Section 4.3.3). Making the same assumptions as before, the rate of nucleation now becomes:

$$R_n = R_i/(1 + A/K_p) \tag{4.55}$$

where $A$ is the total surface area of the particles and $K_p$ is a partition coefficient representing the distribution of $P$-mers between diluent and surface. Like the capture equation of Fitch and Tsai and its extensions, this equation can only be solved if an expression is known for the rate of polymerization, so that the surface area $A$ can be expressed in terms of the time $t$ and number of particles $N$. From the dispersion polymerization model we obtain

$$A = 4\pi r^2 N = (36\pi k^2 R_i^2 t^4 N)^{\frac{1}{3}}$$

As they stand, the equations assume that self-nucleation occurs. The more probable aggregation process is readily accommodated by dividing by $Q$, the average number of oligomers aggregating to form a nucleus. The relationships between parameters defined in equation (4.42) apply also to the equilibrium capture equations, so the final number of particles can be expressed in the form

$$N_f \propto R_i^{\frac{2}{5}} K_p^{\frac{3}{5}} Q^{-\frac{4}{5}} \tag{4.56}$$

Both the solubility $S_p$ and the adsorption coefficient $K_p$ would be expected to fall exponentially with a rising degree of polymerization, the former approximately according to equation (4.12) the latter much less steeply, with empirical parameters which can in principle be determined independently:

$$S_p \simeq e^{-P(\chi-1)}; \qquad K_p = j e^{-aP}$$

We can therefore put

$$N_f \propto R_i^{\frac{2}{5}} Q^{-\frac{4}{5}} j^{\frac{3}{5}} e^{-3aP/5} \tag{4.57}$$

In using the diffusion capture equations above, an inverse power law relationship between the threshold degree of polymerization $P$ and the dispersant concentration $S$ was postulated as a convenient assumption leading to the observed dependence of particle numbers on dispersant. However, there is no *a priori* basis for using such a relationship in other treatments. If the results of the approach using homogeneous nucleation theory are correct, it should be possible to relate $P$ and $Q$ to the solubility characteristics of the polymer, defined by $\chi$ the polymer–solvent interaction parameter, and $\gamma$ the interfacial tension between precipitating polymer and diluent. The dependence of $\gamma$ on dispersant concentration could be determined experimentally. The relevant equations are (see Section 4.3.4)

$$P \propto (\chi - 1 - V_0 H)^{-1}$$
$$V_0 H \simeq 10^{-2} \gamma^{\frac{3}{2}}$$
$$Q \propto r^{*3}/P$$
$$r^* \propto \gamma^{-\frac{1}{2}}$$

Although the substitution of these equations into the expression for $N_f$ gives a very complicated result, it can be shown that the exponential factor is most important at least over small variations in conditions. Taking an average value of 15 as the proportionality factor for $P$ in the first equation, and using the approximation (valid if $\chi - 1 \gg V_0 H$):

$$(\chi - 1 - V_0 H)^{-1} \simeq (\chi - 1)^{-1}[1 + V_0 H/(\chi - 1)]$$

we obtain as a very rough approximation (for the exponential factor only):

$$N_f \propto [2 \cdot 7 e^{\gamma^{\frac{3}{2}}/100(\chi - 1)}]^{-9a/(\chi - 1)}$$

An approximate power law dependence on dispersant concentration is therefore plausible over small ranges of conditions but no single value of the exponent can apply over a very wide range.

Bearing in mind that interfacial tension falls as dispersant concentration rises, we can make the following predictions:

(i) The number of particles increases with rising dispersant concentration (falling $\gamma$).
(ii) The number of particles increases with increasing insolubility of the polymer (rising $\chi$, but exponent changing little since $a$ also increases).
(iii) The number of particles falls with increasing solvency of the medium (falling $\chi$).
(iv) Changes in dispersant concentration have a greater effect if its anchor group is more insoluble and therefore more strongly adsorbed (since the resultant effect on $\gamma$ is greater).
(v) Changes in dispersant concentration have a smaller effect if the polymer is more insoluble (because the value of the exponent falls with rising $\chi$).
(vi) Increasing solvency has a much more powerful effect on reducing particle numbers than lowering the dispersant concentration, since it affects both (iii) and (iv).

All these effects are consistent with experimental observations in many different systems.

## 4.3.5.7. Comparison of the models of oligomer capture

The close relationship of the various equations derived is apparent when they are listed as follows

|  | $dN/dt =$ | where $Z =$ |
|---|---|---|
| Equilibrium (with bulk) | $(R_i/Q)(1 + \pi r^3 NZ)^{-1}$ | $\tfrac{4}{3} S_p^{-1} = \tfrac{4}{3} e^{P(\chi - 1)}$ |
| Equilibrium (with surface) | $(R_i/Q)(1 + \pi r^2 NZ)^{-1}$ | $4 K_p^{-1} = 4 j^{-1} e^{aP}$ |
| Diffusion (Fitch and Tsai) | $(R_i/Q)(1 - \pi r^2 NZ)$ | $(2PD/k_p[M])^{\frac{1}{2}}$ |
| Diffusion (from Fick's Law) | $(R_i/Q)(1 + \pi r NZ)^{1-P}$ | $4D/k_p[M]$ |

In every case the controlling parameter is $P$, the threshold degree of polymerization for nucleation, which links the final number of particles obtained with the solubility of the polymer, the solvency of the medium and the amount and nature of the dispersants. However, the form of the relation is quite different in the diffusion and equilibrium models. The diffusion models include terms for the mean diffusion coefficient and the monomer concentration in their parameters, which are irrelevant to the equilibrium models. The latter are dominated by a molecular weight dependent partition coefficient between diluent and the surface or bulk of the particles.

For the reasons discussed above, the surface equilibrium model probably represents the closest approximation to physical reality of the four. However, its close formal similarity to Fitch and Tsai's equation, both of them based on the total area of the particles, makes the two practically interchangeable for a given value of $Z$ (as defined above), giving absolute values of the same order of magnitude and the same dependence of $N_f$ on $Z$ and the polymerization kinetic parameters. Divergences between the two will only appear when calculations involve the dependence of $Z$ on $P$ and hence on other parameters of the system.

### 4.3.6. Dispersant-limited nucleation

This is essentially the theory which has been applied to particle formation in aqueous emulsion polymerization for many years. In its simplest form, it assumes that every free radical formed in the polymerization produces a new particle as long as sufficient dispersant (surfactant) is available to saturate the total surface area of the particles; as soon as the amount of dispersant becomes depleted to below this level the free radicals formed are absorbed by existing particles and nucleation ceases.

With the additional assumption that every particle grows in volume at the same constant rate, these simple relations follow: The number of particles formed per unit volume, $N_p \propto t R_i$ (time, initiation rate); total surface area of particles at end of nucleation $\propto N_p t^{\frac{2}{3}} \propto S$ (available dispersant).

Eliminating $t$, the number of particles can be expressed as:

$$N_p \propto S^{\frac{3}{5}} R_i^{\frac{2}{5}} \tag{4.58}$$

This is the relationship found by the much more elaborate treatment of Smith and Ewart[3], which takes account of the distribution of size and time of formation among particles, and holds irrespective of whether particles are formed from micelles or from solution.

The model can be adapted to the characteristic kinetic features of non-aqueous dispersion polymerization (Section 4.4.3, equation 4.72) according to which, the total volume of polymer formed is initially proportional to $R_i t^2$:

$$\text{Average surface area per particle} \propto R_i^{\frac{2}{3}} t^{\frac{4}{3}} N_p^{-\frac{2}{3}}$$

$$\therefore \text{Total surface area at end of nucleation} \propto R_i^{\frac{2}{3}} t^{\frac{4}{3}} N_p^{\frac{1}{3}} \propto S$$

Surprisingly, on eliminating $t$, exactly the same relation is obtained as before (equation 4.58) for the number of particles:

$$N_p \propto S^{\frac{3}{5}} R_i^{\frac{2}{5}}$$

In non-aqueous dispersion polymerization, however, the number of particles formed in practice is usually proportional to a much higher power of the dispersant concentration, in the range 1·5 to 3 (Section 4.3.2). Taken in conjunction with the arguments already outlined in Section 4.3.3, above, the experimental data therefore suggest that this is not a valid model for dispersion polymerization in non-aqueous media.

### 4.3.7. Dispersant-limited agglomeration

The simplest theory to explain an inverse relation between particle size and concentration of dispersant assumes that the particles actually observed are formed by the agglomeration of much smaller, primary particles. If in this way, the total surface area of the particles is reduced to an area which can be fully protected by the available dispersant, it should be proportional to the dispersant concentration, and assuming that the particle size distribution remains fairly uniform, this leads to the relationships:

$$\text{radius of particles, } r \propto S^{-1};$$

$$\text{number of particles, } N \propto S^3$$

At least as a rough approximation, this reciprocal relation between particle size and dispersant concentration has often been observed in both non-aqueous dispersion polymerization (Section 4.3.2) and in aqueous emulsion polymerization[66–68,87].

Agglomeration of this type was clearly demonstrated by Fitch and Tsai[56] in the polymerization of very dilute, aqueous solutions of methyl methacrylate. The number of particles rose rapidly at first, but then suddenly fell if less than a certain minimum surfactant concentration was present, and thereafter remained roughly constant (Figure 4.13). Several distinct levels of particle size were observed, whereas above this minimum surfactant level they were uniform in size (primary particles). Up to this minimum value, the numbers approximately fitted the relation $N \propto S^{3.87}$ for initiation by hydrogen peroxide and

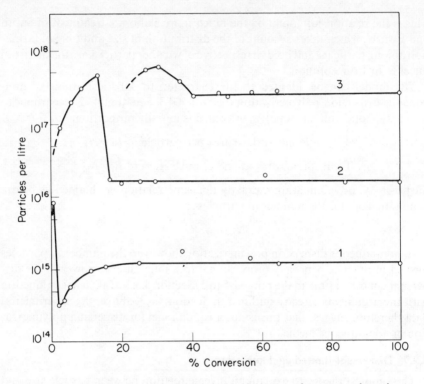

Figure 4.13. Agglomeration of particles during the course of aqueous polymerization of methyl methacrylate[56]: Monomer = 0.036 M; persulphate = 7.35 × $10^{-4}$ M; bisulphite = 1.14 × $10^{-3}$ M; temperature = 30 °C. 1, No additions; 2, as 1, with 2.4 × $10^{-6}$ $Fe^{2+}$ added; 3, as 2, with 7.95 × $10^{-4}$ sodium dodecyl sulphate added. (After Fitch and Tsai in *Polymer Colloids* (Ed. R. M. Fitch), Plenum Press, 1971, p. 90, with permission)

ferrous salt. The exponent observed for a persulphate/bisulphite initiator was about 1.1. Such an initiating system however, would be expected to form stabilizing ionic end groups on the polymer chains in addition to the stabilizing effect of the surfactant.

In non-aqueous dispersion, agglomeration has been shown to occur with some monomers which form semi-fluid particles during polymerization, but the evidence suggests that this type of mechanism is not usually operative in the most thoroughly studied systems using methyl methacrylate. For example, more rigorous measurements over a very wide range of dispersant concentrations, with precautions to maintain the initial solvency of the diluent at a constant value, show proportionality of particle radius to $S^{-0.6}$ rather than $S^{-1}$ as predicted (Figure 4.6, Section 4.3.2). In addition, random agglomeration would be expected to give a wide range of particle sizes (as found by Fitch and Tsai), whereas a fairly uniform particle size is usually observed in dispersion polymerization unless the solvency of the medium is very high or changes substantially, or more dispersant is added. In practice also, the number of particles usually appears to stay essentially constant from a very early stage—if insufficient dispersant is present to stabilize the increasing surface area, the latex undergoes a fairly sudden and gross overall flocculation rather than a more gradual coalescence to form larger particles.

Presumably, an essential requirement for controlled agglomeration rather than overall flocculation is that the dispersant is able to redistribute itself rapidly from the surface 'lost' by agglomeration to protect the remaining surface before further agglomeration ensues. This would be favoured by a relatively loosely-anchored dispersant, low particle size and concentration to slow down the rate of collision, but soft (or monomer-swollen) particles to ensure rapid coalescence after collision. All of these conditions are fulfilled in the system studied by Fitch and Tsai. Clearly also in non-aqueous dispersion polymerization, agglomeration will be much more likely to occur when monomers such as ethyl acrylate or vinyl acetate are used since these form semi-liquid particles under the conditions of polymerization.

Somewhat different forms of agglomeration have been observed in dispersion polymerization where the dispersant is anchored essentially irreversibly and thus cannot redistribute itself. In one example, ultra-fine particles were formed early in the seed stage of a methyl methacrylate polymerization using relatively large amounts of a dispersant with a very insoluble polar anchoring portion also containing copolymerizable groups[64]. These ultra-fine particles agglomerated to form larger particles whose size was almost independent of the amount of dispersant used, presumably because most of the dispersant from the 'lost' surface was buried rather than redistributed.

In the dispersion polymerization of vinyl chloride (see Section 4.4.4) the formation of new particle nuclei appears to continue indefinitely, probably because low molecular weight oligomers have very little tendency to adsorb irreversibly on to existing particles. Monomer absorption is also low and particles grow only very slowly; since the dispersant appears to be irreversibly anchored, existing particles therefore remain stable even if there is insufficient dispersant present to protect the new particles which form. As a result, small aggregates and 'bits' are produced in an otherwise stable latex and become increasingly prevalent, until ultimately flocculation ensues when the older particles grow too large for their anchored dispersant to protect. On the other hand, the polymerizing system will continue to produce small, fully-stabilized particles if dispersant is gradually added to it. In the intermediate situation, when a very limited supply of extra dispersant is fed in, the new particle nuclei appear to be partially stabilized and tend to adhere to existing particles to form larger particles with characteristic protuberances (Figure 4.15, Section 4.4.1).

### 4.3.8. Present status of theories of particle formation

The theoretical approaches which have been described must be regarded as no more than a tentative outline. Qualitatively, they give an adequate explanation of the observed relations between particle size, solubility of the polymer, solvency of the medium and the nature and amount of dispersants. Quantitatively, they are almost entirely untested.

The theory most widely applied to emulsion polymerization—nucleation continuing until insufficient dispersant remains to saturate the growing particle surface—appears to be irrelevant to non-aqueous dispersion polymerization. The evidence indicates that nucleation is normally completed while excess dispersant is still present and supports the view that capture of growing oligomers by particles is the main factor suppressing nucleation. Dispersant therefore merely modifies a process which is already operative in unstabilized precipitation polymerization.

On the other hand, the amount of dispersant available can become the limiting factor at a later stage of particle growth, resulting in secondary agglomeration under certain special circumstances, although gross flocculation is the usual consequence. If secondary agglomeration does occur, an inverse relation between particle size and dispersant concentration follows naturally from the simple theory presented.

The treatment of nucleation, and its suppression by capture of oligomers, as originally described by Fitch and Tsai, has been extended here to obtain relationships corresponding to various polymerization mechanisms and to allow for modification by dispersant. Arguments have been put forward for treating the capture of oligomers not as a diffusion process but as an equilibration moving increasingly in favour of particles as the oligomers grow, and alternative equations have been derived on this basis. However, both approaches are oversimplified: the real process probably involves a gradual transition from equilibration to diffusion as the rate-controlling factor, as each oligomer grows. At very low degrees of polymerization, equations based on diffusion with irreversible capture greatly overestimate the probability of capture, while at high degrees of polymerization equilibration matching the rate of growth and falling solubility of the oligomers would require a rate of capture greater than diffusion allows. A full theoretical treatment of this complex model has not been attempted here. It is quite likely that one or other of the simplified models based on diffusion or equilibration may represent a fair approximation to a practical system, depending on the conditions, solubility characteristics of the polymer, and particularly the value of $P$, the threshold degree of polymerization for nucleation. It may be that equilibration is likely to have a much more important role in hydrocarbon diluents than in the aqueous dispersion systems studied by Fitch and his associates. Such questions can only be resolved by experimental work under carefully defined conditions to maintain constant solvency: the limited data at present available do not permit firm conclusions, particularly in view of the formal similarity of the surface equilibration equation and that of Fitch and Tsai, both depending in a similar way on the total surface area of the particles.

Application of the classical theory of homogeneous nucleation to dispersion polymerization is so far untested in a quantitative sense, but leads to broadly correct conclusions on the effects of solvency and dispersants, when combined with any of the theoretical treatments of the capture of oligomers by particles. It has provided a semi-quantitative basis for a number of important conclusions:

(i) Nuclei probably form by a process of aggregation of oligomers, rather than self-nucleation, since the threshold degree of polymerization for the former may often be quite low ($\sim 10-15$): a longer chain would be needed to coil enough to form a condensed phase.
(ii) The probability of participating in a nucleus rises so steeply with molecular weight, that a single threshold degree of polymerization $P$ above which all oligomers form nuclei is a reasonable approximation in most cases. The number of particles rises as $P$ falls.
(iii) $P$ increases with the solubility of the polymer and falls with rising dispersant concentration (falling interfacial tension), but probably changes only slightly during the course of a typical polymerization.
(iv) The number of oligomers in a nucleus, $Q$, may vary considerably with conditions, rising with increasing dispersant concentration; it probably changes only slightly during the course of a typical dispersion polymerization.
(v) $P$ is usually sufficiently low so that nearly all oligomers form nuclei (or are captured by particles) before their radicals are terminated. In this case, nearly all polymerization takes place within the particles, the concentration of polymer in the diluent phase remains extremely low, and nearly all radicals pass into the particles.
(vi) Some cases may exist where the solubility is marginal: as the degree of polymerization rises, the solubility falls gradually and the probability of capture by particles rises only slowly. $P$ may then be so high that nearly all oligomer radicals are terminated before nucleation or capture; nearly all polymerization takes place in the diluent phase, where an appreciable concentration of polymer may remain in solution; particles grow only slowly since few radicals reach them, and nucleation may continue unabated until a late stage in the polymerization. The balance would tend to shift further in this direction as the initiation rate is raised, enhancing the probability of termination in the diluent phase.

Most of the commoner types of free radical dispersion polymerization appear to conform well to (v), including methyl methacrylate, ethyl acrylate, vinyl acetate, acrylonitrile and their copolymers, in aliphatic hydrocarbons. However, the dispersion polymerization of vinyl chloride and vinylidene chloride appears to fit the model (vi) well. On the other hand, at the much lower initiation rates used in some studies of the bulk polymerization of vinyl chloride, the main locus of polymerization apparently shifts to the particles[88], as might be expected from the reduced probability of termination and longer chain length.

While the qualitative conclusions may be sound, there can be little basis for quantitative testing until much fuller experimental data are available to specify the parameters used in the theory. In particular, there is an almost complete absence of appropriate data on interfacial tension between polymers and hydrocarbon diluents, let alone the way it is affected by the presence of monomer or dispersants. It is known to show some dependence on the molecular weight of lower oligomers[85] but this factor has been completely ignored in the theoretical treatment here. There is also very little quantitative information on solubility relationships for suitable polymer–diluent pairs, or values of $\chi$ which could be used to form an estimate.

On the theoretical side, it must be recognized that the application of bulk properties (free energy of dilution, solubility, interfacial tension) to assemblages of a few molecules is of doubtful validity, although it has worked surprisingly well in other contexts, and there is considerable uncertainty about the value of the pre-exponential factor in the rate equation. At the same time, the Flory–Huggins treatment of polymer solution theory which has been used as a basis for estimating the parameters is admittedly very approximate, and in developing the treatment of nucleation the complexities of polydisperse polymer systems have been largely disregarded. Similar gross simplifying assumptions have been used in dealing with the capture of oligomers. Nevertheless, the basic approach appears sound and may also find application to heterogeneous bulk polymerization and some forms of emulsion polymerization.

Perhaps the least satisfactory aspect is the treatment of dispersants simply through their effect on bulk interfacial tension. In principle, it would be a sounder approach to obtain an expression for the net free energy of formation of an assemblage containing arbitrary numbers of oligomer and dispersant molecules, and then find the maximum corresponding to the activation energy for nucleation, and its dependence on dispersant concentration and oligomer concentration and molecular weight. What can be in little doubt is that the dispersant molecules are involved in the nucleation process. In this connection it is instructive to estimate some molecular magnitudes at the start of a typical dispersion polymerization. For a 5% solution of methyl methacrylate at 80 °C, with 0.2% azo initiator and 0.7% comb stabilizer, the half-life of the oligomer radicals is about 0.1 sec and their average degree of polymerization, 25–30 monomer units. Since they have a high probability of meeting in the precise orientation for termination within a fraction of a second, they obviously undergo extremely frequent collision and their chances of growing to a sufficiently large size for self-nucleation before aggregating with others must be vanishingly small. Collision must be even more frequent with the polymeric backbones of dispersant molecules, which are not only of a somewhat higher degree of polymerization than the oligomer chains but vastly outnumber them for the first 200 sec of polymerization (dispersant molecules $3 \times 10^{17}$ per ml, initiation rate $1.7 \times 10^{15}$ per sec per ml). At this stage the aggregates would be only relatively weakly associated and easily dissociated and re-formed, until the threshold concentration is reached for forming stable nuclei, which clearly incorporate dispersant molecules from the very start. Such a polymerizing system typically forms from $10^{13}$ to $10^{15}$ particles per ml, depending on the solvency of the diluent.

With this picture of the initial stages of particle formation in mind, we can proceed to the main period of particle growth and the kinetics and mechanism of polymerization in dispersion systems.

## 4.4. PARTICLE GROWTH IN RADICAL-INITIATED DISPERSION POLYMERIZATION

### 4.4.1. The mode of polymerization

It has been a marked feature of dispersion polymerization in organic media that essentially no new polymer nuclei are formed after the initial precipitation of polymer particles. Following the nucleation stage, subsequent polymerization is confined to further growth of the polymer particles formed initially. Three basic polymerization mechanisms have been considered as consistent this this observation:

(i) Polymerization in solution, followed by precipitation onto the existing polymer particles;
(ii) Polymerization of monomer adsorbed at the surface of the polymer particles;
(iii) Polymerization of monomer absorbed into the interior of the polymer particles.

All of the evidence from diagnostic tests for the mechanism of dispersion polymerization indicates that mechanism (iii) is the predominant mode of polymerization, although mechanism (i) appears to operate in the dispersion polymerization of vinyl chloride under certain conditions.

The strongest evidence for the polymerization of absorbed monomer within the polymer particles has come from the direct determination of the degree of swelling of preformed polymer particles dispersed in aliphatic hydrocarbons containing monomer under conditions representative of those in an actual dispersion polymerization[89] (Table 4.10). Such studies have shown that methyl

Table 4.10. Degree of swelling of polymer dispersed in cyclohexane by own monomer at 80 °C[89].

| Polymer (% solids) | Monomer (% by wt) | Monomer uptake (g monomer/100 g polymer) |
|---|---|---|
| Poly(methyl methacrylate) (30) | Methyl methacrylate (10) | 9 |
| Poly(methyl methacrylate) (20) | Methyl methacrylate (10) | 8 |
| Poly(vinyl acetate) (30) | Vinyl acetate (10) | 19 |
| Poly(vinyl acetate) (20) | Vinyl acetate (10) | 18 |

methacrylate, at low to moderate concentrations, distributes itself between poly(methyl methacrylate) particles and aliphatic hydrocarbon diluents in approximately equal concentrations. A value near unity for the partition coefficient was also derived from the kinetic data[1]. Similar values have also been reported from determinations on related systems[60].

With vinyl acetate, the partition is somewhat more in favour of the polymer phase (about 2:1) and similar results have been obtained with other monomers, including acrylonitrile. As far as the experimental data go, the distribution of monomer is independent of particle size as predicted by polymer solution theory (Section 4.2.6). In fact, similar values are obtained using polymer in bulk form, providing sufficient time is allowed for equilibration. The levels of monomer absorption determined in these systems ($c.$ 10%) where the monomer is freely soluble in the continuous phase, is much less than found in monomer-swollen particles undergoing aqueous emulsion polymerization. In the latter case, the monomer concentration can be as high as 60–70% by volume in the polymer particles when the aqueous phase is saturated with monomer of low solubility in water[28].

Supporting evidence for the interior of the particles as the main site of polymerization has also emerged from the kinetic studies[1]. For example, the rate of dispersion polymerization is essentially independent of both particle size and the amount of polymeric dispersant used (which controls particle size) over the range in which the polymer particles are fully stable to flocculation (Table 4.11). This could hardly be the case if polymerization took place on the

Table 4.11. Rate of dispersion polymerization of methyl methacrylate at 80 °C using different amounts of graft dispersant[1]. (From Barrett and Thomas, *J. Polym. Sci.* A1, **7**, 2628 (1969), with permission.)

| Graft dispersant (% on monomer) | Rate constant ($k \times 10^3$, sec$^{-1}$) |
|---|---|
| 0.5 | 7.6 |
| 1.0 | 11.6 |
| 2.5 | 10.8 |
| 5.0 | 10.2 |

surface of the particles. In addition, the rate of dispersion polymerization accelerates rapidly as more polymer is formed. The extent of the acceleration runs closely parallel with the extent of acceleration in the corresponding bulk polymerization of monomer at high conversion (i.e. polymer swollen with monomer). For example, at 80 °C the acceleration in rate is very marked for methyl methacrylate but much less so for vinyl acetate[90] (Figure 4.14).

Additional evidence, indicating that the site of dispersion polymerization is within the particle, comes from two further sources. Free radicals formed in a

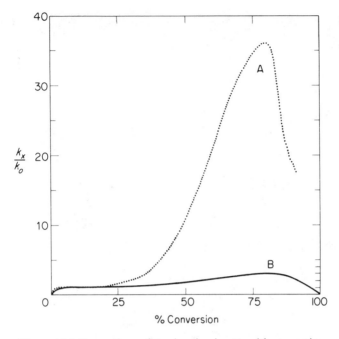

Figure 4.14. Comparison of acceleration in rate with conversion of methyl methacrylate (A) and vinyl acetate (B) in bulk polymerization at 80 °C. ($k_x$, bulk rate; $k_0$, solution rate; direct determination by microcalorimetry[90])

photo-initiated dispersion polymerization of methyl methacrylate have been trapped at temperatures below the polymer glass transition temperature and used to re-initiate the polymerization at a higher temperature[91]. The concentration of trapped free radicals in poly(methyl methacrylate) dispersions was too low for detection by electron-spin resonance but significant levels were detected in poly(acrylonitrile) dispersions. The morphology of the polymer particles formed in dispersion polymerization, as shown by their electron micrographs (Figure 4.15), is also strongly indicative of a micro-bulk polymerization. In the case of the poly(methyl methacrylate), the characteristically smooth, spherical form of the particles is that which would be expected to result from the growth of monomer-swollen particles. In contrast, poly(vinyl chloride) particles which are formed by deposition from solution are unsymmetrical and have a rough and granular surface.

### 4.4.2. Polymerization within a polymer matrix

Free radical addition polymerization is a chain reaction involving three successive stages of initiation, propagation and termination. The first stage of initiation results from the generation of free radicals, usually by the thermal decomposition of an initiator (I) into a pair of free radicals (R·) with a rate

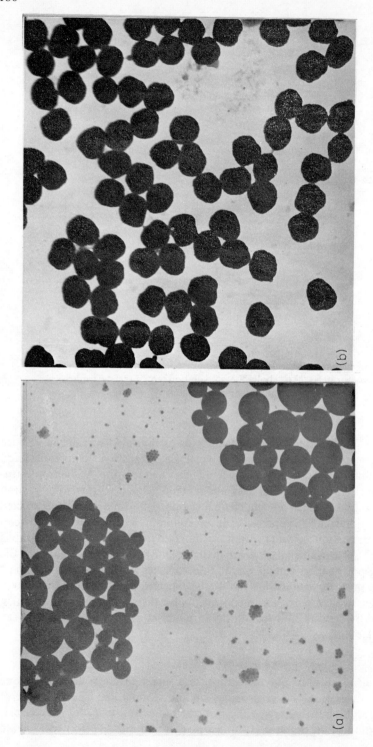

Figure 4.15. Electron micrographs of polymer particles produced by dispersion polymerization: (a), Poly(methyl methacrylate) (×4500); (b), Poly(vinyl chloride) (×9000)

constant, $k_d$, for the decomposition reaction

$$I \xrightarrow{k_d} 2R\cdot \quad (4.59)$$

The radicals produced react rapidly with the unsaturated group of the monomer (M) to yield the chain-initiating species, $M_1\cdot$, with a rate constant, $k_i$

$$R\cdot + M \xrightarrow{k_i} M_1\cdot \quad (4.60)$$

Uusually, the initiation step is much faster than the rate of decomposition of the initiator (equation 4.59) and hence the latter determines the overall rate of initiation, $R_i$

$$R_i = 2fk_d[I] \quad (4.61)$$

where $f$ is the initiator efficiency, i.e. the fraction of radicals formed which initiate polymerization.

Propagation of the chain reaction rapidly ensues by the interaction of the monomer radical ($M_1\cdot$) and the subsequent growing polymer radicals ($M_2\cdot, M_3\cdot, \ldots M_n\cdot$) with monomer, to yield a rapid extension in chain length, with a rate coefficient, $k_p$

$$M_n\cdot + M \xrightarrow{k_p} M_{n+1}\cdot \quad (4.62)$$

Finally, the growing chain is terminated by either a coupling reaction,

$$M_n\cdot + M_m\cdot \xrightarrow{k_{tc}} M_{n+m} \quad (4.63)$$

or a disproportionation reaction

$$M_n\cdot + M_m\cdot \xrightarrow{k_{td}} M_n + M_m \quad (4.64)$$

These can be averaged, using a single termination rate constant, $k_t$. By equating the rate of initiation with the rate of termination in an approximately steady-state condition, and since all the radical species are assumed to be of equal reactivity, the overall radical concentration can be expressed as

$$[R\cdot] + [M_n\cdot] = (R_i/k_t)^{\frac{1}{2}} \quad (4.65)$$

The overall rate of polymerization, $R_p$ defined with respect to the rate of disappearance of single monomer units, is then given by the expression:

$$R_p = -d[M]/dt = k_p[M](R_i/k_t)^{\frac{1}{2}} \quad (4.66)$$

The mean kinetic chain length ($\bar{v}$), that is the average number of monomer units converted to polymer by a single initiating radical, is given by:

$$\bar{v} = k_p[M]/(R_i k_t)^{\frac{1}{2}} \quad (4.67)$$

The molecular weight of polymer produced can be modified by the addition of a chain transfer agent, XA, which competes with monomer in the propagation reaction (equation 4.62), and forms a new free radical for initiation

$$M_n\cdot + XA \xrightarrow{k_{tr}} M_nX + A\cdot$$
$$A\cdot + M \rightarrow M\cdot \tag{4.68}$$

The effect of a chain transfer agent such as a mercaptan, which contains readily abstractable hydrogen atoms, is to reduce the overall molecular weight of the polymer produced. The components of the polymerization—monomer, polymer, initiator or solvent—can also function as transfer agents in a similar manner if they contain reactive groups. For termination by disproportionation, the number average degree of polymerization $\bar{P}_n$, is given by:

$$1/\bar{P}_n = 1/\bar{v} + \sum k_{trx}[X]/[M] \tag{4.69}$$

Radical chain polymerizations frequently exhibit a marked auto-acceleration in rate after the onset of polymerization. This effect, the so-called gel effect (sometimes called the Trommsdorff or Norrish–Smith effect after the workers concerned) is particularly marked in the bulk polymerization of methyl methacrylate but can also be pronounced in solution polymerization with a high initial concentration of monomer[92] (Figure 4.16). Auto-acceleration in rate is also observed in the bulk polymerization of monomers which produce insoluble polymers, such as acrylonitrile or vinyl chloride[93]. The auto-acceleration in rate is associated with a corresponding increase in chain length or overall molecular weight. Similar effects occur also in the polymerization of monomers in the presence of a precipitant for the polymer, such as methyl methacrylate in cyclohexane or styrene in lower aliphatic alcohols.

The gel effect has been shown[94] to result from a decrease in the termination rate as the viscosity of the polymerization medium increases. Although diffusion of monomer is still possible within the increasingly viscous medium, the diffusion of the much larger growing polymer radicals is considerably retarded and makes them much less likely to terminate with each other. Therefore, although the propagation of polymer growth continues largely unhindered under these conditions, the termination rate is considerably reduced. The quantity $k_p/k_t^{\frac{1}{2}}$ (equation 4.66) therefore increases, with a resulting increase in the overall rate of polymerization. For the same reason (equation 4.67) there is a corresponding increase in the chain-length or overall molecular weight of the polymer with increasing conversion. In the extreme case of polymers which precipitate out from the polymerization medium, coiling of the polymer radicals makes the radical ends even less accessible and in some cases trapped radicals have been detected within the polymer matrix[93]. The diminution of termination rate in the polymerization of methyl methacrylate and styrene in poor solvents (but not poor enough to bring about precipitation of the polymer formed at the conversions studied) has been similarly ascribed to hindered radical–radical termination due to the formation of tightly-coiled macroradicals[95].

At higher polymer conversions, even monomer diffusion is hindered in many cases, with a consequent fall in the value of $k_p$. This produces a characteristic

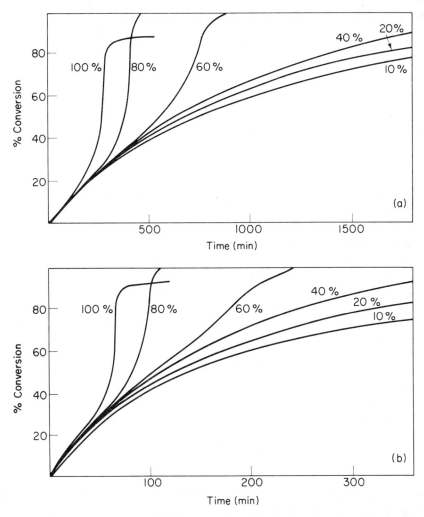

Figure 4.16. Autoacceleration of polymerization of methyl methacrylate in bulk and in solution at various concentrations in benzene at 50 °C (a) and at 70 °C (b)[92]. (From Schulz and Harborth, *Makromol. Chem.*, **1**, 106 (1948), with permission of Hüthig and Wepf Verlag)

pattern of variation in apparent overall rate constant with conversion, resulting in a definite peak rate (Figure 4.18, Section 4.4.4). To a good approximation, the value of $k_p/k_t^{\frac{1}{2}}$ appears to depend only on the monomer/polymer ratio at a given temperature and is little affected by the rate of initiation or molecular weight of the polymer produced[96]. Strong solvents or plasticizers for the polymer behave as if they were extra monomer, displacing the whole rate curve towards higher conversions. Rising temperature has a similar effect and also reduces the magnitude of the auto-acceleration effect.

### 4.4.3. A kinetic model for dispersion polymerization

A simple quantitative model for dispersion polymerization has been evolved from the basic picture of the mode of polymerization described in Section 4.4.1. Once polymer particles have been formed from the initially-homogeneous reaction mixture, they absorb monomer from the diluent phase and polymerization subsequently proceeds within the particles according to the laws of bulk polymerization kinetics. The high viscosity of the monomer-swollen polymer particles greatly hinders radical termination processes and the resulting increase in radical concentration due to the gel effect accelerates the rate of polymerization. The reduction in termination rate also allows the coexistence of many radicals within a single polymer particle. Any polymer radicals initiated in the diluent phase are rapidly swept up by existing particles before they have had time to grow to more than a very few monomer units in length. Consequently, initiation can be considered as taking place as though all the initiator were confined within the particles, even though the types of initiator used are known to be distributed between the polymer particles and the diluent.

A kinetic equation corresponding to this model for dispersion polymerization has been developed in the following manner[1]. If the polymer particles at a given time occupy a volume fraction $V$ of the whole dispersion, and the rate of initiation in the whole dispersion is $R_i$, then the effective initiation rate in the particles will be given by

$$R_{ip} = R_i/V \tag{4.70}$$

If the monomer concentration in the particles is $c_{mp}$, the overall polymerization rate in the particles, $R_{pp}$, will be equivalent to the usual expression for bulk or solution kinetics (equation 4.66 above, Section 4.4.2) in the form

$$R_{pp} = c_{mp}k_p(R_{ip}/k_t)^{\frac{1}{2}}$$

$$= c_{mp}k_p(R_i/k_tV)^{\frac{1}{2}} \tag{4.71}$$

Since essentially all polymerization takes place within the particles in the volume fraction $V$, the overall rate in the whole dispersion is

$$R_p = VR_{pp}$$

$$= c_{mp}k_p(VR_i/k_t)^{\frac{1}{2}} \tag{4.72}$$

In emulsion polymerization, the concentration of monomer in the polymer particles is usually taken as constant up to the stage in the polymerization where the monomer droplets disappear[28]. In dispersion polymerization, since the monomer is completely miscible with the hydrocarbon diluent, the concentration of monomer in the polymer particles depends on its partition coefficient, $\alpha$, between polymer and diluent. Thus, the overall rate of dispersion polymerization, where $c_{md}$ is the monomer concentration in the diluent, is given by

$$R_p = \alpha c_{md}k_p(VR_i/k_t)^{\frac{1}{2}} \tag{4.73}$$

This is the general equation for dispersion polymerization and it takes into account the principal features of the kinetics which have been established[1], such as the proportionality of the polymerization rate to the square root of initiator concentration, the acceleration in rate which follows the increase in the total volume of the polymer particles and the correlation of rate with that in the bulk polymerization of the monomer. It is important to note that the value of $k_p/k_t^{1/2}$ in this expression is not to be taken as constant but varies as the concentration of monomer in the polymer particles changes, in exactly the same manner as in bulk polymerization at high conversions. In principle, the partition coefficient $\alpha$ may also vary with monomer concentration, but the variation is usually not great and a constant value is a sufficiently good approximation for use in most of the kinetic experiments.

The kinetic model proposed differs from the conventional treatment of emulsion polymerization in that there is no enhancement of rate due to the isolation of radicals in separate particles[3]. In dispersion polymerization, factors such as the size of the polymer particles formed, their relatively low absorption of monomer and consequent high internal viscosity and the high rates of initiation used, all contribute to a retardation of cross-termination of radicals within the particles to allow many radicals to coexist in each particle over the whole extent of the polymerization. There is evidence that this situation can also exist in emulsion polymerization but usually only at high conversions in the latter stages of the process[28]. The enhancement of rate observed in dispersion polymerization, in fact, arises from hindered termination within the polymer particles exactly as under bulk conditions and the process can therefore be treated as a micro-bulk polymerization. A similar kinetic model, in which the rate of polymerization is independent of particle size and numbers, has been proposed to account for results obtained in the emulsion polymerization of vinyl acetate[97]. In this case, interchange of radicals between particles may occur due to the ready escape of small mobile radicals produced by chain transfer reactions of this monomer.

There are two limiting cases of the kinetic model proposed which must be considered. The first case is when $\alpha$ and $V$ are small, and $c_{md}$ may be taken as equivalent to the overall monomer concentration, $c_m$. Here $V$ is roughly equal to $c_0 x V_p$, where $c_0$ is the initial monomer concentration, $x$ is fractional conversion and $V_p$ is the volume of polymer per mole of monomer. Then,

$$R_p = \alpha c_0 x^{1/2}(1 - x)(c_0 R_i V_p)^{1/2} k_p/k_t^{1/2} \qquad (4.74)$$

This expression is a good approximation for the behaviour of monomers such as vinyl acetate and methyl methacrylate, and shows two very characteristic features. The rate of conversion, $R_p/c_0$ increases in proportion to the square root of the initial monomer concentration modified by any effects of the altered monomer concentration in the particles on the rate constants, $k_p/k_t^{1/2}$. In addition, the rate of conversion increases in proportion to the square root of conversion over the initial part of the polymerization. This corresponds to an increase in rate directly in proportion to time, and an increase in conversion proportional to the square of time.

The second limiting case is when $\alpha$ is large and nearly all of the monomer passes into the particle at a very early stage. Here, $V$ is approximately equal to $c_0 V_m$, where $V_m$ is the molar volume of monomer and $c_{mp}$ is roughly equal to $(1 - x)/V_m$. Then

$$R_p = (1 - x)(c_0 R_i/V_m)^{\frac{1}{2}} k_p/k_t^{\frac{1}{2}} \tag{4.75}$$

This expression, in kinetic terms, is really the equivalent of suspension polymerization and has the following characteristics. The rate of conversion rises almost at once to a high level which may then change with conversion due to the competing effects of the change in value of the rate constants (due to the gel effect) and depletion of monomer. The rate of conversion $R_p/c_0$ decreases with initial monomer concentration (inversely proportional to its square root). The ratio of monomer to polymer in the particles is defined essentially by the extent of conversion rather than by the overall monomer concentration, and there-

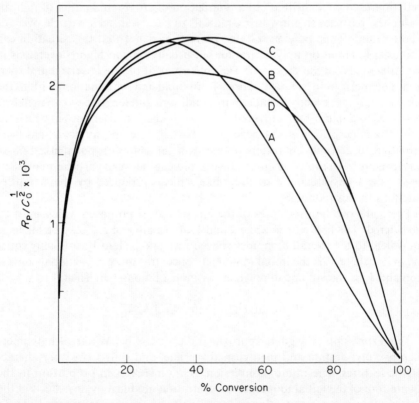

Figure 4.17. Dispersion polymerization of acrylonitrile in cyclohexane[98]. Dispersant 6%; ADIB 0.5%; 80 °C. Acrylonitrile concentrations: A = 1.4, B = 2.8, C = 4.2, D = 5.6 (mol/litre)

fore curves of $R_p/c_0^{\frac{1}{2}}$ plotted against conversion should be almost superimposable, irrespective of any gel effect[98] (Figure 4.17). This form of equation describes the dispersion polymerization of acrylonitrile under certain conditions and would also be expected to apply to other very polar monomers, such as acrylic acid and acrylamide.

### 4.4.4. Results of kinetic studies

Since direct analytical procedures, such as a measure of monomer consumed or polymer formed, are experimentally complicated to apply to heterogeneous polymerizations, the bulk of the kinetic studies has been carried out with a microcalorimetric technique developed for following both homogeneous[99] and heterogeneous polymerization processes[1,100]. The rate of evolution of the heat of polymerization was followed directly with a commercial differential scanning calorimeter (Perkin Elmer DSC-1B) operating under isothermal conditions. Since only small samples (c. 50 mg) of the polymerization mixture were used, ready heat transfer to the thermal sensors is obtained. The method has the advantage that it provides a measure of the rate of polymerization directly and continuously throughout the whole polymerization (Figure 4.1, Section 4.1) and is particularly suited for following the rapid changes in rate associated with heterogeneous micro-bulk polymerizations.

Two types of monomers have been used to establish the kinetics of dispersion polymerization. The first, represented by methyl methacrylate and vinyl acetate, form polymer which is completely soluble in the parent monomer. Consequently, with these monomers dissolved in an aliphatic hydrocarbon diluent (the polymers are, of course, insoluble in aliphatic hydrocarbons) the overall solvency of the diluent mixture for the polymer decreases as monomer is consumed in the dispersion polymerization. The other type, represented by acrylonitrile and vinyl chloride, form polymer insoluble in the parent monomer and so the solvency of the diluent/monomer mixture is essentially constant throughout the polymerization.

*4.4.4.1. Methyl methacrylate*

The kinetics of the dispersion polymerization of methyl methacrylate in organic media cannot be interpreted without taking into account the very large changes in apparent rate constant which takes place during its bulk polymerization. In this case, the gel effect causes a very pronounced decrease in the value of the termination rate constant, $k_t$, with a correspondingly large increase in radical concentration and polymerization rate. At high conversions, the propagation rate constant, $k_p$, also falls in value as the glassy state of the polymer is approached, making even the diffusion of monomer to the macroradicals now difficult. The effect was studied in detail by Schulz and interpreted in terms of diffusion theory[96]. Increases in apparent rate constant of the order of 30- to 50-fold were reported at temperatures around 80 °C. The variation in the acceleration factor, defined as the ratio of apparent first order rate constant to its initial value at low conversion at the same temperature, with conversion

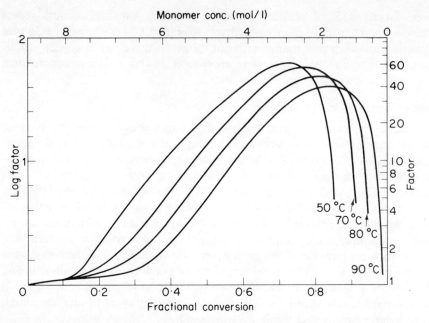

Figure 4.18. Variation of acceleration factor with conversion in bulk polymerization of methyl methacrylate.[1] (From Barrett and Thomas, *J. Polym. Sci.* A1, 7, 2630 (1969), with permission)

obtained in the bulk polymerization of methyl methacrylate is shown in Figure 4.18.

Analysis of the experimental results obtained for the dispersion polymerization of methyl methacrylate, in terms of the simplified equation for the special case ($\alpha$, $V$ small, equation 4.74, Section 4.4.3) showed a similar large increase in apparent rate constant, followed by a decrease at high conversions (Figure 4.19) corresponding to the behaviour in bulk polymerization. Numerical values were obtained both for the partition ratio of monomer in the bulk and continuous phase (c. 1:1) and for the initiator efficiency (c. 0·4) from the rate/conversion data which agreed with the values obtained by direct analytical determinations. Using the values found for these parameters, the extensive studies of the dispersion polymerization of methyl methacrylate which can be carried out with the microcalorimetric method, have shown its behaviour to conform closely to that predicted by the general kinetic model (equation 4.73, Section 4.4.3).

In particular, increase in dispersion polymerization temperature shifts the peak rates to higher conversions and lowers the extent of auto-acceleration in a manner exactly parallel to the behaviour in bulk polymerization (Figure 4.18). The temperature coefficient determined for dispersion polymerization was equivalent to an apparent overall activation energy of 11·4 kcal/mol, 8 kcal/mol lower than the value calculated for the corresponding solution polymerization. The paradoxical effects on the rate of dispersion polymeriza-

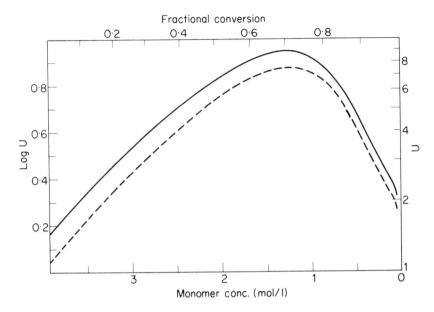

Figure 4.19. Variation of relative rate constant with conversion in dispersion polymerization of methyl methacrylate at 80 °C[1]. [$U = \alpha(2f)^{\frac{1}{2}}.k_p/k_t^{\frac{1}{2}}$ calculated from simplified (———) and fully corrected equation (· · · ·). (From Barrett and Thomas, J. Polym. Sci. A1, 7, 2629 (1969), with permission)

tion where the monomer or the aliphatic diluent is replaced by increasing amounts of a strong solvent, such as n-butyl acetate, can be interpreted in a similar manner. At high reaction temperatures and high initial concentrations of monomer, the rate of polymerization is slowed down because the autoacceleration due to the gel effect is reduced by the plasticizing action of the added solvent. Under conditions in which even the diffusion of monomer is hindered (low monomer content, low reaction temperature) the presence of added solvent may even speed up the reaction.

The model for the dispersion polymerization of methyl methacrylate predicts that the instantaneous molecular weight of the polymer produced should run parallel to the rate of polymerization and should rise to a marked peak value during the course of the polymerization. The peak number average molecular weight ($\overline{M}_n$) should roughly correspond to the overall weight average molecular weight ($\overline{M}_w$) taken over the whole run. This approximation was borne out by comparison of the values of peak $\overline{M}_n$ calculated from the maximum rate with experimental values of overall $\overline{M}_w$ obtained by light scattering (Table 4.12).

### 4.4.4.2. Vinyl acetate

The experimental kinetic studies of the dispersion polymerization of vinyl acetate[101] both by analysis of samples from sealed ampoules and by the continuous recording microcalorimetric procedure have shown that it fits the first

Table 4.12. Estimation of $\overline{M}_w$ from peak conversion rate[1]. (From Barrett and Thomas, *J. Polym. Sci.* A1, **7**, 2641 (1969), with permission.)

| Peak conversion rate (%/min) | Calculated peak $\overline{M}_n$ | Overall $\overline{M}_w$ found |
|---|---|---|
| 1·4 | $1·8 \times 10^6$ | $2·1 \times 10^6$ |
| 1·7 | $2·7 \times 10^6$ | $2·3 \times 10^6$ |
| 1·1 | $1·4 \times 10^6$ | $1·6 \times 10^6$ |

special case ($\alpha$, $V$ small; equation 4.74, Section 4.4.3), taking a constant value for $k_p/k_t^{\frac{1}{2}}$ several times larger than that found in solution polymerization. There is in fact, a small variation of $k_p/k_t^{\frac{1}{2}}$ which closely parallels the small increase and subsequent fall in apparent rate constant found in the bulk polymerization of vinyl acetate at high conversions (the bulk rate only increases 2–3 fold at 70–80 °C) (Figure 4.14, Section 4.4.1). The small gel effect observed with vinyl acetate is no doubt due in large measure to the flexibility of its polymer chain as reflected by its low glass transition temperature.

The experimental data obtained for the dispersion polymerization of vinyl acetate fit well with a roughly constant partition ratio of monomer between disperse polymer and diluent of the order of 2. Other experimental features such as molecular weight [yielding estimates of chain transfer constant, $c_m = 8·4 \times 10^{-4}$ and initiator efficiency, $f = 0·43$ consistent with literature values[102,103]] and temperature coefficient also fit the model described.

#### 4.4.4.3. *Acrylonitrile*

The less extensive kinetic study of the dispersion polymerization of acrylonitrile[104] has shown a somewhat different pattern of behaviour from that of vinyl acetate or methyl methacrylate. For any particular initial monomer concentration, the behaviour at the start of the reaction accords with equation (4.74), Section 4.4.3 (special case: $\alpha$, $V$ small), that is, polymerization rate is proportional to the square root of polymer formed. However, polymerization runs with a wide range of initial monomer concentrations conform to equation (4.75), Section 4.4.3 (special case: $\alpha$ large), in that the rate of conversion is inversely proportional to the square root of initial monomer concentration. These results can be interpreted in terms of the partition of available monomer largely into the polymeric phase as soon as particles form. At first, the amount of monomer in the polymer particles is limited by the amount of polymer formed, and the rate is proportional to the square root of polymer concentration (equation 4.74). However, after a certain conversion is reached scarcely any monomer remains in the diluent and the particle volume fraction is limited only by the total amount of monomer present, yielding kinetic behaviour corresponding to equation (4.75). Superimposed on these effects, of course, is the acceleration due to the gel effect which is comparable in magnitude to that involved in the polymerization of methyl methacrylate.

A direct determination of the monomer partition ratio for poly(acrylonitrile) particles gave a value of 2 in favour of the polymeric phase. This is by no means a large enough value however, to fit the kinetic data found. To explain the effect, it can be tentatively suggested that the swelling of poly(acrylonitrile) by its monomer is only partly reversible. The process can be visualized as starting with polymer particles of a rather tenuous structure heavily swollen with monomer. As polymerization proceeds further, the monomer concentration in the particles falls, the polymer chains become more closely associated through interaction between their polar groups and chain entanglement or even possibly cross-linking ensues. The addition of further monomer at this stage can only swell the polymer and open out its structure to a very limited extent. This interpretation of the behaviour is supported by two experimental observations. The rate of dispersion polymerization of acrylonitrile, carried out in the presence of added preformed poly(acrylonitrile) particles, is markedly different from the corresponding stage of a normal polymerization[104]. This is in sharp contrast to the results obtained in the dispersion polymerization of methyl methacrylate in the presence of added poly(methyl methacrylate) particles which differ little in rate from a polymerization without added particles at the corresponding total polymer content. Additional support for the concept is also provided by the observation that in dispersion polymerizations using a continuous feed of acrylonitrile, uncontrollable reactions occur if the standing monomer concentration is allowed to fall to a very low level and subsequently increased. This can be explained if the polymer particles become partially fixed in a state in which acceleration of polymerization due to the gel effect is very pronounced, i.e. polymeric radicals are immobilized but monomer still has free access. The presence of trapped radicals in poly(acrylonitrile) under similar conditions has been experimentally demonstrated by electron spin resonance[91]. Unfortunately, a fuller analysis of the problem meets with difficulties since with acrylonitrile there is, of course, no corresponding homogeneous bulk polymerization process; the bulk polymerization itself is a precipitation polymerization[93].

*4.4.4.4. Vinyl and vinylidene chloride*

The very limited kinetic study of vinyl chloride dispersion polymerization[104] indicates that the rate and the time/conversion curve at 70 °C differ little from those of the corresponding solution polymerization. Vinylidene chloride also behaves in a similar manner. This is, at first sight, a surprising finding since vinyl chloride, like acrylonitrile, is known to undergo a marked acceleration in rate during its bulk precipitation polymerization[93]. However, the behaviour observed in the dispersion polymerization of vinyl chloride, taken in conjunction with evidence of the protuberant nature of the surface of the particles (Figure 4.15, Section 4.4.1), points to a very different type of polymerization mechanism from those discussed so far, namely solution polymerization followed by precipitation or aggregation with existing particles (see also Section 4.3.4). The reason for this can also probably be attributed to the relatively slight absorption of monomer into the particles, under the reaction conditions, together with the rigid structure of the particles which hinders accretion of oligomer radicals, so reducing growth from within the particles to a low level. Most oligomer radicals therefore, continue polymerizing in solution until they terminate and aggregate with other polymer chains, forming new nuclei which remain separate in the presence of excess polymeric dispersant but accrete to the surface of existing particles if the supply of dispersant is restricted. The growth of the particles thus proceeds in an irregular manner as manifested by their uneven external structure shown in practice.

In spite of the rigid structure of the poly(vinyl chloride) particles, there appears to be little trapping or immobilization of radicals (which would give rise to accelerated polymerization, as in the case of acrylonitrile), probably because facile chain transfer limits the size of the growing radicals so that they readily escape from the polymer matrix. There is some indication of a slight acceleration at lower temperatures, where monomer absorption would be more favoured, the polymer chains would be more rigid and chain transfer would be less. However, definite acceleration only takes place in bulk polymerization where the monomer concentration in the polymer (about 30% at 50 °C), is significantly higher than that operating in dispersion polymerization.

To summarize therefore, dispersion polymerization is a heterogeneous microbulk process whose detailed kinetics depend much on the nature of the monomer and the polymer matrix it forms. Both methyl methacrylate and vinyl acetate polymerize within their monomer-swollen polymer particles although the gel effect is only really pronounced in the former case. Acrylonitrile (and probably similar polar monomers) also shows a marked acceleration in rate due to the gel effect but its detailed kinetics differ from those of the previous species because of the irreversible nature of its swelling process. Finally, vinyl chloride (and also vinylidene chloride) undergoes solution polymerization with precipitation and aggregation of the polymer formed with relatively little interaction between the monomer and the growing polymer particles.

### 4.4.5. Copolymerization parameters

It has long been established, that the chemical composition of a copolymer depends upon the relative reactivity of the monomers concerned towards the corresponding propagating polymer radicals[105]. The monomer reactivity ratios $r_1$ and $r_2$, for a given pair of monomers, are the ratios of the rate constants of the possible propagation reactions below:

$$\sim M_1 \cdot + M_1 \xrightarrow{k_{11}} \sim M_1 M_1 \cdot$$

$$\sim M_1 \cdot + M_2 \xrightarrow{k_{12}} \sim M_1 M_2 \cdot$$

$$\sim M_2 \cdot + M_2 \xrightarrow{k_{22}} \sim M_2 M_2 \cdot$$

$$\sim M_2 \cdot + M_1 \xrightarrow{k_{21}} \sim M_2 M_1 \cdot$$

$$r_1 = k_{11}/k_{12}; \quad r_2 = k_{22}/k_{21} \tag{4.76}$$

Reactivity ratio data for a wide range of monomer pairs have been compiled[106] and can be used directly for predicting copolymer compositions in homogeneous systems. In heterogeneous systems however, the overall reactivity is also influenced by the relative concentrations of monomer at the site of polymerization. Consequently, the effective reactivity of comonomers in dispersion polymerization is considerably modified by their differential absorption by the polymer particles. Qualitatively, the effect is to enhance the apparent reactivity of strongly polar monomers, such as methacrylic acid, which partition mainly into the polymer particles, and to reduce the apparent reactivity of relatively non-polar monomers, such as butyl acrylate or styrene, which reside mainly in the diluent phase. The modification of monomer reactivity resulting from differential absorption of two monomers can be allowed for by the use of a factor $F$, where

$$[M_1/M_2]_p = F[M_1/M_2]_d \qquad (4.77)$$

and the subscripts p, d refer to the ratio of monomers in the particle and diluent respectively. This is equivalent to replacing the reactivity ratios $r_1$ and $r_2$ by new apparent reactivity ratios

$$r'_1 = Fr_1; \qquad r'_2 = r_2/F \qquad (4.78)$$

In a dispersion polymerization with a continuous feed of comonomers, the standing monomer concentrations adjust themselves until the copolymer being formed has a composition corresponding to the comonomer feed composition. This is, of course, provided that the overall polymerization rate is rapid enough to maintain the standing monomer concentration at a low value and that the reactivity ratios and differential absorption factors are not so widely disparate that one of the monomers steadily accumulates. This approach can be used to estimate effective reactivity ratios, and hence $F$, by a measure of the standing monomer concentrations. Alternatively, directly determined values for the differential absorption of monomer mixtures obtained with static systems under equilibrium conditions can be incorporated with reactivity ratio data to obtain the effective reactivity ratios.

So far, it has been tacitly assumed that copolymer is formed in a homogeneous fashion within the particles. This need not necessarily be the case, especially if the monomer composition is deliberately changed during the course of the dispersion polymerization. Usually the second monomer composition will be absorbed by the polymer particles formed by the first composition but if the new polymer composition is incompatible with the preceding one, phase separation may result with the precipitation of microparticles of the second polymer within the first (see Section 5.6.4). Whilst each of the pair of polymers may be either homopolymers or copolymers, the overall process is not a true copolymerization but a blend of polymers prepared in a particularly intimately-mixed form. Even in the absence of actual phase separation, newly-added monomer may tend to polymerize preferentially near the surface of the particles, because many radicals enter the particles from the diluent phase and probably cannot diffuse far before growing too large to move freely within the growing polymer matrix.

### 4.4.6. Control of molecular weight

In the absence of other chain transfer agents, the number average degree of polymerization, $\bar{P}_n$ for termination by disproportionation, is given by

$$1/\bar{P}_n = R_i/R_p + c_m \tag{4.79}$$

where $R_i$ is the overall rate of initiation, $R_p$ the overall rate of polymerization and $c_m$ the chain transfer constant to monomer. Consequently, expressions for $\bar{P}_n$ can readily be derived from the corresponding rate equation for dispersion polymerization (equation 4.73, Section 4.4.3). It follows, therefore that growth in molecular weight will run parallel to the rate of polymerization during dispersion polymerization, provided that the rate of initiation is reasonably constant and chain transfer does not become a limiting factor. Thus, the molecular weight of the polymer formed in a dispersion polymerization using a single, initial charge of reactants will be low initially, rising rapidly to a maximum value as the rate accelerates and finally tailing off as monomer is depleted and the rate falls. Of course, for a process involving continuous addition of monomer with a roughly constant rate of polymerization, the average molecular weight will remain constant with its distribution approaching the ideal for free radical polymerization under constant conditions.

In the dispersion polymerization of vinyl acetate, chain transfer to monomer is the dominating effect in limiting molecular weight. The values determined experimentally, interpreted on the basis of the kinetic model for dispersion polymerization lead to reasonable values for $c_m$ (Section 4.4.4). Similarly, chain transfer will also be an important feature in the dispersion polymerization of vinyl chloride[104]. In the case of methyl methacrylate and related monomers, the very fast polymerization rate due to the gel effect, will also give rise to a correspondingly high molecular weight. The values determined experimentally for poly(methyl methacrylate) dispersions accord well with those calculated theoretically[1] and viscosity average molecular weights up to $2 \times 10^6$ can be readily obtained.

Since the average molecular weight of polymer formed is inversely proportional to the square root of initiator concentration, in principle molecular weight can be controlled by the level of initiator used. However, since this method would usually involve the use of prohibitively high concentrations of initiator (producing very fast polymerization rates) to reduce the molecular weight of acrylic polymers formed in dispersion polymerization, the addition of a suitable chain transfer agent is used in practice. Clearly, bearing in mind the mode of dispersion polymerization, an important factor in the function of the chain transfer agent will be its ability to partition between the polymer particles and the diluent. Generally, the most useful agents for molecular weight control in dispersion polymerizations in aliphatic hydrocarbon diluents are alkyl mercaptans, of moderate chain length, with a balanced partition between the two phases in order to provide an adequate supply of transfer agent at the site of the polymerization within the particle.

### 4.4.7. Continuous processes

The kinetic model for dispersion polymerization can readily be applied to continuous polymerization processes (Section 5.6.1). A steady-state equation can be set up for the 'stirred tank' situation, in which the rate of polymerization (equations 4.73, 4.74; Section 4.4.3) is set equal to the rate of exit of the polymer formed, $-R_{p'}$, i.e.

$$R_p = -R_{p'} = C_0 x/\tau \qquad (4.80)$$

where $C_0$ is the concentration of feed monomer in the system, $x$ is the fractional conversion at the output and $\tau$ is the mean residence time. The effective initiator concentration is also modified by the value of $\tau$ as follows:

$$C_i = i_0 \exp(-\tau k_d) \qquad (4.81)$$

where $i_0$ is the initiator concentration in the feed and $k_d$ its decomposition rate constant. The value of the ratio of the propagation and rate constants involved, $k_p/k_t^{\frac{1}{2}}$, will be a function of the steady state monomer concentration in the particles. The resulting equation, corresponding to the simplified form of equation (4.74), Section 4.4.3, will be:

$$x^{\frac{1}{2}}/(1-x)(k_p/k_t^{\frac{1}{2}})_x = \tau \alpha (C_0 V_p R_i)^{\frac{1}{2}} \qquad (4.82)$$

This expression can either be solved graphically from a knowledge of the dependence of rate constant on conversion or determined directly from a rate/conversion curve of a corresponding polymerization in which all the reactants are added initially. Usually, there will be only a single solution to equation (4.82) with a definite fractional conversion corresponding to a given residence time. In such circumstances, the system will ultimately reach the steady state whatever the starting concentrations in the tank. Of course, this will be reached much more rapidly if the initial concentrations in the tank already approximate to those of steady state conditions.

However, for certain polymerization conditions, more than one steady state value for $x$ can exist. This occurs, for example, when the system exhibits a very rapid acceleration in rate due to a pronounced gel effect, especially when operated near its maximum rate. It can also occur when the concentration of monomer in the feed is high enough to allow solution polymerization to proceed. Under such circumstances, the upper value for $x$ will normally give stable operations, that is, a transient fall in conversion will increase the rate and thus restore steady state conditions. However, the lower value will be unstable since a transient fall in conversion will reduce the polymerization rate still further so that it reaches a very low limiting value, which corresponds to the third solution to equation (4.82). In such cases, therefore, it is important to confine operations to the stable region and to ensure that the initial tank contents are within this range.

Since the rate of dispersion polymerization is zero for $x = 0$ (no particles) and $x = 1$ (no monomer), there will be some conversion corresponding to a maximum rate and usually very high conversions will only be reached by extending the mean residence time. This can be reduced by increasing initiator concentration and by raising temperature, which not only increases the value of the rate constants in the usual way but also shifts the maximum gel effect towards higher conversions.

## 4.5. DISPERSION POLYMERIZATION BY OTHER THAN FREE RADICAL MECHANISMS

There has been little detailed study of the mechanism of particle formation and particle growth in dispersion polymerizations involving other than free radical initiation processes. However, it would be expected that the same general principles evolved should also apply to ionically-initiated systems. Growing oligomers will be formed, aggregation processes will compete against association with polymeric dispersants and the subsequent polymerization will proceed with monomer absorbed within the polymer particles. In some cases, for example in the dispersion polymerization of ethylene oxide derivatives[107], the ionic catalyst will tend to partition largely into the polymer particles producing a situation somewhat analogous to suspension polymerization. One major difference from the free radical processes does arise since in the absence of any mechanism involving the mutual termination of active species there is little basis for expecting a faster overall polymerization rate in dispersion than in the corresponding solution process.

Stampa[108] (see also Section 5.4.1) has observed a faster initial polymerization rate in the anionic dispersion polymerization of α-methyl styrene in heptane compared to the corresponding solution process in benzene (Figure 4.20).

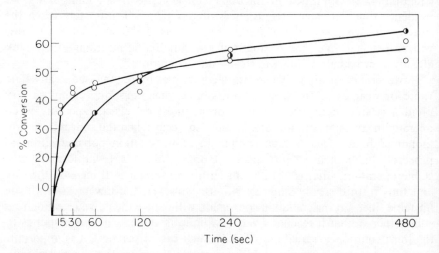

Figure 4.20. Rates of conversion in typical dispersion polymerization of α-methyl styrene[108]: 40% monomer, 1 mol-% BuLi; ●, solution in benzene; ○, dispersion in heptane. (From Stampa, *J. Appl. Polym. Sci.*, **14**, 1230 (1970), with permission)

This difference was ascribed to the presence of an initially high local concentration of monomer in the freshly-precipitated polymer particles with a resulting increase in the propagation rate. After the monomer/polymer equilibrium became established within the particles, the polymerization rate thereafter fell to that of the solution polymerization. In the dispersion process, the yields of polymer obtained and the dependence of molecular weight on initiator concentration were consistent with a bulk-phase polymerization mechanism.

The anionic dispersion polymerization of styrene and methyl methacrylate has been carried out in n-heptane using amphipathic block copolymers as dispersants and lithium alkyl as initiator[109]. Growing oligomers are formed in solution which eventually precipitate out and aggregate in equilibrium with the dispersant to form the growing polymer nuclei. Unlike the corresponding homolytic processes, further additions of initiator during the polymerization are not required since, in the absence of impurities, there is no termination of polymerization and particle growth continues with a steady increase in molecular weight. After precipitation of the growing oligomers at the start of the polymerization, the site of polymer growth is confined to the dispersed particles already formed and unless these particles are disrupted, no new particles are formed even in the presence of a large excess of dispersant.

In practice, it is possible to increase the rate of dispersion polymerizations (by appropriate adjustment of parameters such as reaction temperature and concentration of reactants) to a level quite unacceptable in solution or bulk conditions because of the ease of heat and mass transfer in stirred dispersion systems. Moreover, the greater ease of removal of by-products of condensation reactions carried out in a finely dispersed or emulsified phase (e.g. the elimination of water in the formation of polyesters and polyamides) permits a faster conversion to high molecular weight polymer than in conventional bulk or solution processes, where diffusion to the surface is a rate controlling factor. For example, in a kinetic study of the melt polyamidation of 11-aminoundecanoic acid, a conversion of 75% corresponding to a number average degree of polymerization of 4 was reported[110] after reaction for $5\frac{1}{2}$ h at 185 °C. The same monomer, in a solution of ethylene glycol dispersed in aliphatic hydrocarbon reached a conversion of 99% ($\bar{P}_n = 40$) after 5-h reaction at 160–180 °C[111].

With condensation reactions based on difunctional monomers (such as diacid chloride and diamine[107]), there is also a special feature in the kinetics which results from the geometric rather than the arithmetic progression involved in the growth of the polymer. Consequently, a large proportion of the condensation reaction takes place before the threshold molecular weight for precipitation is reached, resulting in the sudden production of high molecular weight material in high concentration. For this reason, a high concentration of dispersant is required to avoid excessive aggregation in such processes.

### 4.6. REFERENCES

1. Barrett, K. E. J. and Thomas, H. R., *J. Polym. Sci.* A1, **7**, 2621 (1969).
2. Harkins, W. D., *J. Amer. Chem. Soc.*, **69**, 1428 (1947).
3. Smith, W. V. and Ewart, R. H., *J. Chem. Phys.*, **16**, 592 (1948).

4. Roe, C. P., *Ind. Eng. Chem.*, **60**, 20 (1968).
5. Norrish, R. G. W. and Smith, R. R., *Nature (London)*, **150**, 336 (1942).
6. White, T. and Haward, R. N., *J. Chem. Soc.*, 25 (1943).
7. Haward, R. N., *J. Polym. Sci.*, **3**, 10 (1948).
8. Burrell, H., *J. Paint Technol.*, **40**, 197 (1968).
9. Hildebrand, J. H., *J. Amer. Chem. Soc.*, **51**, 66 (1929).
10. Scatchard, G., *Chem. Rev.*, **8**, 321 (1931).
11. Hildebrand, J. H. and Scott, R. L., *The Solubility of Non-electrolytes*, 3rd edition, Reinhold Publishing Corp., New York, 1950.
12. Hildebrand, J. H. and Scott, R. L., *Regular Solutions*, Prentice Hall, Englewood Cliffs, New Jersey, 1962.
13. Flory, P. J., *J. Chem. Phys.*, **10**, 51 (1942).
14. Huggins, M. L., *J. Amer. Chem. Soc.*, **64**, 1712 (1942).
15. Flory, P. J., *Principles of Polymer Chemistry*, Cornell University Press, Ithaca, N.Y., 1953.
16. Small, P. A., *J. Appl. Chem.*, **3**, 71 (1953).
17. Hoy, K. L., *J. Paint Technol.*, **42**, 76 (1970).
18. Burrell, H., *Off. Dig. Fed. Paint Varnish Prod. Clubs*, **27**, 726 (1955).
19. Burrell, H. and Immergut, B. in *Polymer Handbook* (Eds. Brandrup, J. and Immergut, E. H.), Interscience (John Wiley), N.Y., 1966.
20. Gardon, J. L., *J. Paint Technol.*, **38**, 43 (1966).
21. Gardon, J. L. in *Encyclopedia of Polymer Science and Technology*, Vol. 3 (Eds. Mark, H. F., Gaylord, N. G. and Bikales, N. M.), Interscience (John Wiley), N.Y., 1965, pp. 833–862.
22. Crowley, J. D., Teague, G. S. and Lowe, J. W., *J. Paint Technol.*, **38**, 269 (1966).
23. Hansen, C. M., *J. Paint Technol.*, **39**, 104 (1967).
24. Huggins, M. L., *J. Chem. Phys.*, **9**, 440 (1941).
25. Cornet, C. F. and van Ballegooijen, H., *Polymer*, **7**, 293 (1966).
26. Rehner, J., *J. Polym. Sci.*, **46**, 550 (1960).
27. Carpenter, D. K. in *Encyclopedia of Polymer Science and Technology*, Vol. 12 (Eds. Mark, H. F., Gaylord, N. G. and Bikales, N. M.), Interscience (John Wiley), N.Y., 1965, pp. 627–678.
28. Gardon, J. L., *J. Polym. Sci.*, A1, **6**, 2859 (1968).
29. Boyer, R. F. and Spencer, R. S., *J. Polym. Sci.*, **3**, 97 (1948).
30. Maron, S. H. and Nakajima, N., *J. Polym. Sci.*, **42**, 327 (1960).
31. Cotten, G. R., Sirianni, A. F. and Puddington, I. E., *J. Polym. Sci.*, **32**, 115 (1958).
32. Kokes, R. J., Di Pietro, A. R. and Long, F. A., *J. Amer. Chem. Soc.*, **75**, 6319 (1953).
33. Thompson, L. J. and Long, F. A., *J. Amer. Chem. Soc.*, **76**, 5886 (1954).
34. Bristow, G. M. and Watson, W. F., *Trans. Faraday Soc.*, **54**, 1731 (1958).
35. Bristow, G. M. and Watson, W. F., *Trans. Faraday Soc.*, **54**, 1742 (1958).
36. Orofino, T. A. and Flory, P. J., *J. Chem. Phys.*, **26**, 1067 (1957).
37. Fox, T. G., *Polymer*, **3**, 111 (1962).
38. Gerrens, H., Fink, W. and Koehnlein, E., *J. Polym. Sci.*, **C16** (5), 2781 (1967).
39. Doty, P. and Zable, H. S., *J. Polym. Sci.*, **1**, 90 (1946).
40. Schulz, A. R. and Flory, P. J., *J. Amer. Chem. Soc.*, **75**, 3888 (1953).
41. Wolf, B. A., *Adv. Polym. Sci.*, **10**, 109 (1972).
42. Cantow, M. I. R. (Ed.), *Polymer Fractionation*, Academic Press Inc., New York, 1967.
43. Koningsveld, R., *Pure Appl. Chem.*, **20**, 271 (1969).
44. Scott, R. L., *J. Chem. Phys.*, **17**, 268 (1949).
45. Krigbaum, W. R. and Carpenter, D. K., *J. Polym. Sci.*, **14**, 241 (1954).
46. Gibbs, J. W., *Collected Works*, Longmans, Green and Co., New York, 1928.
47. Morton, M., Kaizerman, S. and Altier, M. W., *J. Colloid Sci.*, **9**, 300 (1954).
48. Hibbin, B. C., Reginato, G. and Yearsley, F., ICI Plastics Division, unpublished results, 1969.

49. Hölscher, F., *Dispersionen Synthetischer Hochpolymere*, Part 1, Springer-Verlag, Berlin, 1969, pp. 136–146.
50. Cooke, D. D. and Kerker, M., *J. Colloid Interface Sci.*, **42**, 150 (1973).
51. Cantow, H. J., *Makromol. Chem.*, **70**, 130 (1964).
52. Davidson, J. A., Collins, E. A. and Haller, H. S., *J. Polym. Sci.*, **C35**, 235 (1971).
53. Wales, M., *J. Phys. Chem.*, **66**, 1768 (1962).
54. Nisonoff, A., Messer, W. E. and Howland, L. H., *Anal. Chem.*, **26**, 856 (1954).
55. Kay, D. (Ed.), *Techniques for Electron Microscopy*, Blackwell Scientific Publications, Oxford, 1965.
56. Fitch, R. M. and Tsai, C. H. in *Polymer Colloids* (Ed. Fitch, R. M.), Plenum Press, New York, 1971, pp. 73, 103.
57. Cobbold, A. J. and Gilmour, R. E., *British Polymer J.*, **3**, 249 (1971).
58. Morton, J. and Thompson, M. W., ICI Paints Division, unpublished results, 1964.
59. Tankey, H. W., ICI Paints Division, unpublished results, 1967.
60. Slavnitskaya, N. N., Semchikov, Yu. D., Ryabov, A. V. and Bort, D. N., *Vysokomol. Soed.*, **A12**, 1756 (1970); (translated in *Polym. Sci. USSR*, **12**, 1993).
61. King, J. M., ICI Paints Division, unpublished results, 1965.
62. du Pont, *United States Patent* 3,723,571, 1973.
63. Bueche, F., *J. Colloid Interface Sci.*, **41**, 374 (1972).
64. Pont, J. D., ICI Paints Division, unpublished results, 1972.
65. Osmond, D. W. J., ICI Paints Division, unpublished results, 1964.
66. Gershberg, D., *A.I. Chem. E./I. Chem. E. Symposium, Series No. 3*, p. 4, Institute of Chemical Engineers, London, 1965.
67. Brodnyan, J. G., Cala, J. A., Konen, T. and Kelley, E. L., *J. Colloid Sci.*, **18**, 73 (1963).
68. Gardon, J. L., *British Polymer J.*, **2**, 1 (1970).
69. Shinoda, K., Nakagawa, T., Tamamushi, B. and Isemura, T., *Colloidal Surfactants: Some Physicochemical Properties*, Academic Press, New York and London, 1963, Chapter 1.
70. Dawkins, J. V. in *Block Copolymers* (Eds. Allport, D. C. and Janes, W. H.), Applied Science Publishers Limited, London, 1973, pp. 363, 546.
71. Krause, S., *Macromolecules*, **3**, 84 (1970); *J. Polym. Sci.*, *A2*, **7**, 249 (1969).
72. Graham, N. B., Holden, H. W. and Raymond, F. L., *British Polymer J.*, **2**, 141 (1970).
73. Volmer, M., *Kinetik der Phasenbildung*, T. Steinkopff, Dresden and Leipzig, 1939; reprinted Edward Bros., Ann Arbor, Mich., 1945.
74. Becker, R. and Döring, W., *Ann. Phys.*, **24**, 719 (1935).
75. Dunn, A. S., *Chem. Ind. (London)*, 1406 (1971).
76. Robb, I. D., *J. Polym. Sci.*, *A1*, **7**, 417 (1969).
77. Smoluchowski, M. V., *Z. Physik. Chemie*, **92**, 129 (1917).
78. Glasstone, S., Laidler, K. J. and Eyring, H., *Theory of Rate Processes*, McGraw-Hill, New York, 1941.
79. Von Weimarn, P. P., *Chem. Rev.*, **2**, 217 (1926).
80. La Mer, V. K., *Ind. Eng. Chem.*, **44**, 1270 (1952).
81. Klein, J. and Patat, F., *J. Polym. Sci. Part C*, **16**, 3565 (1968).
82. Bort, D. N. and Vishnevskaya, I. N., *Vysokomol. Soed.*, **13** (9), 1950 (1971).
83. Mandelkern, L., *Crystallization of Polymers*, McGraw-Hill, New York, 1964.
84. Turnbull, D. and Fisher, J. C., *J. Chem. Phys.*, **17**, 71 (1949).
85. Gaines, G. L., *Polymer Eng. and Sci.*, **12** (1), 1 (1972).
86. Gardon, J. L., *J. Polym. Sci.*, *A1*, **6**, 623 (1968).
87. Fitch, R. M. in *Advances in Emulsion Polymerization and Latex Technology*, Lehigh University, Bethlehem, 1973.
88. Ugelstad, J., Fløgstad, H., Hertzberg, T. and Sund, E., *Makromol. Chem.*, **164**, 171 (1973).

89. Barrett, K. E. J., ICI Paints Division, unpublished results, 1966.
90. Wainman, J. E., ICI Paints Division, unpublished results, 1970.
91. Tolman, R. J., ICI Paints Division, unpublished results, 1966.
92. Schulz, G. V. and Harborth, G., *Makromol. Chem.*, **1**, 106 (1948).
93. Bamford, C. H., Barb, W. G., Jenkins, A. D. and Onyon, P. F., *The Kinetics of Vinyl Polymerization by Radical Mechanisms*, Butterworths Scientific Publications, London, 1958.
94. North, A. M., *The Kinetics of Free Radical Polymerization*, Pergamon Press, Oxford, 1966.
95. Cameron, G. G. and Cameron, J., *Polymer*, **14**, 107 (1973).
96. Schulz, G. V., *Z. Phys. Chem. (Frankfurt)*, **8**, 290 (1956).
97. Harriott, P., *J. Polym. Sci.*, $A1$, **9**, 1153 (1971).
98. Barrett, K. E. J. and Thomas, H. R., ICI Paints Division, unpublished results, 1969.
99. Barrett, K. E. J., *J. Appl. Polym. Sci.*, **11**, 1617 (1967).
100. Barrett, K. E. J. and Thomas, H. R., *Br. Polym. J.*, **2**, 45 (1970).
101. Barrett, K. E. J., Thomas, H. R. and Tolman, R. J., *Kinet. Mech. Polyreactions, Int. Symp. Macromol. Chem. Prep.*, **3**, 369 (1969).
102. Clarke, J. T., Howard, R. O. and Stockmayer, W. H., *Makromol. Chem.*, **44/46**, 427 (1961).
103. Bamford, C. H., Jenkins, A. D. and Johnston, R., *Proc. Roy. Soc.*, $A$, **239**, 214 (1957).
104. Barrett, K. E. J. and Thomas, H. R., *Kinet. Mech. Polyreactions, Int. Symp. Macromol. Chem. Prepr.*, **3**, 375 (1969).
105. Alfrey, T., Bohrer, J. J. and Mark, H., *Copolymerization*, Interscience Publishers, New York, 1952.
106. Mark, H., Immergut, B., Immergut, E. H., Young, L. J. and Beynon, K. I. in *Polymer Handbook* (Eds. Brandrup, J. and Immergut, E. H.), Interscience (John Wiley), New York, 1966.
107. Imperial Chemical Industries, *British Patent* 1,095,931 and 1,095,932, 1967.
108. Stampa, G. B., *J. Appl. Polym. Sci.*, **14**, 1227 (1970).
109. Imperial Chemical Industries, *South African Patent* 72/7635, 1973.
110. Colange, J. and Guyot, P., *Compt. Rend.*, **233**, 1604 (1951).
111. Thompson, M. W., ICI Paints Division, unpublished results, 1973.

# CHAPTER 5

# The Preparation of Polymer Dispersions in Organic Liquids

K. E. J. BARRETT AND M. W. THOMPSON

| | |
|---|---|
| 5.1. INDIRECT METHODS | 201 |
| 5.2. THE SCOPE OF DISPERSION POLYMERIZATION | 205 |
| 5.3. FREE-RADICAL ADDITION DISPERSION POLYMERIZATION | 207 |
| 5.3.1. Basic procedures | 207 |
| 5.3.2. Acrylic acids and esters | 209 |
| 5.3.3. Vinyl esters | 212 |
| 5.3.4. Vinyl aromatics | 213 |
| 5.3.5. Acrylonitriles | 214 |
| 5.3.6. Vinyl and vinylidene chloride | 215 |
| 5.3.7. $N$-Vinyl pyrrolidone | 215 |
| 5.4. IONICALLY INITIATED DISPERSION POLYMERIZATION | 216 |
| 5.4.1. Addition polymerization | 216 |
| 5.4.2. Ring-opening polymerization | 218 |
| 5.5. DISPERSIONS OF CONDENSATION POLYMERS | 220 |
| 5.5.1. Condensation dispersion polymerization | 220 |
| 5.5.2. Condensation polymerization of dispersed reactants | 221 |
| 5.6. SPECIAL TECHNIQUES OF DISPERSION POLYMERIZATION | 225 |
| 5.6.1. Continuous dispersion polymerization | 225 |
| 5.6.2. High solids polymer dispersions | 227 |
| 5.6.3. Dispersions of reactive polymers | 230 |
| 5.6.4. Dispersions of microgels and heterogeneous polymer particles | 232 |
| 5.7. SELECTED RECIPES FOR DISPERSION POLYMERIZATION | 234 |
| 5.7.1. Preparation of poly(methyl methacrylate) latex using graft dispersant precursor | 234 |
| 5.7.2. Preparation of poly(methyl methacrylate) latex using preformed graft dispersant | 235 |
| 5.7.3. Preparation of poly(methacrylic acid) latex in chloroform–ethanol mixture using a polyester stabilizer precursor | 237 |
| 5.7.4. Preparation of poly(acrylonitrile) latex in n-hexane | 237 |
| 5.7.5. Preparation of poly(vinyl chloride) latex (semi-technical scale) | 238 |
| 5.7.6. Preparation of a dispersion of poly(t-butyl–styrene–isoprene–styrene) block copolymer | 238 |
| 5.7.7. Preparation of a dispersion of poly(ethylene terephthalate) using direct and reverse emulsification procedures | 239 |
| 5.8. REFERENCES | 240 |

## 5.1. INDIRECT METHODS

Although methods for the preparation of polymer dispersions in water, by the direct polymerization of monomers emulsified or suspended in the medium, have been available for several decades[1], the equivalent direct method for preparing stable polymer dispersions in organic liquids by dispersion polymerization is a comparatively recent development[2,3]. As an alternative to direct polymerization, a number of indirect methods for preparing polymer dispersions in organic liquids have been used, all of which involve the subsequent conversion of polymer, prepared in a variety of ways, to a more or less disperse form in an organic liquid (Table 5.1).

Table 5.1. Indirect methods for the preparation of polymer dispersions in organic liquids.

| Type of polymer | Dispersion method |
| --- | --- |
| 1. Bulk | Comminution in presence of dispersant. |
| | Emulsification of molten polymer, followed by cooling. |
| 2. Solution | Addition of precipitant in presence of dispersant. |
| | Addition of precipitant for one component of graft copolymer ('self-stable' organosol). |
| 3. Aqueous dispersion | Direct 'flushing' or displacement into organic diluent. |
| 4. Particulate powder (ex aqueous dispersion) | Dispersion into organic media in presence of dispersant. |
| | Treatment with 'swelling' solvents (organosol, plastisol). |

Bulk polymer has been converted into fine particle dispersions in organic liquids by various grinding procedures in the presence of a suitable dispersant[4]. In practice, however, the mechanical energy required to reduce the particles to below about 10 μm in size is so great and its efficiency of utilization so low that the heat evolved makes it difficult to control the temperature of the polymer so that it remains in the brittle glassy state.

Above the glass transition temperature of the polymer, the grinding process becomes one of mastication in which subsequent coalescence of the particles formed is more difficult to prevent and chain-fission tends to reduce the molecular weight of the treated polymer and to narrow its molecular weight distribution.

An alternative procedure which has been used, is to emulsify polymer in a molten state into a suitable organic liquid in the presence of a dispersant. The emulsion produced is cooled to yield a corresponding dispersion of polymer[5]. The particle size range of the polymer dispersions produced depends on the efficiency of the emulsification procedure. In practice, the method is limited to polymers of relatively low molecular weight with a low melt viscosity and which have a satisfactory thermal stability in the molten state.

A solution of polymer in an aromatic solvent, such as toluene, in the presence of a suitable dispersant can be converted into a polymer dispersion by the careful addition of an organic liquid in which the polymer is insoluble, e.g. cyclohexane. The addition of a precipitant for one component of a suitable graft copolymer in solution results in the formation of stable micellar aggregates surrounded by solvated polymer chains, to yield the so-called 'self-stable' organosols[6,7]. In this way, stable dispersions of poly(methyl methacrylate-g-isoprene) and poly(methyl methacrylate-g-2-ethyl hexyl acrylate) have been prepared in aliphatic hydrocarbons.

A convenient source of particulate polymer in dispersion is available from the conventional aqueous emulsion or suspension polymerization techniques. The polymers produced by these methods are usually of high molecular weight and fine polymer particles of uniform size can be obtained. A number of techniques have been used, which involve the displacement of the aqueous phase by an organic phase, to yield an equivalent polymer dispersion in an organic liquid. The general principles are essentially the same as the method used for the 'flushing' of pigment pastes from an aqueous into an organic phase[8]. For polymers, the organic liquid is added to an aqueous polymer dispersion in the presence of a dispersant suitable for the particular organic diluent, so that the polymer is gradually dispersed into the organic phase. The aqueous phase remaining can be simply removed by separation in a tap-funnel or by more elaborate procedures such as azeotropic distillation[9].

The displacement methods have generally proved less satisfactory than techniques which involve the redispersion of vacuum-dried emulsion polymer into organic media[10,11]. The surfaces of the dry polymer produced in a powdered form retain a variety of polar groups, such as hydroxyl or carboxyl, which are derived ultimately from the use of ammonium persulphate as the initiator in the emulsion polymerization. These polar groups can be used as the point of attachment for soluble polymers which can stabilize the dispersion by a steric mechanism. Thus, for polymer particles containing anionic groups on their surface, suitable graft copolymers containing complementary cationic groups for attachment, can be used for redispersion. Conversely, graft copolymer dispersants containing anionic groups can be used for emulsion polymers with cationic groups on their surface. The use of a dispersant, based on a backbone of poly(methyl methacrylate-co-ethyl acrylate-co-dimethylaminoethyl methacrylate), for the re-dispersion of a poly(vinyl chloride) powder ('Breon' 121) in white spirit, is illustrated in Figure 5.1. Dispersions of poly(vinylidene chloride) and poly(vinylidene fluoride) have also been prepared in a similar manner. Such methods which utilize polymer particles derived ultimately from aqueous emulsion polymerization are, of course, only feasible with 'hard' polymers below their glass transition temperature.

Dried, particulate polymer powder, derived from emulsion polymerization, is also used in the preparation of 'Organosol' and 'Plastisol' compositions for surface-coating applications. These are pigmented polymer dispersions in organic liquids and in plasticizing solvents, respectively[12]. One of the most widely-used compositions in the surface-coating industry has been the organosol based on poly(vinyl chloride). Organic liquids are added to the polymer powder and the mixture is agitated to break up the agglomerated particles to yield discrete and partially-swollen polymer particles. Complete dissolution of the poly(vinyl chloride) particles is prevented by the presence of crystalline regions in the polymer matrix and stabilization of the dispersion is achieved by the solvation of the outer polymer chains. Cross-linked emulsion polymers or

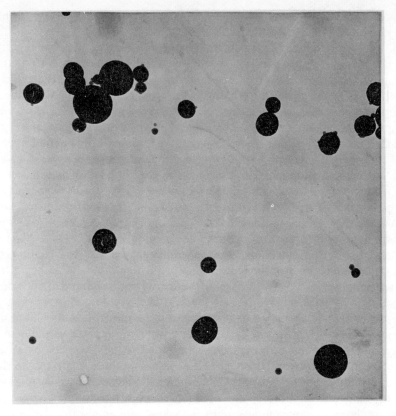

Figure 5.1. Redispersed poly(vinyl chloride) powder, 'Breon' 121. (Magnification × 10,000)

microgels, such as those based on styrene and divinyl benzene or methyl methacrylate and ethylene diacrylate, behave similarly. In this case, complete dissolution is prevented by the cross-linked nature of the interior of the polymer particle and stable dispersions result when the dried copolymer is treated with a swelling solvent[13].

All of the indirect methods described above have been used at various times for the preparation of polymer dispersions in organic media for applications in the surface-coating and other industrial fields, since they have advantages over the more conventional aqueous dispersions or organic solution systems (see Section 1.1). However, the overall properties of the polymer dispersions in organic media obtained by the indirect methods of preparation, have usually fallen below the level required for their widespread application in the surface-coatings field. In particular, these polymer dispersions often have unsatisfactory rheological properties, the overall stability of the dispersions is usually poor and the control of particle size required is frequently difficult to achieve.

Consequently, the development of a direct method for preparing such polymer dispersions, by the dispersion polymerization of monomers in organic media, has overcome many of the defects inherent in the indirect methods of preparation and allowed a more rapid development of surface-coatings based on non-aqueous polymer dispersions (NAD's) particularly as coatings for the automobile industry[14].

## 5.2. THE SCOPE OF DISPERSION POLYMERIZATION

Dispersion polymerization generally involves the polymerization of a monomer dissolved in an organic diluent to produce insoluble polymer dispersed in the continuous phase in the presence of an amphipathic graft or block polymer as the dispersant. The copolymer dispersant can either be prepared separately or its precursor can be added to the dispersion polymerization to undergo grafting reactions simultaneously with the growth of polymer. Clearly, dispersion polymerization is quite different from an aqueous emulsion or suspension polymerization in which the monomer has only a limited solubility in the aqueous phase and also requires surfactant for its stabilization in the emulsified form.

In dispersion polymerization, any mode of polymerization which produces polymer insoluble in the continuous phase can be used. Free radical addition polymerization has been the most widely applied and many of the commonly-used monomers, such as vinyl acetate and methyl methacrylate, are entirely miscible in all proportions with aliphatic diluents allowing a wide range of dispersions of polymers and copolymers to be prepared by the technique (Section 5.3). Ionic initiation can also be used and has been applied to both addition (Section 5.4.1) and ring-opening polymerization (Section 5.4.2). Since the polymerizations are carried out in non-polar media, those involving water or alcohol-sensitive reactants or catalysts whose use would normally be precluded from aqueous emulsion polymerization techniques, can readily be carried out by dispersion polymerization.

Techniques of condensation polymerization can also be utilized, using soluble reactants in the usual way (Section 5.5.1). However, in cases in which the reactants are insoluble in the required organic medium, it is possible to use additional methods to emulsify or disperse the reactants using the same graft copolymer dispersants which are required for the dispersed polymer produced (Section 5.5.2). The disperse nature of the reactants often allows the removal of the by-products of the condensation polymerization at a more rapid rate than the corresponding process under bulk conditions.

Although aliphatic liquids, such as n-heptane or cyclohexane, have been most used as the organic diluent in dispersion polymerizations, more polar liquids such as fluoro- or chloro-substituted hydrocarbons[15], esters and even alcohols can readily be used if required. Normally, the hydrocarbon or mixture of hydrocarbons to be used is selected on the basis of suitable boiling point for the type of polymerization and initiation to be used since the dispersion polymerization is usually carried out under reflux conditions. However, if necessary, the choice can also be based on the requirements of the subsequent applications of the polymer dispersions. In the case of surface coatings, this would require a suitable diluent mixture to achieve the correct rate of evaporation for film formation, or if polymer powders are produced subsequently, a diluent of a suitably low latent heat of evaporation to be distilled off readily with a low heat input.

The kinetic characteristics of free radical addition dispersion polymerization, which enable fast polymerization rates to proceed within the polymer particles independent of their particle size, and the ready removal of the heat of polymerization produced from the mobile dispersion, is particularly suitable for the development of continuous polymerization techniques (Section 5.6.1). In addition, the degree of control over the size of the polymer particles produced by dispersion polymerization which the appropriate choice of the nature, amount and use of a graft copolymer makes possible, can be exploited to prepare bimodal and polymodal distributions of polymer particles which enable solids levels (80–85 %) to be achieved, which are significantly higher than normal (50 %), whilst still maintaining acceptable rheological properties in the final dispersion (Section 5.6.2).

With the use of the appropriate reactive comonomers in dispersion polymerization, dispersions of reactive polymers may be obtained, which undergo subsequent cross-linking reactions (Section 5.6.3). The technique has also allowed the preparation of cross-linked or microgel polymer particles (Section 5.6.4). Heterogeneous polymer particles, which contain separate inclusions of one polymer within the matrix of another, have also been prepared by dispersion polymerization. Other materials, for example inorganic pigments, have also been encapsulated within a dispersed polymer particle by similar means.

Finally, although not strictly polymers in the usual sense, dispersions of metals and inorganic compounds can be prepared in organic diluents by techniques similar to those used for organic polymers but by utilizing decomposition reactions. For example, dispersions of metallic cobalt particles have been prepared by the thermal decomposition of dicobalt octacarbonyl in toluene[16]. In the presence of linear addition polymers of molecular weight > 100,000 (e.g. terpolymers based on methyl methacrylate, ethyl acrylate and vinyl pyrrolidone), stable colloidal dispersions of uniform particle size were formed. Dispersions of nickel could be produced in the same way.

Dispersions of inorganic compounds can, in principle, be prepared by analogous decomposition reactions although the processes involved tend to be chemically complex. For example, tetraethyl silicate dissolved in an aromatic hydrocarbon in the presence of dispersant, should be decomposable by steam to form a dispersion of silica. Similarly, a solution of titanium tetrachloride in aliphatic hydrocarbon, treated in a like manner should also produce a dispersion of titanium dioxide. Alternatively, a 50 % solution of aluminium chlorophosphate ethanolate in methanol can be dispersed in an aliphatic hydrocarbon liquid by means of a 'double-comb' dispersant (Figure 3.13, Section 3.7.4) to produce a fine emulsion[17]. The emulsion is heated to remove methanol and other by-products to produce a fine particle dispersion of poly(aluminium phosphate) in the aliphatic hydrocarbon.

## 5.3. FREE-RADICAL ADDITION DISPERSION POLYMERIZATION

### 5.3.1. Basic procedures

Two main types of procedure have been evolved for dispersion polymerization in organic diluents. The first, a single stage or 'one-shot' process, utilizes an initially homogeneous solution of reactants in an organic liquid which is polymerized to completion in a single stage. The second, a feed process, involves an initial 'seed-stage', in which a polymer dispersion at low solids is produced by a 'one-shot' process and polymerization is completed by the further (usually continuous) addition of reactants to the 'seed-stage'.

In the 'one-shot' process, a solution of monomer, initiator and graft copolymer dispersant in an organic liquid, usually an aliphatic hydrocarbon of suitable boiling point, are heated together with stirring generally under reflux conditions. After a short period under reflux during which solution polymerization takes place, the reaction mixture becomes opalescent due to the formation of very small, 'molecular' particles of polymer precipitated from solution. These primary particles can aggregate with each other to form larger species but are prevented from complete aggregation by their association with the graft copolymer dispersant which provides a barrier of soluble polymer around the particles formed. After the formation of dispersed polymer, monomers such as methyl methacrylate or vinyl acetate, are absorbed into the polymer particles and polymerization then proceeds at an increased rate within the particle (see Section 4.4.1). The onset of the accelerated rate of polymerization is reflected in an increase in the rate of reflux. Dispersion polymerization continues at a fast rate until the monomer concentration (initially at 50%) in the continuous phase has fallen to a low level (c. 2–3%). At this point, the polymerization rate and hence the rate of reflux falls and extra heating is required to complete the polymerization.

The single stage procedure however is not suitable if an accurate control of the particle size (see Section 4.3.2) or molecular weight (Section 4.4.6) of the dispersed polymer is required. Since the monomers used are frequently solvents for their own polymers, the overall solvency of the continuous phase for polymer changes as monomer is consumed in the polymerization. Consequently, the conditions for precipitation of polymer formed in solution change throughout the polymerization. In addition, the degree of association of the graft copolymer dispersant with the dispersed polymer, and hence its effectiveness as a dispersion stabilizer, is reduced by the increased solubility of its 'anchoring' component. These two factors together lead to polymer dispersions of an undesirably wide distribution of particle sizes. The vigorous evolution of the heat of polymerization from the single stage process is also difficult to control in large-scale manufacture and consequently a 'feed' process involving the continuous addition of reactants at a controlled rate, is generally preferred.

In the 'feed' process, the standing monomer concentration can be maintained at the same level so that the overall solvency of the continuous phase remains sensibly constant and the graft copolymer dispersant remains fully effective. In practice, high local concentrations of monomer during the addition process are avoided by diluting the incoming feed of reactants with the returning cooled, condensed organic diluent which has distilled from the reaction mixture. The first stage of the 'feed' process is identical to that of the 'one-shot' process except that a low solids (c. 10%) dispersion of polymer 'seed' of uniform particle size is produced. Controlled addition of a feed consisting of further monomer, initiator and graft dispersant continues particle growth of the polymer 'seed' initially formed to the required solids level. Unless drastic alterations are made either to the solvency of the diluent or to the concentration of graft dispersant during the course of the 'feed' process, little further nucleation of new polymer particles takes place [Figure 5.6(a), Section 5.6.2].

There is a potentially wide selection of organic liquids which can be used as the organic diluent in dispersion polymerization. It is preferable that the monomer or monomers used are readily soluble in the organic diluent and that it is chemically inert. Its boiling point should be below that of the monomers to be polymerized so that the concentration of monomer in the reaction mixture is not depleted under the conditions of reflux. The blanket of condensing vapour of the inert organic diluent also serves as a barrier to the ingress of oxygen into the reaction mixture. The boiling point of the organic diluent should also be close to the optimum half-life of the thermal initiator to be used in the polymerization. It is, of course, essential that the polymer formed is insoluble in the organic diluent and preferably any swelling of the polymer should not be excessive in order to achieve acceptable rheological properties for the polymer dispersion formed. The range of boiling points of the organic diluent is important in subsequent surface-coating applications of the dispersions and usually a range is selected to give a graded evaporation rate in film formation. A low-boiling diluent is required for efficient evaporation when polymer dispersions are converted to polymer powders. Most of the work on dispersion polymerization has been carried out using aliphatic hydrocarbons as the organic diluent with boiling points in the range 30–150 °C.

The initiator for dispersion polymerization is selected to generate a suitable supply of free radicals at the temperature required for gel-phase polymerization. For example, in the dispersion polymerization of methyl methacrylate, gel-phase polymerization proceeds rapidly at 80 °C but with vinyl acetate a more pronounced gel-effect is obtained at 30 °C (see Section 4.4.4). In early systems which involved the use of rubbers etc. as precursors for the graft copolymer dispersants, the free radicals generated, such as those derived from benzoyl peroxide, were also required to provide grafting sites on the precursor by hydrogen abstraction[18]. With systems containing preformed graft dispersant, this requirement is unnecessary and even undesirable and azo-initiators, such as azo-di-isobutyronitrile, are preferred. Their decomposition rates also are virtually independent of the chemical nature of their environment and only depend on the temperature at which they are used.

## 5.3.2. Acrylic acids and esters

The polymerization of methyl methacrylate in petrol to give a stable dispersion of polymer of controlled particle size and molecular weight, was the first system to be studied in detail and has been of major importance in understanding the mechanism of radical-initiated dispersion polymerization in organic liquids. Examples of organic diluents and dispersants which have been used in the dispersion polymerization of methyl methacrylate are given in Table 5.2.

Table 5.2. Examples of organic diluents and dispersants used in dispersion polymerization of methyl methacrylate.

| Organic diluent | Type of dispersant | References |
|---|---|---|
| Aliphatic hydrocarbons | Precursor and preformed graft | 2, 3, 19 |
| Cyclohexane | Graft precursor | 20, 21, 22, 23 |
| Aliphatic hydrocarbons | Poly(vinyl ether) | 24, 25, 26 |
| Aliphatic hydrocarbons | Polyester modified with drying oils | 27 |
| Hexane | Poly(isobutylene-co-isoprene) | 3 |
| Halogenated hydrocarbons | Rubber precursor | 15 |
| Diethyl ether | Preformed graft | 28 |
| Methanol | Methoxy poly(ethylene glycol) preformed graft | 28 |

A broad range of acrylic monomers has now been utilized for the preparation of dispersions of polymers (Table 5.3) in organic media.

Early work on the dispersion polymerization process employed rubbers which took part in grafting reactions during the polymerization, to form the graft copolymer dispersant. Subsequent developments in the technique of dispersion polymerization have followed the improved methods for obtaining graft copolymer dispersant (see Section 3.7). The random grafting process with rubber was soon replaced by the use of synthetic precursors based on, for example, poly(lauryl methacrylate), containing polymerizable double bonds of comparable reactivity to the monomer undergoing dispersion polymerization. In this way, useful graft dispersant was formed in higher yield and a more controllable process was achieved.

The apparatus used (Figure 5.11, Section 5.7) consisted of a glass or stainless steel reactor with either a paddle or anchor stirrer, operating at a few hundred revolutions per minute (stirring rate is not critical). The reactor was fitted with a downward condenser so that the incoming monomer feed was diluted with condensed returning hydrocarbon before reaching the reaction mixture. A reaction temperature of 80–82 °C was maintained by reflux of the hydrocarbon diluent under the exothermic conditions of the polymerization.

The selection of a hydrocarbon medium of the appropriate solvency is important if the required particle size is to be obtained. The aromatic and cyclo-aliphatic content of commercial hydrocarbons should be carefully checked since if these vary from sample to sample, the precipitation conditions

Table 5.3. Examples of polymer dispersions prepared by free radical addition dispersion polymerization.

| Polymer | Organic diluent | Type of dispersant | References |
|---|---|---|---|
| Poly(ethyl acrylate) | Aliphatic hydrocarbon | Precursor and preformed graft | 2, 3, 19 |
| Poly(methyl acrylate) | Aliphatic hydrocarbon | Preformed graft | a |
| Poly(2-ethoxyethyl methacrylate) | Aliphatic hydrocarbon | Precursor and preformed graft | 2, 3, 19 |
| Poly(hydroxyethyl methacrylate) | Aliphatic hydrocarbon | Preformed graft | a |
| Poly(butoxy ethoxy ethyl methacrylate) | Aliphatic hydrocarbon | Preformed graft | a |
| Poly(dimethyl amino ethyl methacrylate) | Aliphatic hydrocarbon | Preformed graft | a |
| Poly(octyl methacrylate) | Methanol | Methoxy poly(ethylene glycol) preformed graft | a |
| Poly(cetyl/stearyl methacrylate) | Methanol | Methoxy poly(ethylene glycol) preformed graft | a |
| Poly(methacrylic acid) | Aliphatic hydrocarbon | Poly(vinyl toluene-co-butadiene) | 3 |
| Poly(methacrylic acid) | Chloroform/ethanol | Methacrylate-terminated polyester | a |
| Poly(acrylamide) | t-Butanol | Preformed graft | a |
| Poly(acrylonitrile) | Hexane | Preformed graft | a |
| Poly(vinyl chloride) | Aliphatic hydrocarbon | Graft precursor | a |
| Poly(ureido-ethyl vinyl ether) | Aliphatic hydrocarbon | Copolymer of long-chain methacrylate, acrylate and N-vinyl pyrrolidone | 29 |
| Poly(chloroprene) | Aliphatic hydrocarbon | Chloroprene/2-ethyl hexyl acrylate graft | 30 |

$^a$ Unpublished work at ICI Paints Division.

in the seed stage will alter and hence modify the particle size of the polymer dispersion produced.

The general method for the dispersion polymerization of acrylic esters, for example methyl methacrylate, in the presence of a precursor for the graft dispersant based on a copolymer of lauryl methacrylate and glycidyl methacrylate reacted with methacrylic acid (random graft precursor, Section 3.7.3) is as follows. (See Section 5.7.1 for a detailed recipe.) A solution of monomer, initiator and graft dispersant precursor in aliphatic hydrocarbon is refluxed with stirring to provide the seed stage dispersion. Chain transfer agent (usually p-octyl mercaptan) is added as a molecular weight control at the completion of the seed stage to ensure equilibration in the heterogeneous dispersion before the main feed of monomer, initiator and further chain transfer agent is added. After the addition of the main feed is completed, the dispersion is refluxed for an additional period of time in order to complete the polymerization.

The method of preparation, involving the use of a monofunctional poly(lauryl methacrylate) graft dispersant precursor (Section 3.7.4), is essentially similar except that since with its use a higher proportion of usable graft copolymer is produced, it can be used in somewhat smaller concentrations (3%) compared with the level (4%) required for the precursor prepared by the random graft process above.

In the method of preparation which utilizes a preformed graft precursor, such as the 'comb'-type graft based on a backbone of poly(methyl methacrylate) with pendant side chains of poly(12-hydroxystearic acid), a different procedure is used (see Section 5.7.2 for a detailed recipe). Unlike the precursor, if all the preformed graft required to stabilize the dispersed polymer is added at the seed stage, the resulting particle size of the seed would be very small. Consequently, the appropriate amount of preformed graft is added at the seed stage for the particle size required. Then, following the addition of mercaptan after the seed stage, an intermediate feed of monomer, initiator, chain transfer agent and the balance of the graft dispersant required, is added before the main monomer feed is commenced.

Attempts to prepare stable dispersions of either poly(acrylic acid) or poly(methacrylic acid) in aliphatic hydrocarbons using a range of graft dispersants have, so far, been largely unsuccessful[31]. Very fine particles of polymer were produced initially but these soon flocculated during subsequent polymerization. Carboxylic acids are known to form hydrogen-bonded dimers in non-polar media[32]. The acrylic acids therefore will behave essentially as difunctional monomers producing cross-linked polymer particles as in the case of microgels (see below, Section 5.6.4). Consequently, the graft dispersant would become firmly attached to the cross-linked polymer particles and eventually lose its stabilizing power by its incorporation within the matrix of the growing polymer particle. However, by the use of more polar organic diluents in which the formation of hydrogen-bonded dimer is less likely, stable dispersions of poly(acrylic acids) have been prepared. The graft dispersant was selected to match the properties of the diluent used. A random graft dispersant based on poly(methyl methacrylate-co-glycidyl methacrylate) reacted with methacrylic acid was used for preparing dispersions in ethyl acetate. A methacrylate-terminated glycol phthalate polyester was also used to prepare dispersions of poly(acrylic acids) in chloroform and chloroform–ethanol mixtures (see Section 5.7.3 for detailed recipe).

The methods available for the preparation of dispersions of copolymers of acrylic esters by dispersion polymerization in organic media are broadly similar to the methods described for homopolymers, using the required mixture of monomers in the appropriate feed. The range of copolymers which have been prepared by these methods, mainly in aliphatic hydrocarbons, is indicated in Table 5.4, which lists a selection of the more important types. Improvements in the design of graft dispersants and their increased effectiveness has enabled a much closer control over the composition and size of the polymer particles to be achieved. The use of dispersant precursors, especially those based on

Table 5.4. Examples of dispersions of acrylic copolymers prepared in aliphatic hydrocarbons by dispersion polymerization (from work at ICI Paints Division).

| Comonomers | Proportions by weight |
|---|---|
| Methyl methacrylate : ethyl acrylate | 35–95 : 65–5 |
| Methyl methacrylate : n-butyl methacrylate | 80 : 20 |
| Methyl methacrylate : n-butyl methacrylate : methacrylic acid : butoxymethyl acrylamide | 63 : 30 : 2 : 5 |
| Methyl methacrylate : n-butyl methacrylate : methacrylic acid : glycidyl methacrylate | 60 : 30 : 5 : 5 |
| Methyl methacrylate : hydroxyethyl methacrylate | 90 : 10 |
| Methyl methacrylate : glycidyl methacrylate | 90–97 : 10–3 |
| Methyl methacrylate : 2-ethoxyethyl methacrylate | 70–92 : 30–8 |
| Methyl methacrylate : 2-ethoxyethyl methacrylate : glycidyl acrylate | 60 : 30 : 10 |
| Methyl methacrylate : acrylic or methacrylic acid | 90–98 : 10–2 |
| Methyl methacrylate : dimethylaminoethyl methacrylate | 90 : 10 |
| Methyl methacrylate : dimethylaminopropyl methacrylamide | 90 : 10 |
| Methyl methacrylate : 2 or 4-vinyl pyridine | 90 : 10 |
| Styrene : acrylic acid | 87 : 13 |
| Styrene : methacrylic acid : ethyl acrylate : glycidyl methacrylate | 43 : 5·5 : 43 : 8·5 |
| Ethyl acrylate : acrylonitrile | 75–95 : 25–5 |

rubber, was particularly hampered by the experimental conditions required for the *in situ* formation of graft copolymer. Grafting of synthetic precursors by hydrogen abstraction with alkoxyl radicals was also incompatible with monomers containing certain reactive groups, such as dimethyl aminoethyl methacrylate. The preparation of dispersions of reactive copolymers (see Section 5.6.3) and microgels (see Section 5.6.4) are described elsewhere.

The compositions of the copolymer dispersions formed are dependent on the relative concentrations of the monomers at the site of polymerization and their relative radical reactivities (see Section 4.4.5 on dispersion copolymerization parameters). Provided that the degrees of absorption of the monomers are comparable and the reactivity ratios are not too disparate, the usual procedures of preparation are adequate. In some cases, these differences can be overcome by appropriate modifications in the feed process. The maintenance of a low standing monomer concentration in the reaction mixture assists the formation of random copolymers by preventing an accumulation of the less reactive monomer in the continuous phase. In some cases, it may even be possible to exploit the modified reactivities of monomers under the heterogeneous polymerization conditions in order to produce copolymers normally difficult to obtain by solution or bulk copolymerization[33].

### 5.3.3. Vinyl esters

The manufacture of dispersions of poly(vinyl acetate) and its copolymers in water by emulsion polymerization is a large, well-established industry having applications in the emulsion paint and adhesives fields. In organic media, dispersion polymerization can be similarly employed to give fluid dispersions of poly(vinyl acetate) of fine particle size up to 70% solids by weight.

Although vinyl acetate is miscible in all proportions with many organic solvents its polymer is insoluble in aliphatic hydrocarbons. Its dispersion polymerization therefore can be carried out in diluents such as n-heptane or cyclohexane. The readiness of vinyl acetate to chain transfer during its polymerization, together with its sensitivity to inhibition, means that an organic diluent free of aromatic compounds or other potentially reactive materials must be used if polymer of reasonably high molecular weight is required. The low glass transition temperature (c. 35 °C) of the polymer, when swollen by monomer, also requires the use of relatively low polymerization temperatures (c. 30 °C) if the maximum gel effect on rate is to be attained.

The dispersion polymerization of vinyl acetate, like methyl methacrylate, can be carried out either as a single-stage polymerization with all the reactants charged initially or by a feed process, following the formation of a seed-stage. In the latter, it is again desirable to dilute the incoming feed of reactants with the returning stream of condensed distillate before it reaches the reaction mixture. In general, aliphatic hydrocarbon diluents of boiling points in the range 50-70 °C and lower temperature initiators, such as isopropyl peroxydicarbonate, have been used, in order to achieve satisfactory rates of polymerization together with polymer of the required molecular weight. A range of both precursor and preformed graft dispersants has been utilized. These have included graft dispersant precursors based on poly(lauryl methacrylate-co-glycidyl methacrylate) reacted with methacrylic acid or containing a crotonate double bond with which the vinyl acetate radical will react. The use of preformed graft dispersants based on a backbone of poly(methyl methacrylate-co-methacrylic acid) with side chains of poly(12-hydroxystearic acid) has also been described elsewhere (Section 3.7.4). Random copolymers based on poly(ethylene-co-vinyl acetate) have also been used as dispersants for the dispersion polymerization of vinyl acetate[34].

A variety of copolymers of vinyl acetate, usually with the comonomer as the minor constituent, have been prepared in dispersion with, for example, methacrylic acid, methyl hydrogen maleate, butoxymethyl acrylamide, vinyl propionate and vinyl versatate[31]. Dispersions of poly(vinyl acetate) prepared in cyclohexane have been converted directly in the disperse state into poly(vinyl alcohol) dispersions by hydrolysis[34]. This was effected by the addition of methanol to the stirred dispersion of poly(vinyl acetate), followed by a solution of sodium methoxide or sulphuric acid in methanol. After stirring for 3 h at 30 °C, a coarse dispersion of poly(vinyl alcohol), 75% hydrolyzed, was obtained. Variations in the procedure produced material up to 98.5% hydrolyzed.

### 5.3.4. Vinyl aromatics

Styrene and its analogues have been the subject of dispersion polymerization by ionic mechanisms (Section 5.4.1) but relatively little has been reported for the corresponding free radical process, probably because of its generally sluggish behaviour in homolytic systems. In addition, although poly(styrene) is not soluble in linear aliphatic hydrocarbons at room temperature it becomes heavily swollen at 80 °C and so in these diluents, the dispersion polymerization must be carried out at lower temperatures in order to obtain fine polymer dispersions.

Dispersions of poly(styrene) in n-hexane (33 % solids) have been made, using an ionically prepared ABA block copolymer based on 2-ethylhexyl methacrylate and styrene as graft dispersant[31]. The rate of the dispersion polymerization, carried out at 30 °C in the presence of isopropyl peroxydicarbonate as initiator, was very slow (reaction time 48 h to obtain 90 % conversion).

Poly(styrene) is insoluble in the lower aliphatic alcohols and successful dispersion polymerizations have been carried out, for example, in a mixture of methanol and ethylene glycol as organic diluent at 80 °C[30]. A graft copolymer precursor was used, based on an ABA block copolymer of ethylene and propylene oxide ('Pluronic F68') with acrylic ester end-groups prepared by an ester interchange reaction catalysed with sodium ethoxide.

The copolymers of styrene are not subject, to the same extent, to the solubility limitations of the homopolymer and dispersions of styrene-based copolymers, e.g. with acrylic acid, ethyl acrylate or acrylonitrile can be prepared in aliphatic hydrocarbons by the usual procedures of dispersion polymerization[31].

### 5.3.5. Acrylonitriles

The dispersion polymerization of acrylonitrile can be carried out successfully in aliphatic hydrocarbon diluents provided the reaction exotherm is controlled by the rate of addition of monomer and the reaction temperature is not allowed to fall due to the formation of the azeotrope of acrylonitrile with hexane or heptane, boiling at about 60 °C (see Section 5.7.4 for a detailed recipe). Control of reaction temperature is essential because the polymerization of acrylonitrile shows a maximum both in rate and in the molecular weight of the polymer produced at about 60 °C[35] and polymerization can easily run out of control if large amounts of monomer are allowed to accumulate at this temperature. If this situation is avoided, the formation of dispersions of poly(acrylonitrile) up to 50 % solids with particle size in the range 0·5–1 μm can be achieved with a fair degree of control although the usage of the 'comb' graft dispersant based on poly(12-hydroxystearic acid) tends to be high due to its strong association with the highly polar poly(acrylonitrile) particle during its growth. Poly(methacrylonitrile) is rather less polar so that stable dispersions may be prepared in aliphatic hydrocarbons using normal amounts of conventional types of graft dispersant[31].

Chain transfer reactions, leading to side reaction grafting can occur during the polymerization of acrylonitrile in the presence of certain polymers. This effect has been exploited to prepare stable dispersions of poly(acrylonitrile) in the presence of polymers which contain no specially-introduced reactive groups or anchor groups to effect stabilization. This has been achieved by using experimental conditions to encourage grafting reactions, e.g. by using a seed stage with a large excess of stabilizing polymer. In this way poly-(acrylonitrile) dispersions were prepared, using ethanol as diluent and poly(N-vinylpyrrolidone) as stabilizer precursor[31].

Dispersions of copolymers of acrylonitrile have been made using rubber as the dispersant precursor. For example, copolymer containing methacrylic acid and ethoxyethyl methacrylate was prepared. Fine particle dispersions of poly(acrylonitrile-co-acrylic acid) have also been prepared in petrol or methyl ethyl ketone, using a 'comb' graft dispersant[31]. The formation of stable dispersions of poly(methyl methacrylate-co-acrylonitrile) in heptane, prepared in the presence of a low molecular weight copolymer based on 2-ethyl hexyl methacrylate and methyl methacrylate and using a non-grafting initiator, has also been described[36].

### 5.3.6. Vinyl and vinylidene chloride

In contrast to the use of methyl methacrylate, vinyl acetate and acrylonitrile, the dispersion polymerization of vinyl chloride (and also that of vinylidene chloride) proceeds entirely by a solution polymerization in aliphatic diluents (Section 4.4.4). Under the conditions used, the polymer formed precipitates out from solution and no further polymerization within the particles ensues. It is therefore necessary to provide a continuous supply of a precursor for the graft dispersant throughout the polymerization since the graft copolymer formed tends either to be buried by the accretion of freshly-formed polymer to the original polymer nuclei or to be used up by the formation of new polymer nuclei.

The dispersion polymerization of vinyl chloride has been carried out in an autoclave under pressure (50–100 lbf/in$^2$), using di-isopropyl peroxydicarbonate as initiator, at 50–70 °C[31] (see Section 5.7.5 for detailed recipe). The graft dispersing agent was formed *in situ* from a soluble copolymer containing an allyl group (e.g. a copolymer based on 2-ethylhexyl acrylate and methacrylic acid, reacted with allyl glycidyl ether) to copolymerize with the vinyl chloride. Usually a straight-chain aliphatic hydrocarbon is present as the aliphatic diluent to limit chain transfer but vinyl chloride in bulk can also be used[37]. As would be expected from a process of particle growth involving continuous accretion of polymer precipitated from solution, the particles of poly(vinyl chloride) formed by dispersion polymerization are only roughly spherical with a coarse micro-structure (Figure 4.15, Section 4.4.1).

Stable dispersions of poly(vinylidene chloride) have not yet been prepared; flocculation took place during dispersion polymerization using a wide range of both preformed and precursor graft surfactants. However, dispersions of its copolymers with methyl methacrylate, ethyl acrylate, vinyl acetate, acrylonitrile and vinyl chloride, containing up to 90% of vinylidene chloride have been made, using preformed graft dispersants of the 'comb' type[31].

### 5.3.7. N-Vinyl pyrrolidone

N-Vinyl pyrrolidone behaves similarly to methyl methacrylate, in its dispersion polymerization in aliphatic hydrocarbons. Both preformed graft dispersants and their precursors may be used for the stabilization of the dispersed polymer[31]. A single stage preparation can be used at low initial monomer concentrations, but above the level of 20 % a continuous feed process is necessary to achieve control of particle size, since the monomer is a good solvent for its polymer. Only aliphatic hydrocarbons have been used as diluent since, with the exception of some esters, the polymer is soluble in most organic liquids.

Dispersions of poly($N$-vinyl pyrrolidone) in aliphatic hydrocarbons show an interesting effect when water is slowly added to the stirred, warm dispersion. Under these conditions, water is uniformly absorbed into the polymer to form a stable emulsion consisting of an aqueous solution of the polymer in petrol. During the addition, the refractive indices of the two phases become equal and a transparent dispersion of quite high solids content is obtained.

## 5.4. IONICALLY INITIATED DISPERSION POLYMERIZATION

### 5.4.1. Addition polymerization

Most of the published work on addition polymerization in dispersion in aliphatic hydrocarbons has been concerned with homolytic polymerization reactions. Dispersion polymerization, however, can also be used with systems involving heterolytic processes. Clearly, the use of an inert aliphatic hydrocarbon as diluent is an advantage in that polymer dispersions may readily be prepared in it using the highly water-sensitive initiators usually involved in ionic polymerizations and whose use is precluded for aqueous emulsion polymerization. Although the acceleration of polymerization rate generally observed in free radical dispersion polymerization due to the diffusion-controlled termination step is not, of course, exhibited by the corresponding ionic addition polymerizations, nevertheless, their inherently fast polymerization rates can provide a useful method for preparing polymer dispersions not obtainable by free radical reactions.

Dispersions of both poly(ethylene) and poly(propylene) have been prepared with the Ziegler–Natta catalysts in aliphatic hydrocarbons, such as pentane or heptane, up to a solids level of 20 to 30%[38]. In this way, poly(ethylene) dispersions of particle size in the range 0·2 to 0·4 μm were prepared, using essentially a block copolymer of ethylene and propylene as the dispersant. The copolymer was first prepared as a coarse dispersion in pentane at a low temperature using a Ziegler–Natta catalyst and subsequently the dispersion was allowed to warm up and eventually dissolve in the diluent. After further dilution, it was used as a solution in the dispersion polymerization of ethylene. Although a detailed analysis of the mode of stabilization of the dispersion was not carried out, it appeared important that the dispersant was used as a solution in the diluent and that it contained polymer segments which associated with the dispersed poly(ethylene) as it was formed. A similar process for the dispersion polymerization of ethylene and other monomers, using a dispersion of a Ziegler–Natta catalyst in aliphatic hydrocarbons, has also been described[39].

Dispersions of poly(propylene) in the presence of poly(octene-1) as a dispersant and a Ziegler–Natta catalyst, have been prepared in kerosene with a solids level of 30 % and a primary particle size in the range 0·2–0·5 μm[38]. The catalyst used was prepared first in refined kerosene from *iso*butylaluminium sesquichloride and titanium tetrachloride and used to polymerize octene-1 at 50 °C. The poly(octene-1) produced was subsequently used as the dispersant in the polymerization of propylene. Diethyl aluminium chloride was treated with oxygen in a solution of poly(octene-1) and purged with nitrogen. A colloidal dispersion of the catalyst containing titanium trichloride was stirred, while propylene and hydrogen was admitted to the reactor over the course of a few hours. After deactivating the catalyst with butanol, the catalyst residues were removed by treatment with an ion-exchange resin. The effect of reaction temperature on the properties of the poly(propylene) produced in dispersion is illustrated in Table 5.5.

Table 5.5. Effect of reaction temperature on the properties of poly(propylene) in dispersion[38]. (From *British Patent* 1,165,840 (1969) with permission of the Controller, H.M. Stationery Office.)

| Polymerization temperature (°C) | Reaction time (h) | Percentage solids | Percentage crystallinity | Intrinsic viscosity | Particle size (μm) | |
|---|---|---|---|---|---|---|
| | | | | | Individual | Clusters |
| 50 | 7·25 | 20·0 | 52 | 3·1 | 0·3 | up to 3 |
| 60 | 5·75 | 20·6 | 58 | 2·7 | 0·3 | up to 2 |
| 70 | 6·67 | 24·4 | 54 | 2·2 | 0·3 | up to 5 |

The first recorded application of an ionic mechanism in a dispersion polymerization was by the anionic polymerization of styrene in heptane using n-butyl lithium as initiator[40]. Various types of rubbers, including poly(butadiene), poly(butadiene-co-styrene) and poly(isoprene) were used to stabilize the dispersions of poly(styrene) formed. Poly(ethylene-co-propylene), atactic poly(propylene) and crepe rubber were also later used for this purpose[41]. The rubbers used have unsaturated groups present in their structure so that the possibility exists for grafting of the growing poly(styrene) to the rubber to take place. The resulting graft copolymer probably functions as the actual dispersant for the polymer formed. The anionic dispersion polymerization of α-methyl styrene has also been described[42] using n-butyl lithium and hexamethyl phosphoramide as promoter. A poly(vinyl alkyl ether) was used as the interfacial agent. In the dispersion polymerization of styrene, using poly(butadienyl) lithium as initiator, an AB graft dispersant can also result and when used in a relatively large amount gave a dispersion of poly(styrene) of particle size in the range 0·4–1·3 μm. The use of AB and ABA block copolymers, based on styrene and t-butyl styrene, has been reported[43] as dispersing agents in the anionic polymerization of styrene in aliphatic hydrocarbons.

The use of butyl lithium and similar anionic initiators for the polymerization of olefins and vinyl aromatic monomers to give soluble 'living' polymers has been described[44]. The addition of a second monomer, which produces a polymer insoluble in the continuous phase, to the 'living' polymer, enables an AB block copolymer to be formed, which precipitates out to yield a very fine polymer dispersion stabilized with the soluble component of the copolymer. Similarly, an AB block copolymer in solution can be treated with a third monomeric component whose polymer is insoluble in the continuous phase, to produce an ABC block copolymer dispersion. By this method, dispersions of block copolymers of controlled composition, structure and molecular weight can readily be obtained. For example, t-butyl styrene was treated in n-heptane with a solution of sec-butyl lithium in n-hexane and polymerization allowed to proceed to completion at 50 °C. This was followed by the second component forming soluble copolymer, isoprene, which again was allowed to polymerize to completion and finally styrene was polymerized to yield a milk-white dispersion of the ABC block copolymer of particle size 0·1 µm. Gel permeation chromatography indicated a block copolymer with the following molecular weights for its components: poly(t-butyl styrene), 15,000; poly(isoprene), 15,000; poly(styrene), 70,000 (see Section 5.7.6 for detailed recipe).

Other types of ionic dispersion polymerization have included the polymerization of isocyanates[45], e.g. the polymerization of phenyl isocyanate using butyl lithium as catalyst and hexane as diluent, but relatively little has so far been published in this field.

### 5.4.2. Ring-opening polymerization

The technique of dispersion polymerization in organic media has also been applied to a range of ring-opening polymerizations of heterocyclic ring compounds using ionic modes of initiation. These have included the polymerization of cyclic ethers (epoxides, oxacyclobutanes), cyclic esters (lactones), cyclic amides (lactams) and cyclic acetals (trioxane), (Table 5.6).

For the dispersion polymerization of cyclic ethers, polymeric dispersants have been used which are based on soluble polymers containing reactive groups capable of copolymerization with the main polymerizing species or groups, such as hydroxyl or carboxyl, which can initiate or terminate the main polymer chains[45]. Various preformed graft copolymer stabilizers have also been used[46]. More recently, the use of swollen poly(epoxide) microgels to stabilize the polymer dispersions formed have also been described[47].

The copolymerizable group most generally used in the precursor for stabilization in the polymerization of oxygen ring systems has been itself an epoxide group, usually in the form of glycidyl methacrylate randomly copolymerized with lauryl methacrylate. A minimum molecular weight of 50,000 was desirable to achieve completely satisfactory stabilization. The anhydride group, in the form of random copolymers of vinyl stearate and maleic anhydride was also utilized as a copolymerizable group.

Lewis acids, particularly boron trifluoride etherate, were preferred for oxygen ring-opening polymerizations but other catalysts, for example aluminium trialkyls, have also been used. Boron trifluoride etherate was particularly useful

Table 5.6. Typical heterocyclic monomers used for ring-opening dispersion polymerization.

---

Cyclic ethers

$\underset{CH_2-CHR}{\overset{O}{\triangle}}$  epoxides (oxiranes)

$\begin{array}{c} O\text{——}CH_2 \\ | \qquad | \\ CH_2-CR_2 \end{array}$ oxacyclobutanes (oxetanes)

Cyclic esters

$\underset{CO-R}{\overset{O}{\triangle}}$ lactones (R = —(CH$_2$)$_5$—, ε-caprolactone)

Cyclic amides

$\underset{CO-NH}{\overset{R}{\triangle}}$ lactams (R = —(CH$_2$)$_5$—, ε-caprolactam)

Cyclic acetals

$\overparen{(CH_2O)_n}$  (n = 3, trioxane)

---

as a catalyst with the glycidyl methacrylate copolymers used as polymeric dispersants since they readily form a soluble pink complex with several times the stoichiometric equivalent of boron trifluoride etherate required for the 1:1 reaction with the glycidyl groups. This feature was advantageous as it enabled a fully homogeneous catalyst system to be prepared in aliphatic hydrocarbon solvents. More efficient stabilization of the polymer dispersion and a finer particle size was obtained by pre-reacting the catalyst with the stabilizer in this way. The glycidyl groups in this case, almost certainly initiated polymerization as well as taking part in the copolymerization, so that the graft copolymer required for stabilization was formed in sufficient quantity before much insoluble polymer precipitated out. In this way, dispersions of poly(ethylene oxide), poly(glycidyl acetate), poly(phenyl glycidyl ether), poly(epichlorhydrin), poly(styrene oxide) and poly(3,3-bis(chloromethyl)oxacyclobutane) have been prepared in a range of aliphatic diluents.

Various preformed graft stabilizers, e.g. poly(lauryl methacrylate-g-vinyl pyrrolidone), have also been used in oxygen ring-opening polymerizations and dispersions of poly(glycidyl acetate) were prepared in this way, using boron trifluoride etherate as catalyst. In some cases, it was possible to prepare the graft copolymer directly in the polymerizing system. Thus, glycidyl stearate can be polymerized in aliphatic hydrocarbon solvents using boron trifluoride etherate to form a soluble polymer, followed by the addition of phenyl glycidyl ether (with the addition of more catalyst if required) to yield both stabilizing graft copolymer and polymer dispersion *in situ*.

The dispersion polymerization of ε-caprolactone has been carried out in heptane with dibutyl zinc as catalyst[48]. Polymeric dispersants, such as poly(lauryl methacrylate) and poly(vinyl chloride-co-lauryl methacrylate) were used to stabilize the polymer dispersions obtained which were of a uniform particle size of about 1 μm. It is not clear in this case whether graft copolymer was formed with the polymerizing species. It is conceivable that either chlorine or ester groups in the polymeric dispersant could interact at a few points with the polymerizing lactone to form sufficient graft copolymer to achieve stabilization. A similar method was also used to produce dispersions of copolymers of ε-caprolactone with ethylene oxide and other epoxides[49]. Catalysts used included phosphorous pentafluoride and boron trifluoride etherate. The dispersion polymerization of $\beta$-propiolactone has also been carried out in cyclohexane in the presence of boron trifluoride etherate with a copolymer of lauryl methacrylate and glycidyl methacrylate as a graft copolymer stabilizer precursor[45]. The dispersion polymerization of lactams in the presence of synthetic rubber polymers in solution in aliphatic hydrocarbons has also been described[50]. Presumably, grafting reactions take place with the soluble polymer. For example, ε-caprolactam was treated with sodium ε-caprolactam and toluene diisocyanate as activator to yield a dispersion of ε-caprolactam polymer in aliphatic hydrocarbon in the presence of poly(butadiene). Dispersions of ε-caprolactam polymer have also been prepared in a mixture of aliphatic and aromatic hydrocarbons by treatment with sodium in the presence of a random copolymer precursor based on lauryl methacrylate and N-methacryloyl caprolactam[45].

Dispersions of trioxane polymer have been prepared in cyclohexane using boron trifluoride etherate as catalyst and a stabilizer precursor based on lauryl methacrylate and glycidyl methacrylate[46]. Formaldehyde polymers were prepared directly by the passage of formaldehyde into n-hexane at $-76\,°C$ in the presence of a random copolymer of lauryl methacrylate and diethylaminoethyl methacrylate as polymer dispersion stabilizer.

## 5.5. DISPERSIONS OF CONDENSATION POLYMERS

### 5.5.1. Condensation dispersion polymerization

The first recorded example of the preparation of dispersions of condensation polymers in organic media involved mechanically grinding solid resin condensates in aliphatic hydrocarbons in the presence of dissolved rubber[51]. Much later, condensation polymerizations were carried out with the reactants dispersed in aliphatic hydrocarbon in the presence of a swelling agent, such as tetramethylene sulphone to provide a polymer dispersion with some measure of stability in the particulate state[52]. The condensation reaction of a tetracarboxylic acid dianhydride with a diamine has also been used in acetone at low temperatures to produce polyamide acids in a granular form[53,54]. Again, the limited dispersion stability achieved was obtained by swelling the polymer formed with relatively polar liquids, such as tetrahydrofuran, dioxane and acetone. However, these methods in general only provided relatively crude and ill-defined polymer particles of poor dispersion stability.

Stable dispersions of condensation polymers can be produced by the direct dispersion polymerization of soluble reactants in organic liquids in the presence of a polymeric dispersant or its precursor[45,46]. The process is analogous to that used for free radical addition dispersion polymerization with the important difference that in condensation polymerization, the growth of polymer proceeds by a stepwise union of the reactants leading to a slow, multiple build-up in molecular weight and a sudden, large requirement for polymeric surfactant when the condensation polymer becomes insoluble in the organic diluent. In general, the method is limited to the narrow range of condensation polymer reactants which are soluble in organic media. The most widely studied reaction has been the condensation of diacid chlorides with diols in ester solvents involving the elimination of hydrogen chloride with a tertiary amine as acid acceptor.

For example, the condensation polymerization of bis(4-hydroxyphenyl)-2,2-propane with a mixture of terephthaloyl and isophthaloyl dichloride was carried out in a mixed solvent of methylene chloride and ethyl acetate in the presence of a polymeric dispersant based on a graft copolymer of lauryl methacrylate, methyl methacrylate and methacrylic acid. After the addition of 2,6-lutidine as acid scavenger, a vigorous exothermic reaction ensued with the separation of the base hydrochloride as a coarse crystalline precipitate and the formation of a stabilized dispersion of polyester.

### 5.5.2. Condensation polymerization of dispersed reactants

In free radical addition polymerization, most of the readily available monomers are soluble in hydrocarbons and only infrequently, as for example in the case of the preparation of a dispersion of poly(ureidoethyl vinyl ether)[29], is it usual to use an emulsion of a monomer for subsequent polymerization in dispersion. However, condensation polymerization frequently involves the use of reactants which are insoluble in aliphatic hydrocarbons and it is then necessary to use emulsions or dispersions of one or more of their reactants for subsequent polymerization in dispersion. Clearly in such cases, the graft copolymer used as the dispersing agent must be designed so that it can provide a stable emulsion or dispersion of the reactants as well as acting as a dispersant for the polymer produced throughout the polymerization.

A number of variants of this basic method have been described. Nicks and Osborne have directly emulsified a liquid reactant into fine droplets in the organic diluent with the use of a suitable dispersant[55]. Solid materials can be melted prior to emulsification provided that the melt viscosity is sufficiently low and that no polymerization takes place before emulsification is completed. If necessary, solutions of the reactants can be emulsified in the organic diluent and the solvent used subsequently separated from the organic diluent by distillation during the course of the polymerization[56]. Alternatively, solid reactants can be dispersed directly into the organic diluent by the orthodox mechanical methods (ball-mill, sand-mill) which are used for preparing dispersions of pigments for paint manufacture[57].

In contrast to dispersion polymerization with soluble reactants where particles are formed and grow during the course of the polymerization, in emulsification processes the particle size of the condensation polymer is largely determined by the size of the initial emulsion droplets. Their size depends not only on the level and nature of the polymeric dispersant used but also on the overall efficiency of the emulsification process. The emulsifying agent can either be dissolved in the organic diluent or in the liquid reactant. The latter method is known as 'reverse emulsification'. Both methods are capable of producing droplets of colloidal dimensions provided that the viscosity of the liquid to be emulsified is low enough to allow it to be broken down to the required particle size by the shearing forces available. The polymeric dispersing agents used have been of two types. Block copolymers consisting of one component soluble in the organic diluent and the other component soluble in or attached to the droplet of reactant have been used. However, 'comb-like' graft copolymers are more efficient and the so-called 'double-combs' (Figure 3.13, Section 3.7.4) consist of a polymeric backbone with side chains which are soluble in the reactant droplet and other side chains which are soluble in the organic diluent[55].

The organic liquid used as a diluent in the condensation polymerization in dispersion must, of course, not contain groups which interfere with the polymerization reaction. This generally excludes the use of alcohols, acids, esters and the like, except where the reaction is carried out at low temperatures. In general, the most suitable inert liquids have been aliphatic or aromatic hydrocarbons. The choice of a diluent of suitable boiling point allows a range of reaction temperatures to be used and the hydrocarbons are also suitable for supporting the azeotropic removal of the by-products of the condensation reaction (usually water or alcohol). Although convenient, removal of by-products by azeotropic distillation is not essential and can be carried out by fractional distillation, as in the formation of poly(p-oxybenzoate) from p-acetyloxybenzoic acid where acetic acid is eliminated. Condensation polymerizations are usually equilibrium reactions and consequently the rate of condensation depends largely on the rate of removal of the by-product of the reaction. In general, the greater ease of removal of the by-product from the reaction of a fine dispersion in a medium of low viscosity compared with bulk polymerization, enables the heterogeneous polymerization to be carried out at some 50–70 °C below the normal bulk polymerization temperature (see Section 4.5).

The variants in the basic method of condensation polymerization in dispersion in hydrocarbons are illustrated by the following examples. An emulsion of bis(hydroxyethyl) terephthalate in a high-boiling (170–190 °C) petrol was converted to poly(ethylene terephthalate) by the azeotropic distillation of ethylene glycol. A graft polymer dispersant was used consisting of a poly(methyl methacrylate) backbone with pendant poly(12-hydroxystearic acid) soluble side chains. Emulsification was achieved either by dissolving the graft dispersant in the petrol and stirring added molten monomer or by melting the monomer with the graft dispersant and adding petrol with rapid stirring ('reverse emulsification') (see Section 5.7.7 for detailed recipes). The dispersion of poly-(ethylene terephthalate) obtained is illustrated in Figure 5.2.

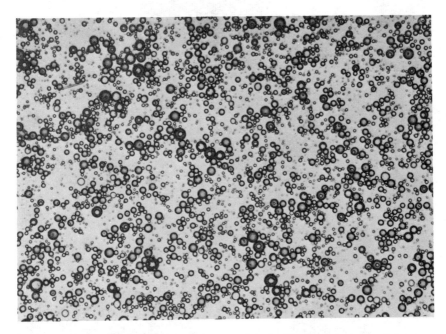

Figure 5.2. Dispersion of poly(ethylene terephthalate) in petrol prepared from an emulsion of bis(hydroxyethyl) terephthalate. (Magnification × 225)

A solution of dimethyl terephthalate in petrol was reacted with an emulsion of an aqueous solution of hexamethylene diamine in the same diluent. Methanol and water were removed by azeotropic distillation to yield a dispersion of poly(hexamethylene terephthalamide). An emulsion of molten adipic acid was also condensed with an emulsion of an aqueous solution of hexamethylene diamine, to give a dispersion of poly(hexamethylene adipamide). By the same general methods, a range of polyamides and both saturated and unsaturated linear polyesters can be prepared. Dispersions of solid reactants can also be used, e.g. nylon-66 salt (hexamethylene diamine, adipic acid salt) dispersed in petrol was directly converted to nylon-66 by the azeotropic removal of water. Nylon-69 salt (hexamethylene diamine, azelaic acid salt) and 11-amino-undecanoic acid have also been used for the preparation of the corresponding polyamides by the same method (Figure 5.3). A dispersion of nylon-6:9/6 in petrol, prepared from hexamethylene diamine and a mixture of azelaic and adipic acid, is illustrated in Figure 5.4.

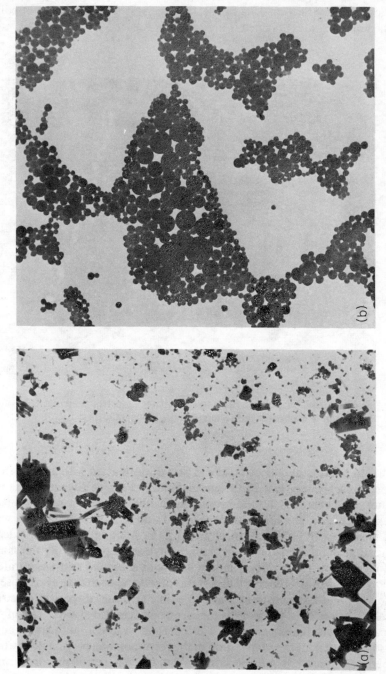

Figure 5.3. Dispersion of 11-amino-undecanoic acid in petrol (b.p. 160–180 °C) (a) and corresponding dispersion of poly(undecanoamide) (b) obtained by removal of water of reaction. (Magnification ×4500)

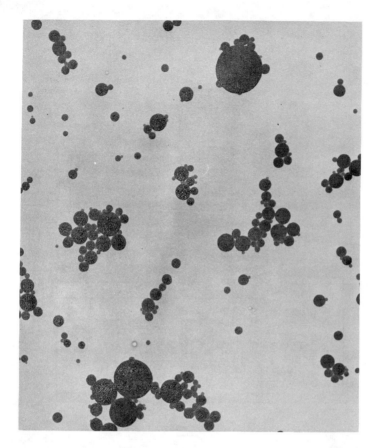

Figure 5.4. Dispersion of nylon-6:9/6 prepared in petrol. (Magnification × 5000)

## 5.6. SPECIAL TECHNIQUES OF DISPERSION POLYMERIZATION

### 5.6.1. Continuous dispersion polymerization

The kinetic features of free radical dispersion polymerization using soluble reactants, described in detail in an earlier chapter (Section 4.4), make it a particularly suitable process for its use in continuous polymerization methods. Unlike aqueous emulsion polymerization, the initial reactants are homogeneous, the fast rate of polymerization is not dependent on the particle size of the polymer formed and the particle size required can readily be controlled by the use of an appropriate level of dispersant. A process for the continuous dispersion polymerization of methyl methacrylate alone and with other acrylic monomers has been described[58]. This process used a stirred reactor in conjunction with an evaporator to produce polymer powder directly from the polymer dispersion formed.

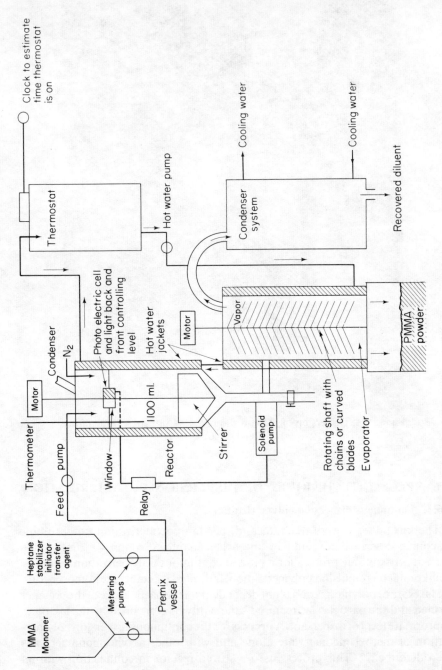

Figure 5.5. Outline diagram of apparatus for continuous dispersion polymerization

A continuous feed of monomer, initiator, chain transfer agent and graft copolymer dispersant in heptane was added at the top of the stirred reactor so that the liquid feed was rapidly distributed throughout the polymer dispersion already present in the reactor. This ensured that monomer was rapidly absorbed into the polymer particles to give fast micro-bulk polymerization accelerated by the gel effect. The dispersion polymerization was carried out with an azo initiator at 90–95 °C, just below the boiling point of heptane and enabled residence times of 10–20 min duration with polymer conversions of 97–99% to be achieved. With methyl methacrylate, polymer dispersions of 60% (w/w) solids were obtained. The polymer produced had a high and uniform molecular weight with a particle size distribution in the range 0·1–5 µm.

An outline diagram of the process is shown in Figure 5.5. The polymer dispersion produced was withdrawn from the base of the reactor at a rate controlled by a photo-electric cell which maintained the level of the dispersion in the reactor. An injection pump was required to feed the polymer dispersion obtained into the evaporator since during the early stages of the evaporation process the dispersion passes through a critical packing value when the dispersion has low mobility. The dispersion was evenly distributed on to the heated walls of the evaporator by a rotating assembly and the heptane was removed by a stream of inert gas and collected. The polymer powder which deposited on the walls of the evaporator was also removed by the rotating assembly and collected at the base of the reactor. The heat of polymerization produced (13 kcal/mol for methyl methacrylate) can be used to assist evaporation of the diluent and, in theory, sufficient self-heating is available at a 60% level of monomer in the feed to evaporate all of the heptane used. A laboratory-scale plant was operated for up to 30 h continuously to produce about 50 lb of dry polymer powder and its rate of polymer production was some 15 times greater than a corresponding solution process using the same level of reactants.

The continuous polymerization of vinyl chloride in dispersion has also been described[59]. Separate feeds of methanol, benzene containing peroxide catalyst and vinyl chloride were passed under pressure into a 4·4-m tubular reactor at 55 °C and the reaction mixture circulated every 6 sec. Excess reaction mixture was allowed to overflow into pressurized accumulators from which polymer was separated as a fine powder. The polymer formed however, was of low molecular weight and no polymeric dispersing agent was apparently used for its stabilization.

### 5.6.2. High solids polymer dispersions

The requirements of different particle sizes to obtain high packing fractions for assemblies of spherical particles have been described by Yerazunis and co-workers[60] and by McGeary[61]. In general, high packing fractions can be obtained with a bimodal distribution of spherical particles if the ratios of the diameters are greater than 12:1 and 70% of the total volume is occupied by the larger particles. Very high packing fractions, up to 95% of the theoretical density, may also be obtained with a tetramodal distribution of spherical

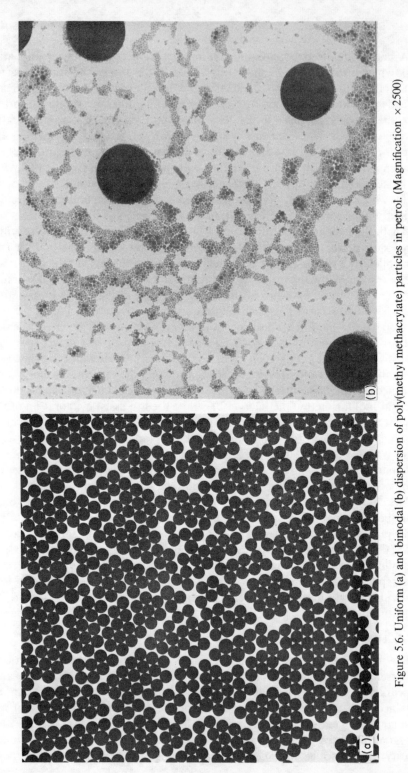

Figure 5.6. Uniform (a) and bimodal (b) dispersion of poly(methyl methacrylate) particles in petrol. (Magnification ×2500)

particles. Dispersion polymerization provides a means to produce spherical polymer particles of the required particle sizes and can therefore, be used for the preparation of high solids dispersions[62].

To obtain a bimodal distribution of particle sizes, a polymer dispersion of uniform particle size (5–6 µm) was first prepared in the usual way. Further graft copolymer dispersant was then added so that after the resumption of the monomer feed, fresh nucleation of polymer seed was induced. Particle growth of the two species was then continued, eventually producing polymer particles of two different size ranges. These were of 0·1–0·3 µm and 6–7 µm in size respectively, with volume ratios of 30/70 and a total solids level of 85% (Figure 5.6). Such bimodal dispersions have unusual rheological properties in that their viscosity decreases with an increasing shear force to a constant limiting value (Figure 5.7). This behaviour can be ascribed to the redistribution of the particle sizes from a state of rest when the larger particles sediment preferentially under gravitational forces to optimum packing arrangements produced by shear forces of sufficient magnitude (see also Chapter 6).

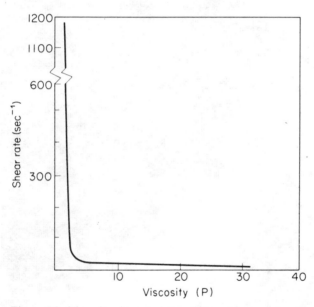

Figure 5.7. Viscosity characteristics of bimodal dispersion of poly(methyl methacrylate) in petrol (78·4% solids) measured by Weissenberg Rheogoniometer

Polymodal polymer dispersions were prepared by a two-stage process. In dispersion polymerization, the particle size of the polymer precipitated initially in the early stages of the polymerization depends in the first place on the solvency of the medium for the polymer produced. The most convenient method for increasing solvency is by the addition of monomer in which the polymer to be produced is soluble. By this means, the solvency of the medium gradually falls as monomer is consumed in the polymerization. The first stage of the dispersion polymerization was carried out with up to 55–60% of monomer in the reaction mixture, so that the solvency of the diluent favoured the formation of large particles by delaying the onset of precipitation. The higher solvency of the diluent also affects the solubility of the insoluble component of the graft copolymer dispersant so that 'anchoring' to the dispersed polymer and hence the overall stabilization becomes less effective. By this means, a 60% solids polymer dispersion of a wide particle size range from 0.1–20 μm was obtained in roughly four size groups. Using this dispersion as the 'seed stage', further continuous addition of monomer, transfer agent, initiator and graft dispersant resulted in a polymer latex of 83% (w/w) solids (Figure 5.8). Above this solids value, the dispersion rapidly became dilatant.

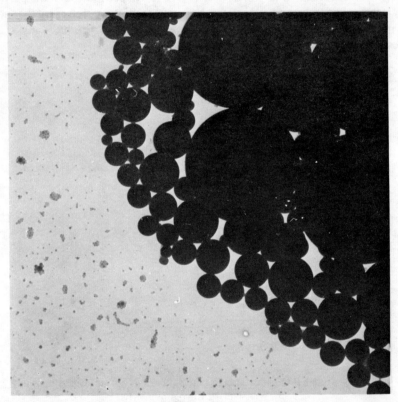

Figure 5.8. Polymodal dispersion of poly(methyl methacrylate) particles in petrol. (Magnification × 5000)

### 5.6.3. Dispersions of reactive polymers

In the surface coatings field, solutions of polymers containing various co-reactive groups are often used to produce cross-linked, thermosetting film-forming compositions. Since in solution, the reactive groups present are free to undergo cross-linking reactions during storage, it is frequently necessary in practice either to separate the reactive species into 'two-pack' systems which are mixed before use or to add a catalyst for the cross-linking reaction to achieve cure when required. With polymer dispersions containing reactive groups, however, the dispersed state makes it physically possible either to keep the reactive groups in a given particle apart from each other[63,64] or to use complementary reactive groups in different particles[63,65]. In general, the latter systems do not produce as high a degree of cross-linking after reaction as do dispersions in which the co-reactive groups are present in the same particle. The use of chemically inert diluents, such as aliphatic hydrocarbons, allows the incorporation of reactive groups into the particles which can remain in a protected state until activated by heat or by coalescing solvents during film-formation. Polymer dispersions, in which a reactive plasticizer is incorporated into the polymer particles, have also been described[66]. Here, the film produced by heating is softened by the plasticizer and its reactive groups cross-link with complementary groups in the main polymer. For this purpose, melamine–formaldehyde resins can be incorporated into dispersion polymers by dissolving the resin in the monomer feed. Dispersed polymer particles containing reactive groups in their surface layers can also react with oligomers or monomers containing reactive groups in solution[67,68] or be activated by catalysts in solution[66,69]. The use of reactive groups incorporated in the graft copolymer dispersant associated with the dispersed polymer have also been reported[70].

A typical cross-linking reaction used for thermosetting systems is of the acid-catalysed melamine–formaldehyde type[71], using butoxymethyl acrylamide and methacrylic acid. In dispersion polymerization, methacrylic acid and butoxymethyl acrylamide have been incorporated into the same polymer particle by the following means[63]. A stable dispersion of a copolymer of methyl methacrylate with a small quantity of methacrylic acid was prepared in aliphatic hydrocarbon as a seed stage and dispersion polymerization was continued to give a copolymer of methyl methacrylate, butyl methacrylate and methacrylic acid. Finally, the polymerization was completed with butoxymethyl acrylamide replacing methacrylic acid in the monomer feed. In this way, the reactive groups are prevented from initiating cross-linking reactions at the temperature of polymerization (85 °C) by their physical separation in the semi-solid particle. When the two reactive monomers were introduced together in the same monomer feed, internally cross-linked particles were produced.

An alternative method of preventing premature cross-linking during the preparation of polymer dispersions containing co-reactive groups is to carry out the dispersion polymerization with a low free monomer level ($<10\%$) with the reactive monomer concentration kept below $2\%$[64]. This can be illustrated by the preparation of dispersions of acrylic copolymers containing methacrylic acid and glycidyl methacrylate as the reactive species which cross-link by an epoxide–carboxylic acid reaction. The copolymers produced in dispersion were completely soluble in a mixture of ethylene dichloride and ethanol (95:5) but the corresponding films produced by heating at 150 °C for 30 min had a low solvent (xylol) extractability, indicating that extensive cross-linking had occurred. The copolymer dispersions themselves showed no significant increase in intrinsic viscosity during storage for 6 months.

Cross-linking processes have also been used which involve the interaction of reactive groups incorporated in the surface layers of the dispersion polymer with complementary groups dissolved in the continuous phase of the dispersion. Using dispersions of polyesters or polyamides with hydroxyl, carboxylic acid or amino groups incorporated in their surface layers and melamine/formaldehyde or urea/formaldehyde resin dissolved in the continuous phase, it is possible to produce films which cross-link on heating[55]. Alternatively, reactive groups may be incorporated in the graft copolymer dispersant which can interact with reactive polymers in solution, such as poly(butadiene)[70] or melamine/formaldehyde resins[67,72]. Catalysts dissolved in the continuous phase can also be used to initiate cross-linking reactions between reactive groups in the surface layers of the dispersion polymer when film-formation takes place. In this way, $p$-toluene sulphonic acid was used as a catalyst for the curing reaction between $N$-butoxymethyl methacrylamide and 1,4-butane diol methacrylate incorporated in a polymer dispersed in petrol[69].

### 5.6.4. Dispersions of microgels and heterogeneous polymer particles

Stable dispersions of cross-linked or microgel polymer particles in organic diluents can readily be prepared by dispersion polymerization techniques in the required particle size range. Addition polymer dispersions were prepared by the incorporation of co-reactive components, such as methacrylic acid and glycidyl methacrylate, into the acrylic monomer feed[73]. It is important that the cross-linking reaction takes place after the main polymer particle has been formed otherwise the dispersion becomes unstable. This is largely due to the loss in effectiveness of the graft copolymer dispersant due to its incorporation into the interior of the cross-linked polymer matrix during growth. A similar effect was observed in the dispersion polymerization of acrylic acid (Section 5.3.2) and acrylonitrile (Section 5.3.5). In the formation of poly(ethyl acrylate) microgel dispersions, the cross-linking reaction between the epoxide and carboxylic acid was effected in a final stage by the addition of diazobicyclooctane as catalyst.

Microgels based on dispersions of condensation polymers have also been prepared[73]. When the reactants of the condensation polymerization are emulsified in the continuous phase, little change occurs in the volume or surface area of the droplet during polymerization so that the graft copolymer dispersant is retained at the surface of the particle. Consequently, in this case, polymer formation and cross-linking can be carried out simultaneously without loss of dispersion stability. In this way, dispersions of cross-linked polyesters based on adipic acid and a mixture of ethylene glycol and glycerol have been prepared in both aromatic and aliphatic hydrocarbon diluents.

The preparation of heterogeneous polymer dispersions, with inclusions of discrete particles of one polymer within the matrix of another, has been demonstrated by the use of a sequential dispersion polymerization technique[74]. Inclusions of poly(methyl methacrylate) within poly(ethyl acrylate) particles were prepared by the addition of a feed of methyl methacrylate, initiator, transfer agent and dispersant into a preformed dispersion of poly(ethyl acrylate) in petrol. Monomer was absorbed into the poly(ethyl acrylate) particles and a separate phase was formed when the molecular weight of the resulting polymer reached a value high enough to make it incompatible with the main polymer matrix, which in this case was soft and rubbery at the polymerization temperature. Electron microscopy showed that several particles of poly(methyl methacrylate) were incorporated within the poly(ethyl acrylate) matrix (Figure 5.9).

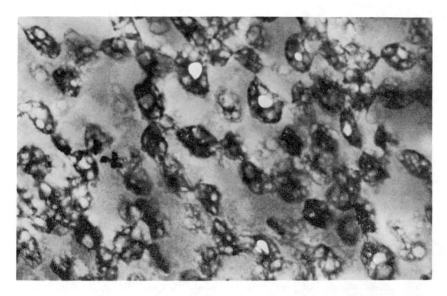

Figure 5.9. Cross section of poly(ethyl acrylate) particles with inclusions of poly(methyl methacrylate). (Magnification ×28,000.) (Particles embedded in Araldite and stained with dodecatungstophosphoric acid to show poly(ethyl acrylate) as dark areas)

The compound polymer particles exhibited two glass transition temperatures, as determined by differential scanning calorimetry. The values obtained, −13 °C and 94 °C, corresponded to the values of the two homopolymers concerned. By a reversal of the above technique, inclusions of poly(ethyl acrylate) in poly(methyl methacrylate) were also prepared.

In general, the particle size, number and amount of the included component is dependent on the level of cross-linking and the rigidity of the polymer matrix. This can be modified by the incorporation of suitable amounts of cross-linking agents in the main polymer composition. In this way, inclusions of poly(methyl methacrylate), poly(acrylonitrile) and poly(styrene) have been incorporated into lightly cross-linked samples of poly(ethyl acrylate). Poly(methyl methacrylate) inclusions have also been prepared in lightly cross-linked polyesters, e.g. those based on adipic acid, diethyl glycol and glycerol.

Finally, techniques for the preparation of dispersions of heterogeneous polymers incorporating inorganic pigment particles in organic liquids by dispersion polymerization have also been described. The first essential step in their preparation was to precipitate a layer of polymer over the surface of the pigment particle to serve as the site of the subsequent polymer growth. In effect, the pigment particle coated with polymer can be used as the 'seed stage' for the subsequent dispersion polymerization. The initial coating of polymer was prepared by dispersing the pigment into a solution of polymer by the usual mechanical procedure (ball-mill) in the presence of a suitable graft dispersant and subsequently adding a non-solvent for the dissolved polymer so that it was precipitated onto the surface of the pigment[75]. Dispersion polymerization was then used to build up a shell of polymer around the pigment particles (Figure 5.10). In this way, dispersions of rutile titanium dioxide encapsulated with

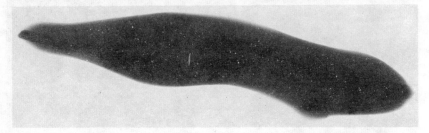

Figure 5.10. Encapsulation of iron oxide pigment particle within poly(methyl methacrylate). (Magnification × 70,000)

polymers based on both poly(methyl methacrylate), poly(vinyl acetate) and poly(vinyl chloride) were obtained and it was possible to produce a wide range of polymer-encapsulated pigments, including red iron oxide, thio indigo red and vegetable black.

## 5.7. SELECTED RECIPES FOR DISPERSION POLYMERIZATION

### 5.7.1. Preparation of poly(methyl methacrylate) latex using graft dispersant precursor

The preparation of a poly(methyl methacrylate-co-methacrylic acid) (98:2 by wt) latex using a graft dispersant precursor (Section 3.9.1) based on poly-(lauryl methacrylate) is described below.

#### 5.7.1.1. Reactants

*Seed stage* (wt-% of total): Methyl methacrylate, 7·28; methacrylic acid, 0·13; azo-di-isobutyronitrile, 0·10; graft dispersant precursor (32%, Section 3.9.1), 6·49; petroleum (boiling range 70–90 °C, sulphonatable hydrocarbon 5–9%), 38·83; petroleum (boiling range 230–250 °C, sulphonatable hydrocarbon <3%), 3·43.

*First feed* (wt-% of total): n-Octyl mercaptan (10% by wt solution in petroleum, boiling range 230–250 °C), 0·32.

*Main monomer feed* (wt-% of total): Methyl methacrylate, 41·38; methacrylic acid, 0·84; azo-di-isobutyronitrile, 0·10; n-octyl mercaptan (10%), 1·10.

*5.7.1.2. Procedure*

The type of apparatus required is illustrated in Figure 5.11. The components of the seed stage are added to the reaction vessel, stirring is commenced and the contents heated to reflux (80–82 °C). After the completion of the seed stage (45 min), the first addition of mercaptan is made, followed by the main monomer feed over the course of 3 h. The main feed is added initially at half the required rate and then raised over 30 min to the maximum rate. A slow initial feed rate in the main feed is preferred so as to avoid a build-up of unreacted monomer in the reaction mixture. The monomer feed is mixed with the returning cooled distillate before addition to reduce the possibility of temporary local concentrations of monomer in the reaction mixture. The flow rate of the returning distillate should be at least 1·5–2 times the monomer feed rate. After the completion of the main monomer feed, the batch is held at reflux for a further 30 min to ensure completion of reaction. Final solids content, 52–55%; reduced viscosity [0·5% solution (w/v) in ethylene dichloride–ethanol, 95:5, at 25 °C], 0·39–0·43; particle size (electron micrograph), peak at 0·2 μm, very few >0·4 μm.

### 5.7.2. Preparation of poly(methyl methacrylate) latex using preformed graft dispersant

The preparation of a poly(methyl methacrylate-co-methacrylic acid) (98:2 by wt) latex using a preformed 'comb' graft dispersant (Section 3.9.2) is described below.

*5.7.2.1. Reactants*

*Seed stage* (wt-% of total): Methyl methacrylate, 2·14; methacrylic acid, 0·03; azo-di-isobutyronitrile, 0·10; preformed graft dispersant solution (Section 3.9.2), 1·03; petroleum (boiling range 70–90 °C, sulphonatable hydrocarbon 5–9%), 36·52; petroleum (boiling range 230–250 °C, sulphonatable hydrocarbon <3%), 4·77.

*First feed* (wt-% of total): n-Octyl mercaptan (10% by wt solution in petroleum, boiling range 230–250 °C), 0·32.

*First monomer feed* (wt-% of total): Methyl methacrylate, 23·45; methacrylic acid, 0·48; azo-di-isobutyronitrile, 0·05; n-octyl mercaptan (10%), 0·36; preformed graft dispersant solution (Section 3.9.2), 6·41.

*Second monomer feed* (wt-% of total): Methyl methacrylate, 23·45; methacrylic acid, 0·48; azo-di-isobutyronitrile, 0·05; n-octyl mercaptan (10%), 0·36.

*5.7.2.2. Procedure*

The same general experimental procedure as 5.7.1 is used. The seed stage components are charged and the clear mixture raised to reflux (80 ± 1 °C)

236

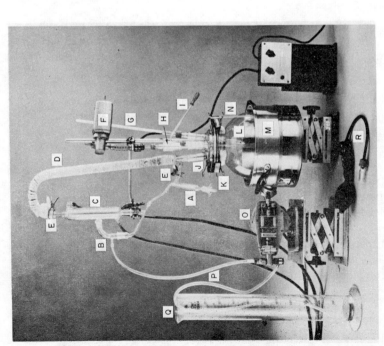
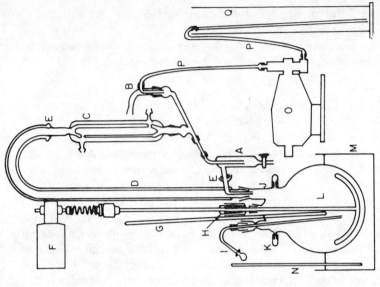

Figure 5.11. Apparatus used for dispersion polymerization on the laboratory scale: A, Trap to prevent escape of vapour and to ensure adequate mixing of diluted feed. Traces of water are also retained here; B, Feed entry-point and outlet to atmosphere; C, Double-surface water condenser; D, Lagged vapour outlet with reflux return tube at lower end designed to prevent diluted feed draining down sides of lid and flask; E, Spherical ground glass joints to reduce rigidity of assembly; F, Stirrer motor with variable speed control, operated electrically or by compressed air; G, Flask thermometer; H, Glass stirrer-gland filled with glycerine; I, Sampling tube; J, Flange lid fitted with appropriate sockets; K, Metal clamp for flange (the use of a neoprene gasket is optional); L, Wide-necked flange flask; M, Oil bath (a steam-bath may be used if preferred); N Oil bath thermometer; O. Metering pump for feed, with micrometer control: P. PVC tubing; O. Graduated cylinder for feed

and maintained at this temperature for 20 min after the initial whitening. At the completion of the seed stage, mercaptan is added and the first monomer feed metered in over the course of $1\frac{1}{2}$ h with a reduced initial feed rate. The second monomer feed is added at the normal rate throughout, over the same period. After the final addition is made, heating under reflux is continued for 30 min to complete the reaction. Final solids content, 52–54%; reduced viscosity (as 5.7.1), 0·40–0·42; particle size, 0·1 μm.

### 5.7.3. Preparation of poly(methacrylic acid) latex in chloroform–ethanol mixture using a polyester stabilizer precursor

*5.7.3.1. Reactants*

*Polyester stabilizer precursor* (wt-%): Ethylene glycol, 31·51; phthalic anhydride, 51·82; methacrylic acid, 6·02; trichloroethylene, 9·94; toluene sulphonic acid, 0·71.

*Poly(methacrylic acid) dispersion* (wt-%): Methacrylic acid, 5·82; chloroform, 75·58; ethanol, 11·63; polyester stabilizer precursor (1 above), 5·81; azo-di-isobutyronitrile, 1·16.

*5.7.3.2. Procedure*

*Preparation of polyester stabilizer precursor*: The reaction mixture ('precursor' above) was heated under nitrogen in a stirred flask fitted with a reflux condenser and a Dean and Stark take-off to remove the water of esterification. After 6-h heating, the reaction temperature had risen from 107 °C (initial reflux temperature) to 148 °C. After removal of 100 ml of aqueous distillate the acid value was reduced to 47 mgKOH/g (corresponding to an estimated mean end-group molecular weight of 1200).

*Preparation of poly(methacrylic acid) dispersion*: The methacrylate-terminated glycol phthalate polyester was used as a stabilizer-precursor in a 'one-shot' recipe ('dispersion' above) using a chloroform/ethanol mixture as diluent. After refluxing the stirred reaction mixture for several hours, a slightly viscous dispersion of fine particle size was formed, having a total solids content of 9·0% (82% yield).

### 5.7.4. Preparation of poly(acrylonitrile) latex in n-hexane

*5.7.4.1. Reactants*

*Seed stage* (g): Acrylonitrile, 100; n-hexane, 1000; comb graft stabilizer solution (50% solids, Section 3.9.2), 25; di-(4-t-butyl cyclohexyl) peroxydicarbonate, ('Perkadox Y16', AKZO (Novadel) Limited), 0·2.

*First monomer feed* (g): Acrylonitrile, 400; comb graft stabilizer solution, 100; di-(4-t-butyl cyclohexyl) peroxydicarbonate, 0·8.

*Second monomer feed* (g): Acrylonitrile, 500; comb graft stabilizer solution, 100; di-(4-t-butyl cyclohexyl) peroxydicarbonate, 0·8.

### 5.7.4.2. Procedure

The seed stage components were added to a 5-litre flask, fitted with a stirrer, a thermometer and an 'up and over' condenser recycle system (Figure 5.11). The charge was heated to recycle the n-hexane/acrylonitrile azeotrope boiling at a temperature of 61–62 °C. After heating for $\frac{1}{2}$ h, the charge became opalescent and the dispersion formed. After a further $\frac{1}{2}$ h under reflux, the first monomer feed was added *via* the returning cold hexane over 2 h. This was followed by the second feed, introduced in the same manner and also added over 2 h.

After the addition of the second feed was complete, recycling was continued until the temperature of the reaction mixture rose from 62 to 69 °C (b.p. of n-hexane), indicating that the acrylonitrile was completely polymerized. The latex produced (at 52% solids) was stable and had a particle size of less than 0·5 µm.

### 5.7.5. Preparation of poly(vinyl chloride) latex (semi-technical scale)

#### 5.7.5.1. Reactants

*First charge* (kg): Hydrocarbon diluent (low odour white spirit), 22·99; di-isopropyl peroxydicarbonate ('Perkadox IPP'), 0·138; graft dispersant precursor solution (30% solids, Section 3.9.3), 3·06.

*Monomer charge* (kg): Vinyl chloride, 46·00.

*Dispersant feed* (kg): Graft dispersant precursor solution (30% solids, Section 3.9.3), 4·60.

#### 5.7.5.2. Procedure

The components of the first charge were added to a pressurized autoclave (159 kg water rim capacity) fitted with stirrer. Oxygen was removed by reducing the pressure within the autoclave to less than 60 mmHg and restoring the pressure to atmospheric with nitrogen. This procedure was repeated. Monomer was added, via a pressurized burette, to give an initial pressure of 15 to 20 lbf/in$^2$. Heating was applied and the batch temperature raised to 50 °C as quickly as possible (pressure 80 to 90 lbf/in$^2$). The dispersant feed was started and added at a uniform rate over 100 min. The batch was maintained at 50 °C for 6 h (the maximum exotherm required circulating water at a temperature of 30 °C). After 6 h, the pressure was less than 10 lbf/in$^2$ with solids at 60%, corresponding to a conversion of more than 90%. Residual monomer was removed under reduced pressure.

### 5.7.6. Preparation of a dispersion of a poly(t-butyl styrene–isoprene–styrene) block copolymer

#### 5.7.6.1. Reactants

*First monomer charge* (g): n-Heptane, 680; t-butyl styrene, 48; s-butyl lithium, 0·2.

*Second monomer charge* (g): Isoprene, 48.

*Third monomer charge* (g): Styrene, 227; tetrahydrofuran, 0·3.

## 5.7.6.2. Procedure

A glass reactor was used fitted with stirrer, an inlet port for the introduction of monomers and initiator and gas inlet ports (Figure 3.14, Section 3.9.4). The apparatus was washed with chloroform, rinsed with n-heptane, evacuated and purged with nitrogen until dry. A positive pressure of nitrogen (approximately 20 mmHg above atmospheric) was maintained throughout the preparation.

The reactor was charged with n-heptane and t-butyl styrene and a molar solution of s-butyl lithium in n-hexane added slowly until a faint pink colouration was produced. A further quantity of the same solution (containing $10^{-2}$ mol s-butyl lithium per mol of t-butyl styrene) was added and polymerization allowed to proceed at 50 °C until completed.

After $1\frac{1}{2}$ h, isoprene was added, and the reaction temperature maintained at 50 °C for a further $4\frac{1}{2}$ h. After this time, the reactor contained a colourless, clear mobile solution. Half the charge of styrene was added to the reactor with stirring, followed after 15 min by the rest of the styrene, together with the tetrahydrofuran. The reactor contents became opalescent and a milk-like dispersion was then formed. Stirring was continued for 15 h after the first addition of styrene. After this period, the contents of the reactor were allowed to stand in contact with air. A stable dispersion of particle size 0·1 μm was obtained.

### 5.7.7. Preparation of a dispersion of poly(ethylene terephthalate) using direct and reverse emulsification procedures

#### 5.7.7.1. Reactants

Bis(hydroxyethyl) terephthalate, 280 g; high-boiling petrol (b.p. 250 °C), 420 g; graft dispersant solution (40% in white spirit), 49.

#### 5.7.7.2. Procedure

*Direct emulsification*: Bis(hydroxyethyl) terephthalate was prepared by the ester interchange reaction of dimethyl terephthalate and ethylene glycol and contained approximately 7% free ethylene glycol and 0·06% calcium acetate. The graft dispersant used was based on poly(12-hydroxystearic acid), methyl methacrylate and glycidyl methacrylate in the weight ratio 100:90:10.

The charge was heated with gentle stirring in a 2-litre flask until the monomer became molten (c. 140 °C). The stirring rate was increased to 300 rev/min to achieve emulsification and the emulsion brought to reflux. The lower boiling petrol (from the graft dispersant solution) was distilled off through a Dean and Stark apparatus. Ethylene glycol began to distil over at c. 190 °C and heating was continued until the final reflux temperature of 250 °C was reached (c. $2\frac{1}{2}$ h). The bulk of the ethylene glycol (c. 80%) was removed in the first half-hour. A dispersion of a solids content of 31% and with a particle size in the range 2 to 20 μm was obtained. The dispersion could be readily filtered to obtain the solid material.

*Reverse emulsification*: A mixture of the monomer and graft dispersant solution was heated with gentle stirring until molten. Heating was continued until the temperature reached 180–190 °C and the rate of stirring was increased but care was taken that the material thrown up on the flask walls did not solidify. At this stage, the white spirit from the graft dispersant solution distilled over together with a small amount of glycol. The melt of monomer and graft dispersant was heated for a further 10 min and the stirring rate then increased to 300 rev/min. High-boiling petrol (420 g) was added from a dropping funnel to the molten reactants at a slow rate (10 ml/min) initially. The emulsion formed gradually became thicker and inversion took place after *c*. 150 ml of petrol had been added. After inversion, the rate of addition of petrol was increased and was complete after *c*. 45 min. During this period the temperature was maintained at *c*. 190 °C and the stirring rate gradually increased. After the addition of petrol was complete, the ethylene glycol formed was removed by azeotropic distillation over $2\frac{1}{2}$ h. A dispersion similar to the previous example was obtained, but with a particle size in the range 1–10 μm.

## 5.8. REFERENCES

1. Bovey, F. A., Kolthoff, I. M., Medalia, A. I. and Meehan, E. J., *Emulsion Polymerization*, Interscience Publishers Inc., New York, 1955.
2. Imperial Chemical Industries, *British Patent*, 893,429 (1962).
3. Rohm and Haas, *British Patent*, 934,038 (1963).
4. FMC Corporation, *German Patent*, 1,946,688 (1970).
5. National Distillers and Chemicals, *United States Patent*, 3,586,654 (1971).
6. British Rubber Producers' Research Association, *British Patent*, 797,346 (1958).
7. du Pont, *German Patent*, 2,103,849 (1971).
8. Moilliet, J. L., Collie, B. and Black, W., *Surface Activity*, E. and F. N. Spon Limited, London, 1961.
9. du Pont, *United States Patent*, 3,637,569 (1972).
10. Pennsalt Chemicals, *British Patent*, 1,049,088 (1966); *British Patent*, 1,141,225 (1969).
11. Imperial Chemical Industries, *British Patent*, 1,206,398 (1970).
12. Sarvetnick, H. A. (Ed.), *Plastisols and Organosols*, Van Nostrand Reichold Co., New York, 1972.
13. Fitch, R. M., *J. Elastoplastics*, 3, 146 (1971).
14. Baylis, R. L., *Trans. Inst. Metal Finishing*, 50, 80 (1972).
15. Rohm and Haas, *British Patent*, 992,636 (1965).
16. Hess, P. H. and Parker, H. P., *J. Appl. Polym. Sci.*, 10, 1915 (1966).
17. Imperial Chemical Industries, *South African Patent*, 72/3550 (1972).
18. Waite, F. A., *J. Oil Col. Chem. Assoc.*, 54, 342 (1971).
19. Imperial Chemical Industries, *British Patent*, 941,305 (1963).
20. Rohm and Haas, *United States Patent*, 3,232,903 (1966).
21. Imperial Chemical Industries, *British Patent*, 1,052,241 (1966).
22. Imperial Chemical Industries, *British Patent*, 1,096,912 (1967).
23. Imperial Chemical Industries, *British Patent*, 1,104,403 (1968).
24. Rohm and Haas, *British Patent*, 1,009,004 (1965).
25. Union Carbide, *French Patent*, 1,531,022 (1968).
26. Badische Anilin- & Soda-Fabrik, *Belgian Patent*, 712,716 (1968).
27. Rohm and Haas, *British Patent*, 1,002,493 (1965).
28. Imperial Chemical Industries, *British Patent*, 1,122,397 (1968).
29. Rohm and Haas, *British Patent*, 956,453 (1964).

30. du Pont, *British Patent*, 1,272,890 (1972).
31. Imperial Chemical Industries, Paints Division, unpublished work.
32. Pimentel, G. C. and McClellan, A. L., *The Hydrogen Bond*, W. H. Freeman and Company, San Francisco, 1960.
33. Imperial Chemical Industries, *British Patent*, 987,751 (1965).
34. Union Carbide, *British Patent*, 1,199,651 (1970).
35. Bamford, C. H., Barb, W. G., Jenkins, A. D. and Onyon, P. F., *The Kinetics of Vinyl Polymerization by Radical Mechanisms*, Butterworths Scientific Publications, London, 1958.
36. Ahn, Y., *J. Korean Chem. Soc.*, **14**, 29 (1970); *Chem. Abs.*, **73**, 89215b (1970).
37. Imperial Chemical Industries, *British Patent*, 1,178,233 (1970).
38. Hercules, *British Patent*, 1,165,840 (1969).
39. Imperial Chemical Industries, *Belgian Patent*, 669,261 (1966).
40. Firestone Tire and Rubber, *British Patent*, 1,007,476 (1965).
41. Firestone Tire and Rubber, *British Patent*, 1,008,188 (1965).
42. Stampa, G. B., *J. Appl. Polym. Sci.*, **14**, 1227 (1970).
43. Mobil Oil, *Netherlands Patent*, 06366 (1972).
44. Imperial Chemical Industries, *South African Patent*, 72/7635 (1973).
45. Imperial Chemical Industries, *British Patent*, 1,095,931 (1967).
46. Imperial Chemical Industries, *British Patent*, 1,095,932 (1967).
47. Hercules, *United States Patent*, 3,634,303 (1972).
48. Union Carbide, *United States Patent*, 3,632,669 (1972).
49. Union Carbide, *United States Patent*, 3,646,170 (1972).
50. Firestone Tire and Rubber, *British Patent*, 1,008,001 (1965).
51. British Thomson-Houston, *British Patent*, 385,970 (1932).
52. Rohm and Haas, *British Patent*, 1,151,518 (1969).
53. du Pont, *British Patent*, 1,147,515 (1969).
54. du Pont, *United States Patent*, 3,441,532 (1969).
55. Imperial Chemical Industries, *German Patent*, 2,215,732 (1972).
56. Imperial Chemical Industries, *British Patent Application*, 28837 (1973).
57. Imperial Chemical Industries, *British Patent Application*, 17250 (1972).
58. Imperial Chemical Industries, *British Patent*, 1,234,395 (1971).
59. Gulf Oil of Canada, *British Patent*, 1,220,777 (1971).
60. Yerazunis, S., Cornell, S. W. and Winter, B., *Nature (London)*, **207**, 835 (1965).
61. McGeary, R. K., *J. Amer. Ceram. Soc.*, **44**, 513 (1961).
62. Imperial Chemical Industries, *British Patent*, 1,157,630 (1969).
63. Imperial Chemical Industries, *British Patent*, 1,095,288 (1967).
64. Imperial Chemical Industries, *British Patent*, 1,156,012 (1969).
65. BALM Paints, *British Patent*, 1,248,972 (1971).
66. Imperial Chemical Industries, *British Patent*, 980,633 (1965).
67. Cook Paint & Varnish, *British Patent*, 1,134,977 (1968).
68. Badische Anilin- & Soda-Fabrik, *British Patent*, 1,240,317 (1971).
69. Badische Anilin- & Soda-Fabrik, *British Patent*, 1,240,320 (1971).
70. Celanese Coatings, *British Patent*, 1,243,274 (1971).
71. Walker, J. F., *Formaldehyde*, Reichold, New York, 1964.
72. Imperial Chemical Industries, *German Patent*, 2,158,821 (1972).
73. Imperial Chemical Industries, *British Patent*, 1,242,054 (1971).
74. Imperial Chemical Industries, *German Patent*, 2,140,135 (1972).
75. Imperial Chemical Industries, *British Patent*, 1,156,653 (1969).

# CHAPTER 6

# *The Properties of Polymer Dispersions Prepared in Organic Liquids*

D. W. J. OSMOND AND I. WAGSTAFF

| | |
|---|---|
| 6.1. Rheology | 243 |
| 6.2. Sedimentation and Aggregation | 249 |
| 6.3. Film Formation | 250 |
|    6.3.1. The nature of the problem | 251 |
|    6.3.2. Case A | 253 |
|    6.3.3. Case B1 | 255 |
|    6.3.4. Case B2 | 256 |
|    6.3.5. Case C | 258 |
|    6.3.6. Case D | 260 |
|    6.3.7. Case E | 261 |
|    6.3.8. Case F | 262 |
| 6.4. Ultra-colloidal Polymer Dispersions | 263 |
| 6.5. References | 270 |

## 6.1. RHEOLOGY

(For a general account of the rheology of dispersed systems, see Reference 1.)

The rheological behaviour of particulate dispersions in liquids has been the subject of considerable theoretical and practical investigation since the pioneering work of Einstein[2]. Quite different types of behaviour can be distinguished for the three main concentration regimes considered.

Dispersions with a phase volume of only a few percent, as considered by Einstein, are so dilute that the transfer of momentum between colliding particles during flow is negligible. The differences observed between the viscometric behaviour of the dispersion and of its continuous phase alone is due only to the perturbation of the normal (Newtonian) flow of the latter. Under these circumstances, the viscosity ($\eta$) of the dispersion is proportional only to the viscosity ($\eta_0$) of the continuous phase and the volume fraction ($\phi$) of the particles present and quite independent of their size or size distribution, that is:

$$\eta/\eta_0 = 1 + \alpha_0 \phi \qquad (6.1)$$

where the Einstein coefficient, $\alpha_0$, has the value of 2·5 in this ideal case.

In dispersions of an intermediate range of particulate concentrations, that is from four to forty percent phase volume but still well below critical packing, a significant number of collisions between particles can now occur. However, over the usual range of shear rates applied, the dispersions still show essentially

Newtonian behaviour. Many attempts have been made to extend the Einstein relationship (equation 6.1) to this range of higher disperse phase volumes. These have usually taken the form:

$$\eta/\eta_0 = 1 + 2{\cdot}5\phi + k\phi^2 \tag{6.2}$$

with various values for the factor $k$. This work has already been extensively reviewed[3]. The methods used have ranged from essentially empirical curve-fitting operations of the experimental data to more refined approaches, such as that of Vand[4], which took account of two-body collisions and gave a value for $k$ of 7·35. More recent work by Mason and his co-workers, based on direct optical observations of the two-body collisions, gave values for $k$ of 9·15[5] and 10·5[6], depending on the assumptions made in the treatment.

Such expressions, however, are limited in their application to dispersions of a volume fraction less than 0·15. Thomas[7] after a critical analysis of the published experimental data on the suspension viscosities of uniform spherical particles, minimized the inertial and non-Newtonian effects by extrapolation techniques and produced an expression of the form:

$$\eta/\eta_0 = 1 + 2{\cdot}5\phi + 10{\cdot}5\phi^2 + A \exp B\phi \tag{6.3}$$

This gave a satisfactory fit for disperse phase volumes up to 0·6 with $A = 0{\cdot}00273$ and $B = 16{\cdot}6$. The micro-rheological approach has been extended by Lee[8], who incorporated the effect of three-body collisions of spheres under shear to give an expression:

$$\eta/\eta_0 = \left\{\frac{1}{1-\phi}\right\}(2{\cdot}5 + 1{\cdot}92\phi + 7{\cdot}74\phi^2) \tag{6.4}$$

which has given a moderately good fit at the higher volume fractions.

In general, all of these expressions apply to dispersions of rigid, spherical particles in which the colloidal stability is virtually ideal and the volume occupied by the stabilizer system quite negligible. This will be the case for large particles, but for particles of sub-micron size an effective stabilizing sheath will produce a significant increase in the hydrodynamic drag of the particles and methods have been developed[9,10] for measuring this property. The effect due to the presence of stabilizer can be expressed in two ways: as an increase in the disperse phase volume or the Einstein coefficient ($\alpha_0$) by a factor $f$; alternatively, an increase in the particle diameter ($D$) by a distance $2\Delta$ can be used. Thus equation (6.2) becomes

$$\eta/\eta_0 = 1 + \alpha_0 f\phi + kf^2\phi^2 \tag{6.5}$$

It was shown[9] that

$$\alpha_0 f = \alpha_0 + (6\alpha_0\Delta/D) \tag{6.6}$$

where $\alpha_0 f$ is now the effective Einstein coefficient.

The hydrodynamic effect of a stabilizing barrier of poly(12-hydroxystearic acid) attached to poly(methyl methacrylate) particles dispersed in aliphatic hydrocarbon was found[11] to be equivalent to an increase of 6·2 nm in the radius of the particle core and the way in which it becomes significant for particles less than 1 μm in size is shown in Figure 6.1.

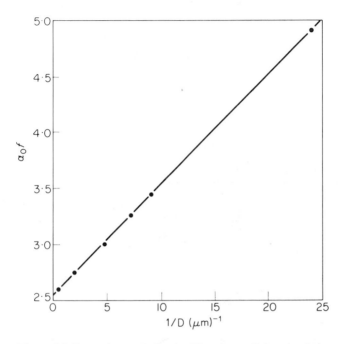

Figure 6.1. Dependence of effective Einstein coefficient ($\alpha_0 f$) on particle diameter, $D^{11}$. (From Barsted et al., Trans. Faraday Soc., **67**, 3598 (1971), with permission)

For dispersions of the higher concentration regime and which are approaching the region of critical packing, significant departures from Newtonian behaviour are observed with shear rates in the range 1 to 5000 sec$^{-1}$. In some examples, a decrease in viscosity takes place with increasing shear rate ('shear thinning'); in others, increasing shear rate produces an increase in viscosity ('shear thickening'). The latter behaviour is frequently found where the dispersions are virtually ideal and have Einstein coefficients close to the ideal value for spheres of 2·5. Other high phase volume dispersions may exhibit thixotropy due to flocculation of the particles and form a high viscosity 'structure' at rest. This structure is readily destroyed by shear forces but more or less slowly reforms when the shearing force is removed.

Krieger and co-workers have shown that shear thinning occurs with dispersions of cross-linked poly(styrene) particles (particle diameters 0·15–0·43 μm, Einstein coefficients 2·64–2·68) in water[12] benzyl alcohol and m-cresol[13]. Wagstaff and Waters[14] found both shear thinning and shear thickening, over different ranges of shear rate, to occur in dispersions of poly(methyl methacrylate) in petrol (particle diameters 0·04–2·0 µm and Einstein coefficients 2·54–5·0, depending on particle size). In the latter case, when an allowance for the effect of the stabilizing barrier was made, an Einstein coefficient of 2·56 was obtained, indicating that the particles were virtually spherical (Figure 6.1).

The rheological behaviour of the sterically-stabilized polymer dispersions in aliphatic hydrocarbon under shear is illustrated in Figures 6.2 and 6.3. Typically, as the shear rate was increased, the apparent viscosity first decreased, passed through a broad, flat minimum and then increased abruptly above some critical shear rate. This abrupt increase in viscosity at high shear rates was usually associated with the appearance of a significant force normal to the direction of shear. The shear thinning observed with these polymer dispersions was also distinctive compared with some other systems in that the equilibrium viscosity value was attained almost instantaneously at all shear rates.

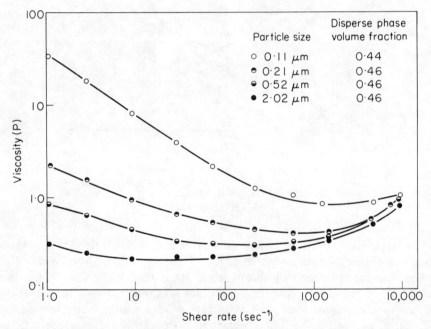

Figure 6.2. Rheological behaviour of poly(methyl methacrylate) particles in size range 0·11–2·02 µm in petrol[14]

The general pattern of behaviour described showed certain characteristic modifications with changes both in particle size and phase volume. At a constant phase volume, the shear thinning was more pronounced and the minimum viscosity was higher with very fine particles (Figure 6.2). At the same time, the shear thickening was more pronounced with the coarser particles. On the other hand, for a given particle size, a reduction in phase volume led to a significant reduction in viscosity at very low shear rates, whilst the onset of shear thickening was delayed to higher shear rates (Figure 6.3). In general, the effects observed were more pronounced with mono-disperse systems; the use of blends of particles of different sizes or broad distributions of particle size tended to flatten out the characteristic U-shaped curve and to give less rapid changes in viscosity with shear.

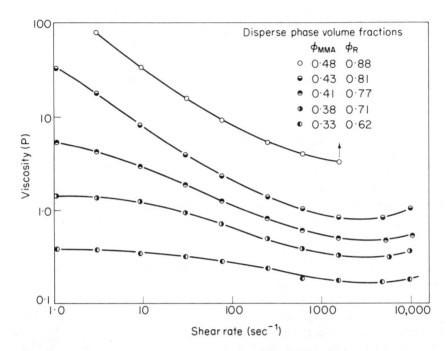

Figure 6.3. Rheological behaviour of poly(methyl methacrylate) particles (0·11 μm) in petrol over range of disperse phase volume fractions[14]

An explanation of the shear thinning behaviour observed with near-ideal stable dispersions of rigid spheres has been suggested by Krieger and Dougherty[15]. It is known that the transient formation of doublets, due to a combination of Brownian motion and shear forces, can cause a greater dissipation of energy than the individual particles[4–6]. It was suggested therefore, that the number of doublets formed will be reduced by increasing the shear rate and hence reducing the shear stress developed at a given shear rate. It was shown[12,16] that if the shear stress was corrected by a factor $a^3/kT$ (where $a$ = particle radius), then the reduced viscosity/reduced shear stress curves for different dispersion media and different particle sizes could be superimposed.

For poly(methyl methacrylate) dispersions in petrol however, this correction factor over-compensates for the effect of particle size; the apparent viscosity at a particular reduced shear stress increases with particle size instead of being independent as predicted by theory. On the other hand, when the results are examined on the basis of the combined volume ($\phi_R$) of the polymer core and the stabilizing sheath rather than the polymer core alone ($\phi_{MMA}$), then significant shear thinning is found only in those dispersions for which the combined phase volume is similar or greater than that corresponding to random close-packing (i.e. $\phi_R \geq 0.64$, Figure 6.3).

In those cases in which the log viscosity/shear rate curves tend to infinity as the shear rates tend to zero, it is found that the packing of the particles plus sheaths is closer even than that corresponding to hexagonal close packing (i.e. $\phi > 0.72$). Under these circumstances, it is obvious that, at rest, the fringes of the stabilizer sheaths must be interpenetrated as this is energetically preferable to the compression of the sheaths (Section 2.3). Those examples which appear to show significant shear thinning, but which extrapolate to a finite viscosity at zero shear rate, are usually found to correspond to packing fractions of the particle plus sheath in the region of that for random close packing, but below that for hexagonal packing. In this case, there is probably little or no interpenetration of the stabilizer sheaths either at rest or at low shear rates.

Presumably, as higher and higher shear rates are applied, particles are pushed past one another, causing distortion and even disentanglement of the stabilizer sheaths so that their hydrodynamic drag or the level of their interpenetration is reduced. Although, as described previously, the interpenetration of two interacting sheaths is energetically preferable to the compression of each individual sheath, the difference in energy between the two configurations is small, so that in turn, only small applied stresses would be required to rearrange the sheaths into the non-interpenetrated form, with an associated reduction in viscosity.

It is worth noting that there is no detectable hysteresis in the shear thinning behaviour, i.e. this system is not thixotropic. In addition, the fact that the shear thinning occurs over a shear rate range from $1\ \text{sec}^{-1}$ to $1000\ \text{sec}^{-1}$, suggests that the process involved has a relaxation time in the region of a few milliseconds, which is consistent with a mechanism involving rearrangement of polymer molecules.

Even when the results from a range of fine particle size latices having a common stabilizing barrier have been normalized, so that the viscosities can be compared at the same combined phase volume, there still remains a particle size dependence (Figure 6.2, above). Thus, the finer the particle size, the greater the shear thinning, or rather the greater the shear rate or shear stress required to reach the minimum in the viscosity shear rate curve. For a given shear stress, the shear force acting on a particle and deforming the stabilizer sheath must be inversely proportional to some power of the particle diameter and so it is not surprising that the normalized viscosity/shear rate relationship shows a particle size dependence.

Although the results described above have been exclusively derived from uncharged, sterically-stabilized particles, it is clear that this type of behaviour is similar to the electro-viscous effect[17]. Here, long range coulombic interactions between charged particles in aqueous media of low ion content, repel one another at ranges which are comparable to their mean spacing in the latex to produce non-Newtonian rheology and pronounced shear thinning.

The poly(methyl methacrylate) dispersions in petrol represent a clear example of stable dispersions of spherical particles exhibiting shear thickening at the higher phase volumes of the rigid particle cores and at high shear rates. This effect was shown by the dispersions having particle diameters in the range 0·12–2·0 μm (the largest particle size examined). Finer particle diameter dispersions than 0·1 μm could not be prepared at adequately high disperse-phase volumes because the dispersions had intractably high viscosities due to the proportionately larger volume occupied by the stabilizer barrier. There was also a tendency for higher disperse-phase volumes and higher shear rates to be required to produce shear thickening in the finer-particle-size dispersions.

The mechanisms proposed for such behaviour have usually involved, in various ways, modifications to the packing of particles under the influence of shear. It is easy to visualize an increase in the size and number of shear-induced aggregates with increasing shear rates resulting in particle jamming. Alternatively, for one particle to move smoothly past another, it will be necessary, at high disperse-phase volumes, for there to be some closer packing of the surrounding particle array in order to make sufficient space available. This would be expected to require longer times for the larger particle sizes and for the higher phase volumes, while less time will actually be available as the shear rate increases. Such explanations however, are only qualitative speculations at present and indicate the need for a more rigorous analysis of the problem.

## 6.2. SEDIMENTATION AND AGGREGATION

It is important to distinguish between the stability of dispersions in practice and the ideal or thermodynamic concept of stability. An ideal stable dispersion is one in which the particles undergo no mutual attraction but only short-range, elastic repulsions. They behave, in fact, rather like the molecules of an ideal gas. Simple calculations using the Stokes' Law sedimentation rate and the Brownian diffusion rate make it quite clear that only under the most exceptional circumstances (usually a near identity in density between the particles and the liquid medium) will such ideal systems be free from settlement under gravity.

An example of extreme stability in practice is that of the often-quoted red gold sols in water, which do not settle after many months or even years of storage under normal laboratory conditions. In this case, the stability does not arise from the situation in which the Brownian diffusion rate in an upward direction is fast enough to compensate for gravitational settlement. For this to be so, the particle size would have to be of atomic dimensions. In fact, it has been shown[18] that the stability of such systems is associated with the convective stirring which inevitably occurs when samples are left in uncontrolled laboratory conditions; storage of gold sols in a controlled temperature environment results in complete settlement over a period of a week or so.

Since the particles of an ideal dispersion can, by definition, slip past one another very easily, settlement continues until a close-packed array of particles is achieved in the sediment. This close-packed array is mechanically quite strong and, as will be clear from the work described in Section 6.1 above, is highly dilatant. As a result, while such clay-like sediments are readily re-dispersed by gentle agitation over very long periods of time, they are extremely difficult to re-disperse rapidly by vigorous mechanical agitation. The formation of such an intractable sediment after prolonged storage constitutes an 'unstable' system in the eyes of the practical paint formulator or other technological user. For these applications it is preferable that the dispersions either do not settle at all, or at least settle to a soft and easily re-incorporated, open-textured sediment. This type of behaviour requires a critical degree of instability or non-ideality in the behaviour of the dispersion. When particles come together, they undergo an attraction of the order of a few $kT$, so that they flocculate rather than move freely past one another to pack into a close sediment; at the same time the strength of the attraction must be sufficiently small to enable the floc structure to be readily redispersed by mild agitation.

The structure of flocs formed from dispersions in which the particles form contacts of finite attractive energy, has been discussed by Vold and others[19]. Interestingly, it is found that the stronger the attraction between the particles (i.e. the *more* energy that would be available if the system formed a compact floc), the more open-textured is the floc which is actually formed. Negligible rearrangement of the primary floc structure is possible in these cases because of the large activation energies associated with such processes. Thus, the phase volume of the most open-textured flocs may be as low as a few percent, compared with a value of about 60% for the phase volume of the closely-compacted sediment from an ideal stable dispersion.

In the case of ideal stable dispersions of soft deformable particles and which settle slowly to form compact sediments, problems can arise especially when a very large volume of a dispersion is stored. Under these circumstances, the thickness of the sedimentation layer may be such that the gravitational loads on the lower portions of the sediment are sufficient to distort the particles, and in effect, bring about the earliest stages of irreversible aggregation/film formation described below (Section 6.3). Similar effects are of course, readily produced when high '$G$' loadings are applied to such dispersions by centrifugation.

## 6.3. FILM FORMATION

The principal industrial outlet for polymer dispersions in organic media has, so far, been in the formulation of surface coating compositions. Consequently, a knowledge of the detailed structure of the films produced from polymer dispersions and an understanding of the mechanism by which the films are formed is of considerable relevance to their practical application.

The introduction of surface coatings based on aqueous polymer dispersions—the so-called 'emulsion' paints—has led to an intensive study of the mode of film formation from aqueous latices over the past decade or so[20]. However, much of the work described so far has lacked a sufficiently broad analysis of the problem of film formation from disperse polymer systems in general terms. The more recent development of sterically-stabilized polymer dispersions in non-aqueous media has also introduced additional features not usually considered in many of the aqueous systems. The account of film formation given here will inevitably concentrate on the problems involved with non-aqueous polymer dispersions but many of the features described will also be relevant to aqueous polymer latices as well.

### 6.3.1. The nature of the problem

In principle, the problem before us is extremely simple; to consider a dilute dispersion of fine polymer particles suspended in some liquid and to discuss the sequence of events which takes place when this liquid evaporates. In practice, the problem proves to be one of daunting complexity, because what in fact does happen, depending on circumstances, may be modified by almost every conceivable attribute of the particles, of the diluent and of the conditions of film formation.

Some of the more obvious possibilities are listed in Table 6.1. The diluent liquid may be totally volatile, or may be a mixture of volatile and essentially non-volatile fractions. In the latter case, the non-volatile fraction may be larger or smaller than the total volume of the voids between the particles when they are close-packed. In either case, this non-volatile fraction may be inert with respect to the polymer particles under all circumstances; it may be inert at the temperature at which the diluent evaporates, but becomes a solvent or modifier at some elevated temperature; or it may commence to dissolve the particles strongly at the temperature of film formation, as soon as substantial quantities of the volatile non-solvent fraction of the diluent has evaporated.

Table 6.1. Properties of the polymer dispersion relevant to film formation

| Component | Properties |
|---|---|
| Liquid diluent | Volatility, solvency, presence of dissolved components |
| Polymer particles | Size, degree of flocculation, hardness |
| Stabilizer | Size of soluble polymer barrier, degree of attachment to particle, compatibility of soluble polymer with dissolved components |

The above implies that the non-volatile portion of the diluent is essentially liquid, but it may, of course, contain a dissolved solid of high or low mechanical strength, good or bad adhesion to the polymer particles and so on, as well as having all of the other possible characteristics listed above. (Of course, only in very special cases is such a solid likely to be molecularly compatible with the polymer itself, when, by definition, the solid was initially soluble in an environment in which the polymer was insoluble.)

The disperse particles themselves may be completely deflocculated at all concentrations throughout the process of film formation; they may be more or less severely flocculated at all stages; or they can make the transition from one state to the other—more importantly, from a stable dispersion to a flocculated one—as the diluent evaporates and the particle concentration rises.

The particles themselves may be hard, soft but rubbery, or soft and plastic. They may be fairly large, moderate or very small in absolute size (that is by colloidal standards), say of the order of μm, tenths of a μm, or tens of nm in diameter. Irrespective of the absolute size of the particles, they may carry stabilizer sheaths of dimensions which are negligible, significant, or very large compared to the size of the particles themselves. The stabilizer molecules which comprise these sheaths may be bound irreversibly to the particle surface, or be in equilibrium adsorption with the particle surface throughout the whole of the film formation process, or make a transition from one state to the other during the process of film formation (more importantly, from initially strongly-bonded, to more weakly-bonded states as evaporation of the diluent proceeds). The soluble groups of the stabilizer may cover the whole range of compatibility with any non-volatile component dissolved in the diluent. Further, the soluble groups (in the absence of the volatile fraction of the diluent) may be weak, viscous liquids, or relatively strong solids, tough or brittle.

In the early work on the structure of aqueous latex films, the systems considered were primarily those of particles which were capable of plastic deformation under sufficiently large applied stress, dispersed in a single component (predominantly aqueous) environment, and stabilized by low molecular weight, usually anionic, surfactants. Not surprisingly therefore, the attention of most workers[20] was concentrated primarily upon the source and quantification of the force which deformed the polymer particles from spheres to close-packed polygons in this key stage of film formation (see Figure 6.5, below); the fate of the stabilizer and the subsequent history of the interface between the surfaces of the close-packed deformed particles, although discussed to some extent were on the whole relegated to a place of secondary importance. Equally, low priority was generally given both to the earlier stages of film formation before particle deformation and to the analysis of systems of relatively undeformable particles.

In contrast, in the case of non-aqueous systems used for industrial paints, the polymer particles are likely to be hard and not amenable to easy plastic deformation at room temperature and the diluent may be highly complex and of varying solvency. In addition, the need for steric stabilization (usually by means of a block or graft copolymer) means that the stabilizing component constitutes a significant proportion of the final film composition, so that the fate of this relatively high molecular weight polymer, and its influence on the final film properties, cannot be ignored in this case.

While a complete analysis of all conceivable types of film formation process is quite impracticable here, some general understanding of the processes involved can be obtained by considering a number of ideal cases arranged in a logical order (Table 6.2). The field falls into two distinct parts; on the one

Table 6.2. Combination of polymer dispersion properties considered in general cases for film formation

| Case | Particles | Medium | Dispersion | Stabilization |
| --- | --- | --- | --- | --- |
| A | Hard/plastic | Major inert, non-volatile component | Deflocculated | Real stabilizer |
| B1 | Plastic | Volatile | Deflocculated | Ideal[a] |
| B2 | Plastic | Volatile | Deflocculated | Real steric stabilizer |
| C | Hard | Volatile | Deflocculated | Ideal[a] |
| D | Hard | Volatile | Flocculated | — |
| E | Hard | Volatile | Deflocculated | Real steric stabilizer |
| F | Hard | Major strong solvent, non-volatile component | Deflocculated | Real steric stabilizer |

[a] i.e. stable but stabilizer is either 'volume-less' or disappears on film formation.

hand those systems in which (like the aqueous cases described in the bulk of the literature) the particles are capable of plastic deformation at the temperature at which the diluent evaporates, and, on the other, those in which the particles are rigid and essentially non-deformable at this stage.

### 6.3.2. Case A

This case concerns deflocculated particles (hard or plastic) in a medium containing a large volume of non-volatile component which is incompatible with the particles. The dispersion consists of deflocculated particles in a diluent containing a large volume (that is a significantly greater volume than the total volume of the voids between the particles when randomly close-packed) of inert, non-volatile material. On evaporation of the volatile fraction of the diluent, although the particles will move closer together, the process will stop before close-packing of the well-dispersed particles occurs. The final film will therefore consist of a concentrated dispersion of the particles in the non-volatile component of the diluent (Figure 6.4). If the residual diluent is a liquid, the final film will be a mud or paste; if a solid, a solid film of essentially composite material, having a continuous phase derived from the soluble film former in which are embedded the dispersed polymer particles.

Modern composite polymer technology[21,22], has shown that materials of this heterogeneous type have overall properties which are not usually the simple weighted average of those of the two components involved, but depend critically upon the interactions between them. Widely different properties can be obtained, depending upon the nature of the continuous binder phase, the nature of the dispersed polymer particle phase, the degree of interfacial binding and the size and phase volume of the particles in the disperse phase.

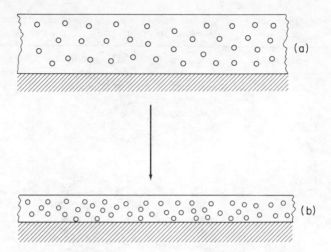

Figure 6.4. 'Film' formation from a dispersion of deflocculated particles in a medium containing a large volume of non-volatile and incompatible diluent. Dilute stable dispersion, (a); 'Film' of concentrated stable dispersion, (b)

At one time, it was customary to think of heterogeneous systems of this sort as quite different from the results of true film formation from polymer dispersions. However, it is now clear that materials showing at least some degree of heterogeneity of this type are a common end-product of films produced from sterically-stabilized polymer dispersions. On this view, Case A systems, in which the non-volatile component is a film-forming polymer compatible with the stabilizing polymer, merely represents an extreme case of the Classes B discussed below.

A special type of Class A system, which in the past was of some importance in the formulation of paints for stoving industrial applications (Section 7.3.2), and which has no direct parallel in the aqueous field, was formulated in the following way. The non-volatile component of the diluent phase was a plasticizer whose rate of diffusion into the particles at room temperature (even in the absence of any significant amount of diluent) is trivially slow. However, on raising the temperature of the system to above the $T_g$ (or crystalline melting temperature, $T_m$) of the polymer particles, there is a rapid onset of attack by the plasticizer.

Clearly, after the evaporation of the inert component of the diluent, the resultant paste of the particles in the 'potential' plasticizer is closely akin to a conventional 'plastisol' paste[23]. When these are heated, the particles of such pastes imbibe the plasticizer while simultaneously deforming and swelling to fill the spaces previously occupied by the plasticizer without any first-order volume change. The result is a film of close-packed, polygonal, plasticized polymer particles (Figure 6.5). The properties of such films and the extent of further modification in their structure depends to some extent upon their

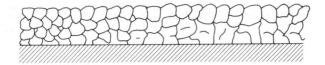

Figure 6.5. Formation of film of close-packed polygons from polymer dispersion in a medium containing a non-volatile, plasticizing component

subsequent history, but more importantly, on the properties of any stabilizing material at the particle surfaces. This point is discussed at some length below.

### 6.3.3. Case B1

These are deflocculated, deformable particles in a totally volatile diluent (idealized case ignoring the presence of real stabilizer).

Consider a system of particles which, in some miraculous fashion, remain totally deflocculated at all concentrations, although the particles have no stabilizing material associated with their surface. Further, let us assume that these particles are plastic or ductile, i.e. they behave elastically (at least over short time spans) up to some modest stress, after which they undergo permanent deformation. As the diluent evaporates from such a system, the particles will move closer and closer together, but remain in random motion until critical packing is reached. Beyond this point, further evaporation of the diluent must result in the formation of a very steeply curved liquid surface as the free surface withdraws into the capillary channels between the close-packed colloidal particles. The radii of curvature thus produced are so small that for all of the liquids (aqueous or non-aqueous) likely to be encountered in practice, provided that the liquid wets the particle surfaces with an energy which is not trivial compared to its own internal cohesion (this will always be true of a stable dispersion) then very large tri-axial tensile stresses will be developed in the interstitial liquid. These in turn, will generate very large compressive forces on the particles throughout the whole film.

Much of the early work of film formation from aqueous latices was concerned with the quantification of this capillary compressive stress and its comparison with the forces required to produce plastic flow of the particles. Equations were generated relating the surface tension of liquid with the particle size, the modulus and the yield strength of the particles etc.[24-27]. Although such calculations are of interest, their details need not concern us greatly here. It is clear that the volatile diluent will continue to evaporate (although possibly at a slower rate and with an apparently larger latent heat of evaporation than from a bulk surface). This evaporation of the diluent can lead to only two possible situations; either the capillary stress generated is sufficiently large to cause plastic flow of the particles, in which case the final film will consist of particles packed in an essentially void-free, polygonal structure, or; on the other hand, the particles are so stiff and strong that they are capable of withstanding the maximum capillary compressions to which they are subjected, and porous, friable films of packed, essentially spherical, particles will be produced. (See Case C, Section 6.3.5, below.)

In the idealized (i.e. stabilizer-free) system considered here, the deformation of the particles results in the close contact of clean polymer surfaces. Even at this stage, the contact adhesion between such surfaces will be high, especially for moderately polar polymers such as those normally encountered in non-aqueous dispersion systems. However, as Voyutskii[28] has pointed out, when compatible polymers, which are capable of plastic or ductile deformation, are brought into contact in the absence of significant surface contamination, they slowly undergo molecular diffusion across their boundary, so that ultimately the boundary effectively disappears. Hence, in the case of the idealized system we are discussing here, a totally integrated, monolithic film, having properties essentially similar to those cast from a polymer solution should be obtained, although in practice the ageing-time required may be of very long duration.

It will already be clear that in many ways both this type of system and the plastisol system discussed as a special class of Case A (Section 6.3.2, above), are essentially similar.

In the case of conventional plastisol pastes and of aqueous latices stabilized solely by low molecular weight surfactants, there is either no stabilizer as such or the stabilizer (in the absence of the aqueous phase) may have very significant solubility in, and rates of migration through, the polymer phase. Under these circumstances, it is possible to conceive, certainly after sufficiently prolonged ageing, very considerable auto-adhesion of the Voyutskii type. However, as soon as significant amounts of polymeric material, soluble in the continuous phase, are introduced, whether in the form of steric stabilizers in non-aqueous systems or in the form of colloidal thickeners in the case of many practical aqueous latex systems, the ultimate fate of the stabilizer can no longer be ignored. The behaviour of such systems is discussed in Case B2 immediately below.

### 6.3.4. Case B2

This case concerns deflocculated, deformable, particles in a totally volatile environment with steric stabilization. Clearly, the early stages of film formation from such systems is essentially similar to Case B1 above. However, when the particles first start to deform as a result of the rising capillary compression, the loads are carried at the points of contact between adjacent particles and very high local stresses are built up. Further, the deformation process results in a modest increase in the surface area of the particles. At the same time, the rising concentration of soluble polymer in the small amount of the continuous phase remaining, reduces the equilibrium spacing of the chains on the particle surfaces, as described above in Section 2.3.6. The net result of all of these processes is generally the local displacement of stabilizer (provided that it has some lateral mobility on the surface) from the points of first contact between the particles so that direct particle–particle contact is made (Figure 6.6).

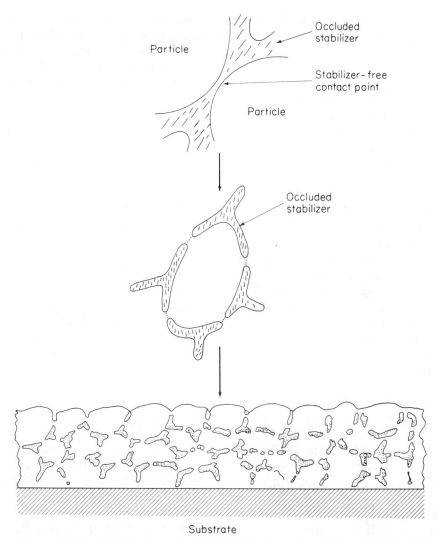

Figure 6.6. Mode of film formation from dispersions of particles stabilized with polymeric dispersants

However, all of the experimental evidence[29] would suggest that, in the case of a stabilizer which is firmly attached to the particle surface, the total area of virgin surface which is generated in this way is always quite small compared to the total surface area of the particles. The residual area remains covered with closely-packed stabilizing molecules, even after the evaporation of all of the diluent. These layers of opposing stabilizer molecules can be regarded as giant laminar micelles sandwiched between the faces of polygonally-packed polymer particles, and the whole film consists of two interpenetrating polymer matrices (Figure 6.6).

It is interesting to note that where the anchoring of the stabilizing chain to the particle surface is good, and the mechanical properties of the solvent-free polymer chains are suitable, the optical and mechanical properties of such films (despite their micro-heterogeneity) are extremely satisfactory[29]. Indeed, such structures clearly show some resemblance to modern heterophase, rubber-modified bulk plastics such as ABS (based on acrylonitrile, butadiene and styrene). Even those cases in which the residual stabilizing chains undergo no further chemical or physical change which renders them less soluble in the original continuous phase than before the formation of the film, the resistance to re-peptization of the film in the original diluent is often surprisingly high. As Cousens[30] pointed out, this is probably associated with the very low cross-section area (as a fraction of the total surface area of the film) of these laminar micellar channels, so that the rate of diffusion through them is extremely slow. Indeed, in many cases it would appear that the dominant mode of transport of the diluent is by diffusion through the insoluble polymer of the original particle phase. It thus follows that, if the polymer of the original particle phase is very insoluble in the original diluent (and hence the original diluent diffuses very slowly through it), the micellar 'glue' (see Section 6.4, below) which holds the greater part of the particle surfaces together can be well protected from attack by the diluent. At the same time, the primary adhesive contacts between virgin surfaces at the first points of particle/particle interaction provide permanent bonds which are insensitive to diluent and which produce films of reasonable integrity, even on prolonged exposure to diluent.

### 6.3.5. Case C

Dispersions of deflocculated hard particles in totally volatile diluent form Case C systems (ideal systems ignoring stabilizer). Let us consider a stable deflocculated dispersion of hard particles, either 'self-stable' or having a stabilizer sheath which is trivially small in size compared to that of the particles, suspended in a simple diluent which is totally volatile (for the effect of finitely-sized sheaths, see Case E, Section 6.3.7, below). During the early stages of evaporation of the diluent, the concentration of particles will rise and the particles will jostle closer to one another until the stage of critical packing is reached, as in the cases considered above. At this stage, providing the particles are truly non-deformable and the diluent continues to evaporate, then cavities must appear in the voids between the closely-packed particles (Figure 6.7).

There are several possible routes by which the diluent in the voids may be replaced by air to form these cavities. Where the particles are very coarse (far coarser in fact than any under discussion here), the dipping of the liquid surface below the top line of the packed particles may not result in such a steep curvature of the liquid meniscus that any significant modification of the evaporation process occurs; the bulk of the liquid can evaporate freely from the complex but continuous passages between the particles and the only capillary stress to which the particles are subjected comes from the liquid lenses which will remain around their points of contact.

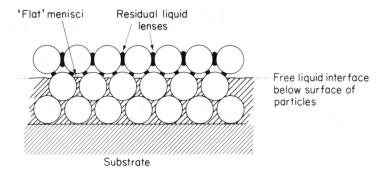

Figure 6.7. Loss of diluent from large, hard, close-packed spherical particles

No polymer dispersion of practical interest however, has particles so coarse that their interstices, when closely packed, are not well within capillary dimensions, so that in all practical cases, as evaporation first causes the free liquid surface to dip between the top layer of particles, substantial curvature of the liquid surface and hence substantial capillary compressions will then operate as described for cases B1 and B2 above. If however, the film consists of well-packed, rigid particles, its volume cannot be reduced even by the very high compressive forces in the capillaries; the diluent continues to evaporate from the cavities between the particles and is replaced by air until eventually a dry film of close-packed, hard particles is achieved. Assuming only trivial adhesion at the points of contact between such particles, the film will have only negligible mechanical strength and is indeed, essentially a dry white powder.

Returning to the question of the exact way in which the diluent evaporates, there are essentially two types of behaviour. Either evaporation occurs from the free liquid surface, which gradually retreats further and further within the bed of the particles, or the free surface retreats into the film less quickly than the rate at which the diluent is evaporating and hence cavitation occurs in cusps deep within the wetted part of the film. Whether the cavities are formed in the body of the liquid in the cusps (presumably around some fortuitous nucleus) or by the liquid de-wetting from the surface of one of the particles forming the cusp, the initial radii of curvature seem likely to be much smaller than those of the menisci of the free surface retreating through the mass of the particles. Hence, in a simple homogeneous model system, a cavitation mechanism of this sort is implausible. However, in more complex non-homogeneous systems, in which some sort of skin forms over the surface of the drying film and in which substantial amounts of diluent may be lost by molecular diffusion through the polymer particles themselves, such behaviour might well be observed. Indeed, where capillary compression results in particle deformation rather than cavitation, this type of mechanism may be quite important. Visual observation of the drying of dispersions of hard particles clearly indicates that the whitening (corresponding to air replacing diluent in the cusps) spreads from a few, apparently favourable, starting points (where presumably a local radius of curvature was larger than average) and slowly spreads over the film until the whole is gradually engulfed.

It is interesting to note, in view of the discussion in the earlier literature concerning the relative roles of capillary compression in the presence of the evaporating diluent phase and the high particle/air interfacial tension in providing the driving force for film coalescence, that in the case of many such dry, powdery films of hard particles cast from non-aqueous dispersions, on raising the temperature to just above the $T_g$ of the polymer particles, there is rapid collapse, with the elimination of the interstitial air, to form a continuous film. This product appears to be in no way significantly different from that which is obtained if the diluent is evaporated at a similar temperature. (The ease of control of the boiling point of the diluent in non-aqueous systems makes this point experimentally easy to establish.)

The very high temperatures which are frequently required to consolidate some of the powdery films that are obtained from aqueous systems from which the water has evaporated below the minimum filming temperature (which in practice is very close to the effective $T_g$ of the particles), may well, therefore, be due to causes other than an intrinsic inadequacy of the air/polymer interfacial tension as a driving force. For example, the water-soluble polymers which may be present in aqueous latex formulations are of necessity highly polar and therefore, in the absence of water, usually form hard and horny solids of high melting point, which may well completely enclose the polymer particles within strong skins capable of resisting subsequent deformation and integration. In contrast, the stabilizers normally used in organic environments of low polarity are themselves soft, oily liquids or low melting point solids which offer comparatively little resistance to the integration of the particles.

It is obvious that in the presence of a stabilizer firmly anchored to the particle surface, the complete contact between the stabilized surfaces, with the elimination of their mutual free interface, will be almost as effective in reducing the free energy of the system as the coming together of naked polymer particle surfaces, so that similar effects will be observed on treating dry powdery films, even in the presence of stabilizer.

### 6.3.6. Case D

These are dispersions of hard particles in totally volatile diluents in which the particles flocculate as the diluent evaporates. Consider a system such as in the previous case, but in which the particles become flocculated at moderately high volume concentrations. As Eveson[31] has pointed out, this situation is likely in the case of ionically-stabilized aqueous dispersions, because of the steady rise in concentration of flocculating electrolyte which occurs as the diluent evaporates. When flocculation takes place, particles will cease to jostle and will pack more closely one against the other with further loss of diluent so that the film effectively sets at a lower disperse phase volume than that for critical packing. Diluent will continue to evaporate from between the particles as described above, but now, when severe capillary compressions are applied, the film, as it is not critically packed, can undergo a further reduction in volume. Such shrinkage in a film of fixed area, sets up tensile stresses which sooner or later exceed the local strength of the film at some weak spot or other, and the whole of the stress in that region is released in the form of a major crack or fissure. Many simple conventional organosols (Section 5.1) 'mud-crack' in this way, and it is characteristic of such systems that the amount of non-volatile diluent required to eliminate this defect is very large compared to the theoretical voids-content of a close-packed film. Further, the final film is often rough in texture and on heating does not (unlike Case C above) convert to smooth and transparent polymer fragments.

### 6.3.7. Case E

Case E concerns deflocculated hard particles in a totally volatile diluent with a stabilizing sheath of significant size compared with that of the particle. Consider a system similar to that in Case C above, comprising stable dispersions of hard particles, but in which the size of the stabilizing sheath surrounding each particle has dimensions which are significant compared to those of the particle itself. The early stages of the evaporation of the diluent will be similar to those described above; the particles will jostle into a close-packed array, but in this case close-packing implies packing of the particle plus the stabilizer sheath. On evaporation of the residual diluent, the tenuous, diluent-swollen stabilizer sheaths slowly shrink. Consequently, additional space for a further reduction in film volume appears. However, as the stabilizer jackets begin to shrink, the particles cannot easily and rapidly slide past one another to take up this small but significant volume change, partly perhaps because of incipient flocculation, but probably more importantly because of the interpenetration of the outer fringes of the sheaths with an associated, very steep, rise in the effective viscosity of the medium. Hence the capillary stress, which is being applied simultaneously, causes local failure and mud-cracking. This type of mud-cracking however, due to the volume change in the finite stabilizer barrier, is clearly different in several respects from the mud-cracking associated with flocculated systems, as described in Case D above. Firstly, the severity of the mud-cracking effect (in terms of the fraction of the total surface area of the final film which is affected) is closely related to the molecular weight of the stabilizer used and the particle size of the polymer. Indeed, it is obvious that as the segment concentration in the solvated stabilizer barrier is probably of the order of 5 to 20%, while in the fully collapsed barrier it is of course close to 100%, then the volume shrinkage of the film on stabilizer collapse is approximately 10-times the weight percentage of the soluble group used to stabilize the particles.

For the three main particle size ranges considered here, i.e. several $\mu$m, few tenths of a $\mu$m and a few tens of nm, the typical amounts of soluble groups used are a few tenths of a percent, a few percent, and a few tens of a percent on polymer respectively, so that the expected volume contraction leading to mud-cracking should be of the order of 1%, 10% and 100% of the polymer core volume respectively. Empirical observations of the behaviour of these latices have been broadly in line with these values, especially for the coarser particles. It is also true for this type of system, that by increasing the non-volatile fraction of the diluent to a value just sufficient to fill the voids of the close-packed particles plus sheaths, the mud-cracking defect is usually eliminated.

The dry, white, mud-cracked fragments of the films from this system (as those from Case C above, but unlike the mud-cracked fragments from the products of Case D) when raised to a temperature above the softening point of the polymer, coalesce to form transparent and continuous film fragments. However, in this case, the contractile stress which follows the elimination of the air voids is absorbed by an 'all-round' shrinkage of the fragment. As a result, any mud-cracks which are present before heat treatment are substantially increased in area after stoving is carried out. In contrast, with Case C systems, the similar substantial diminution in volume is generally taken up in only a single dimension, normal to the substrate. This is apparently because the auto-adhesion between the particle contacts which form before film shrinkage occurs (the particles have to be soft and sticky before they can flow and deform) are sufficiently uniform and strong to absorb this contractile stress without localized breakdown leading to mud-cracking. This implies that not only are the particles converted from spheres to polygons, but they are also substantially flattened to 'bun-shaped' polygons.

In the Cases A to E described above, the main type of film-formation processes have been considered. Of course, there are many other additional special cases but only one is discussed in detail here—that of the so-called 'panel-coalescing' recipes (Section 7.3.3) which is of particular practical importance in automotive paints.

### 6.3.8. Case F: 'Panel coalescing' systems

Consider a rather dilute dispersion of hard, film-forming particles, dispersed in a diluent which, typically, contains at least two components—a rather freely volatile non-solvent and a less (but still finitely) volatile strong solvent. There may be other components present, such as non-volatile plasticizing solvents for the particles. When such a composition is applied by spraying, the highly volatile non-solvent diluent evaporates very rapidly, and the relatively large volume of strong solvent which remains immediately attacks the particles. In many cases, the strong solvent appears to partition itself between the particle and diluent phase on storage, so that the particles are to some extent already plasticized before the volatile diluent evaporates. As a result, the attack on the particles by the rest of the strong solvent in the continuous phase is much expedited. If the level of coalescing solvent is very high and the stabilizer is very labile on the surface of the particles, the composition on the panel may become virtually a concentrated solution. Ideally, the free stabilizer will be dispersed in the form of inverted micelles (see Section 6.4) throughout the body of the solution; less favourably it may be rejected in the form of a 'sweat layer' to the upper or lower surfaces of the film. Subsequent ageing at room temperature or forced drying eliminates the modestly-volatile strong solvent in exactly the same way as from a conventional solution lacquer film, to yield a product which is more nearly homogeneous than any usually produced from polymer dispersions.

More usually, the level of strong coalescing solvent used is such, together with the very high levels of stabilizer anchoring required to maintain stability in strong solvent environments (Section 3.5.3), that the particles behave, not so much as if they had gone into virtually complete solution, but rather as though they were merely very readily deformable particles (as discussed in Case B2, above). In this case therefore, film formation occurs essentially by particle deformation with minor areas of virgin, primary adhesive contacts and major areas of secondary, stabilizer-cemented contacts. On prolonged air drying or force drying, each swollen particle loses its coalescing solvent, shrinking in volume and becoming harder as it does so. However, the individual particles retain their original identity during this process. It would appear that the significant shrinkage stresses which must inevitably be set up can be accommodated (as in a lacquer film) by contraction and deformation normal to the surface only. That is, the primary and secondary adhesive contacts appear to be sufficiently strong and the particles—at least throughout the greater part of the evaporative process—are sufficiently plasticized so that the shrinkage stress never exceeds the strength of the inter-particle bonds. Careful examination of their crazing tendencies however show that these films, like conventional lacquers, do contain residual stress, especially if rapidly dried at temperatures just below the $T_g$ of the unswollen polymer. Nevertheless, despite the special problems involved, films of this type, i.e. those in which relatively hard particles are transiently plasticized by a modestly volatile strong solvent, so that at the instant of film formation, they fall essentially into the class of Case B2 above, while eventually yielding films of $T_g$'s substantially above ambient temperature, do in fact represent the most commonly used and generally useful approach to formulating industrial paints from non-aqueous dispersions.

## 6.4. ULTRA-COLLOIDAL POLYMER DISPERSIONS

By appropriate modifications of the dispersion polymerization process, it is possible to prepare polymer dispersions of a very small particle size, of the order of a few tens of nm in diameter. Such dispersions are usually transparent, although having obvious optical peculiarities, and they form gelatinous solid materials (presumably by close-packing, see Section 6.3) at around 40% total solids. Applying the values found for the amount of stabilizer required to cover a unit surface area of the particle (Section 3.3), one would expect that about 20% of soluble group, calculated on insoluble particle core, would be required at this particle size, and this is indeed found to be experimentally of the correct order. Assuming a normal ASB (anchor/soluble group mass balance, see Section 3.6, above) of about 1:1, this implies that some 40% of the total particle is actually stabilizer and about 60% ungrafted, insoluble polymer. Allowing for the fact that there is generally at least an order of magnitude difference between the molecular weight of the anchor group of the stabilizer and that of the ungrafted polymer, it follows that there are at least as many molecules of anchor polymer as there are ungrafted polymer molecules in the typical particle. Clearly, any significant further reductions in particle size moves one into a region in which the particle is primarily composed of stabilizer molecules, which may or may not be contaminated with small numbers of ungrafted polymer chains. Micellar dispersions of this type can indeed be prepared[32] (see also Section 3.6, above). It has also been demonstrated by molecular weight measurement (vapour pressure osmometry) that ABA block copolymers with A segments of poly(12-hydroxystearic acid) and B segments of a copolymer derived from diphenylol propane and epichlorhydrin form aggregates of two or three units in hemi-solvent media[33]. It is convenient to discuss at this point the characteristics of these micellar dispersions and the way in which their properties are modified by the composition and properties of the graft copolymers from which they are formed.

It is fundamental to the understanding of the behaviour of an amphipathic block or graft copolymer in a hemi-solvent (that is, a solvent for only one component of the block or graft copolymer) to realize that, in such an environment, the soluble chain is expanded essentially to the same dimensions and to the same effective segment density as it would be in isolation in the same solvent. Similarly, the insoluble chain is collapsed as it would be in isolation in the non-solvent. The fact that the two chains are covalently linked together at one point only slightly perturbs the characteristic behaviour of the individual chains.

As has been described above (Section 2.3), two soluble chains moving into one another's vicinity, undergo perturbations which result in a mutual repulsion. On the other hand, insoluble polymer chains tend, on contact, to attract one another. This latter effect is even more marked than might at first be expected, because for low molecular weight polymers typical of those used as the anchoring component of graft copolymer dispersants in practical systems, the length and stiffness of the chain are such that the degree of 'self-coiling' (i.e. the extent of intra-molecular segment–segment association) which can be achieved is actually surprisingly small. For example, the ratio of length:diameter of a typical poly(methyl methacrylate) anchor chain of a few thousand molecular weight, is only of the order of 10:1. Such a molecule can be visualized as having the flexibility of, for example, a 1-ft length of thick walled, large (1 in) diameter pressure tubing. Under these circumstances, it is clear that when two or more chains come together, the number of *inter*-molecular segment–segment associations which they can achieve is much greater than the sum of their individual *intra*-molecular associations. As a result, the driving force towards aggregation of the insoluble polymer chains is very strong. Equally, any configuration of the polymer in the core of the particle in which much intertangling of the chain occurs is strongly favoured. It seems obvious that the large energy gain from aggregation on the one hand and from the large repulsion of the soluble chains on the other will result in the build-up of individual molecules of graft copolymer into organized aggregates or micelles [Figure 6.8(a)]. This growth is only stopped when the particle core is surrounded by a continuous layer of soluble polymer which then repels further individual stabilizer molecules in exactly the same way that repulsion between soluble groups limits the build-up of stabilizer on the surface of the particle.

Taking appropriate values for the area of a particle surface stabilized by soluble polymer chains, the molecular weight of the anchor chain of the dispersant and the number of soluble polymer groups attached to it, together with certain simple assumptions concerning the density of the particle core, it is possible to calculate the number of dispersant molecules required to produce a continuous layer of soluble polymer on the surface of the particle[32]. This number clearly corresponds to the micellization number for that particular amphipathic block or graft copolymer. The appropriate equations, together with representative worked examples for an AB block copolymer (Example A) and a 'comb'-type graft copolymer (Example B) are given below:

$$n = \frac{36\pi}{(0\cdot6023)^2}\left(\frac{M_B}{\rho}\right)^2 \frac{1}{C^3 x^3} \tag{6.7}$$

where

$n$ = micellization number
$\rho$ = density of particle core (assumed)
$M_B$ = molecular weight of insoluble chains
$C$ = surface area (Å$^2$) stabilized per soluble chain
$x$ = number of soluble chains attached to each insoluble chain of dispersant

$$r = \left(\frac{3nM_B}{4\pi\rho \times 0\cdot6023}\right)^{\frac{1}{3}} \tag{6.8}$$

where $r$ = micelle core radius (Å).

Example A: *AB block copolymer* [Figure 6.8(a)]

$$\rho = 1; \quad M_B = 5000; \quad C = 400 \text{ Å}^2; \quad x = 1$$

corresponding to one soluble chain of about 2000 in molecular weight attached to an anchor chain of 5000 molecular weight.

$$n = 122$$

$$r = 63 \text{ Å} \ (\equiv \text{core diameter of } 12\cdot6 \text{ nm}).$$

Example B: *'Comb'-type graft copolymer* [Figure 6.8(b)]

$$\rho = 1; \quad M_B = 5000; \quad C = 400 \text{ Å}^2; \quad x = 2\cdot5$$

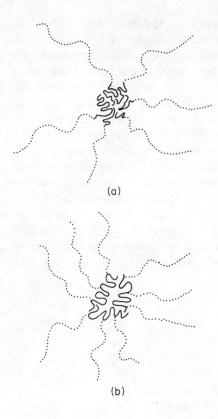

Figure 6.8. Schematic representation of aggregates of six AB block copolymer molecules (a) and three 'comb'-like graft copolymer molecules with three soluble chains per insoluble chain (b): (····) soluble chain, (——) insoluble chain

corresponding to two or three soluble chains each of about 2000 in molecular weight attached to one anchor chain of 5000 molecular weight.

$$n = 8$$

$$r = 25 \text{ Å } (\equiv \text{ core diameter of 5 nm}).$$

The disposition of six molecules of the AB block dispersant (Example A) and that of three molecules of the 'comb'-type graft dispersant (Example B) are schematically illustrated in Figure 6.8. As will be evident from these illustrations and the worked examples, an increase in the number of chains associated with a single anchor chain results in a sharp fall in the value of the micellization number. It can also be seen that for such comb graft copolymers, the micellization number is actually very small, often less than 10. An inspection of the form of the equations (6.7) and (6.8) indicates that by changing the anchor/soluble group balance towards a higher soluble-group content, thus causing an increase in the area stabilized by the soluble chain relative to the increase in core surface area produced by the attached insoluble chain, also causes the micellization number to fall.

It is instructive to consider two limiting cases of the micellization number/core radius as calculated above. Firstly, the point at which the micellization number passes through unity and thereafter becomes fractional. Secondly, the case at which the micellization number is so large that the core size required to achieve it is larger than the physical size of the insoluble polymer chains comprising the core (the upper micellization limit). This latter limit of course, depends upon the absolute molecular weight and other properties of the insoluble component of the graft copolymer, and is not a unique value like a micellization number of unity. A suitable method for calculating this limit is given by equation (6.9), below.

$$n_{max} = \frac{4\pi\rho r_{max}^2 \times 0.6023}{3M_B} \tag{6.9}$$

where $r_{max} = L_m$, the length of insoluble polymer chain in an AB block copolymer or $L_m/2x$, for a graft or ABA block copolymer ($L_m \approx 1.1q$ Å where $q$ = the number of atoms in the backbone of the insoluble polymer chain).

Micellization numbers of less than unity clearly indicate that the soluble component of a single molecule of the amphipathic graft copolymer is sufficiently large to envelop and protect the insoluble polymer chain and form a stable entity, without having to associate with even one other molecule. Equally, such polymers have little affinity for bulk polymer surfaces and therefore do not micellize in the ordinary sense and do not either adsorb at particle surfaces nor act as stabilizers. Micellization numbers of one or less can arise in two fundamentally different ways. Firstly, from molecules containing one (or a very few) soluble polymer groups which are very large compared to the insoluble polymer. Such molecules, as illustrated in Figure 6.9(a), below, remain highly organized in hemi-solvents and are essentially equivalent to monomolecular micelles. On the other hand, graft copolymer molecules of extreme 'comb'-like configurations, but in which the ASB is normal, may also have fractional micellization numbers because of the geometric restrictions. These make it impossible to arrange the molecule in a configuration in which all of the soluble polymer is in the continuous phase, whilst at the same time all of the insoluble polymer is out of solution. In a sense, such molecules exceed the upper micellization limit when the micellization number is still fractional. When fractional micellization numbers occur with such molecules, situations result in which either the insoluble polymer is forced into an extended form, to allow all of the soluble polymer to exist in the continuous phase, or, a core is formed from all of the insoluble polymer at the expense of burying significant quantities of the soluble polymer within it. For molecules of this type, the transition from a micelle-forming, comb-graft copolymer, to material with a fractional micellization number is essentially the transition from an amphipathic graft copolymer of comb configuration to a random copolymer of two components of differing polarity.

When the energy of solution of the more soluble component exceeds the energy of precipitation of the less soluble component, so that the molecule adopts an extended configuration, then clearly we have a soluble copolymer, while the reverse energy balance corresponds to an insoluble copolymer which

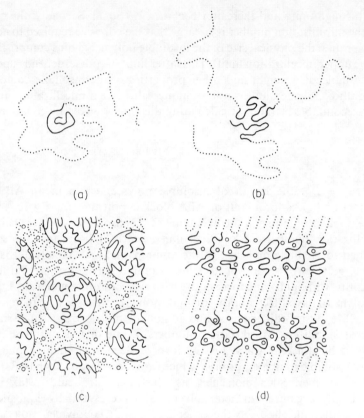

Figure 6.9. Schematic representation of types of aggregates formed with increasing concentrations of amphipathic block copolymers. (a), monomolecular micelle; (b), multimolecular micelle; (c), organized cylinders of rod-like micelles; (d), organized lamellar structure; (· · · ·) soluble chain, (——) insoluble chain. (From Sadron, *Angew. Chem. Internat. Edn.*, **2**, 248 (1963), with permission)

precipitates out of solution to form a bulk insoluble phase. As in the case of the organized monomolecular micelle, normally neither of these configurations can provide high stability for dispersed particles. In the case of the soluble polymer there is usually only weak adsorption (as discussed in the case of anchoring of soluble polymer chains above); if the polymer is insoluble it is obviously incapable of providing a repulsive barrier.

Where the micellization number is large compared to unity and also exceeds the upper critical value for the formation of spherical micelles (i.e. without the incorporation of some of the soluble components into the micelle core), this must result, not from an extreme multiplicity of soluble groups as discussed above, but from low soluble-group content. Such systems form highly organized, non-spherical micelles in hemi-solvents. This situation is particularly favoured in concentrated solutions in which the area stabilized by each soluble chain is limited by the amount of solvent available. Indeed molecules which are capable of forming normal spherical micelles in dilute solution, often exceed the critical micellization number at high concentrations.

The solution properties of such AB and ABA block copolymers have been widely studied[34-36]. In cases where the critical micellization number is only slightly exceeded, the aggregates rearrange into long rod-like micelles [Figure 6.9(c)]. Under more extreme conditions, however, when even these cylindrical assemblies have greater surface areas than can be satisfied by the soluble groups present, then a laminar structure is formed, as in Figure 6.9(d). In the ultimate limit, phase inversion to rods and finally to inverted spherical micelles can take place. Since block copolymers of this type have organized structures on a macroscopic scale, the products are not fluid, and indeed often display surprising mechanical strength, quite unlike even fairly concentrated solutions of spherical micelles. These various possibilities are summarized in Table 6.3.

Table 6.3. Criteria for aggregation behaviour of amphipathic graft copolymers

| Type of aggregate | Micellization number, $n$ | Core radius, $r$ | No. soluble chains per insoluble chain, $x$ | Anchor-soluble group balance (ASB) |
|---|---|---|---|---|
| Spherical 'real' micelles | $>2$ | $<r_{critical}$ | $\not> 1$ | $A:S \simeq 1:1$ |
| Rod-like or laminar micelles | $>2$ | $>r_{critical}$ | $\not> 1$ | $A:S > 1:1$ |
| Spherical 'monomolecular' micelles | $<2$ | $<r_{critical}$ | $\not> 1$ | $A:S < 1:1$ |
| Random copolymer | $\ll 2$ | $>r_{critical}$ | $\gg 1$ | $A:S < 1:1$ (soluble) $A:S > 1:1$ (insoluble) |

Although it is often convenient to discuss the formation of micelles in terms of the association of individual molecules pre-existing in a totally hemi-solvent environment, this situation is never of course, actually realized in practice. For this to take place would require the initial concentration of the graft copolymer in the hemi-solvent, to be at a value well below that of the critical micelle concentration (CMC). Such initial solutions would have to be so dilute (a small fraction of a per cent) that the degree of subsequent concentration required to produce a reasonably high solids polymer dispersion would be enormous and quite impracticable. In practice therefore, it is usual to start with a fairly concentrated solution of the graft copolymer in a common solvent and to add an excess of hemi-solvent. The common solvent used initially is then removed by distillation. Under these conditions, micellization invariably occurs in circumstances in which the polymer forming the micelle core—although insoluble in the continuous phase—is nevertheless significantly swollen by the residue of the common solvent. This swelling assists maximum entanglement of the insoluble polymer. However, although the polymer forming the core of the micelle is of quite low molecular weight, and both the absolute physical size of the core and the number of molecules involved is small, yet nevertheless, the core may still exhibit to some extent, the properties of the bulk polymer of that composition. In the case of polymers with a glass transition temperature ($T_g$) far above room temperature, such as poly(methyl methacrylate), even the tiny fragment of polymer which constitutes the core of the micelle may be below its $T_g$ in the hemi-solvent, and therefore be quite rigid. In an ordinary micelle, although the association of the anchor groups within the core is energetically highly favoured, yet nevertheless, this energy of association is not indefinitely large compared to $kT$ and no additional activation energy is involved in the break up of the micelle. Hence a finite equilibrium (and also a finite CMC) can exist. However, when the core of the micelle is below its $T_g$, although the equilibrium energies may be similar to those of a normal micelle, the activation energy required to disentangle an anchor chain from the rigid core may be very large compared to the equilibrium energy of association. Under these circumstances, the rate of equilibration becomes vanishingly small, and the aggregate formed will tend to have a permanent, metastable identity and should perhaps be regarded as an ultra-colloidal particle rather than as a true micelle in equilibrium. In some senses, particles such as those with rigid cores may also be regarded as larger versions of the monomolecular micelles in that they are self-stable entities which do not interact with other particle surfaces, and are incapable of any stabilizing action. However, a change in the environmental conditions (e.g. by raising the temperature or adding limited-swelling solvents to the continuous phase) can lead to the point at which the core of the micro-particle becomes soft, causing it to revert to a normal, equilibrium micellar state. As has been described elsewhere (Section 3.6), this phenomenon is of considerable practical importance in the preparation of stable polymer dispersions at low temperatures when a micellizing graft copolymer is used as the stabilizer for the dispersion.

Due to the permanence of the core structure of micro-particles which have been prepared as spherical entities at high dilutions, they can be concentrated (as long as the core retains its rigid structure) to far beyond the point at which anchoring micelles would have rearranged to the rod-like or laminar structures described above[32]. However, on warming or on the addition of solvent, transformation to the non-spherical equilibrium state rapidly occurs.

## 6.5. REFERENCES

1. Sherman, P., *Industrial Rheology*, Academic Press, London and New York, 1970, Chapter 3.
2. Einstein, A., *Ann. Phys.*, **17**, 459 (1905); **19**, 271 (1906); **34**, 591 (1911).
3. Rutgers, I. R., *Rheol. Acta*, **2**, 202, 305 (1962).
4. Vand, V., *J. Phys. Chem.*, **52**, 277 (1948).
5. Goldsmith, H. L. and Mason, S. G. in *Rheology*, Vol. IV (Ed. Eirich, F. R.), Academic Press, New York, 1967.

6. Manley, R. St. J. and Mason, S. G., *Can. J. Chem.*, **32**, 763 (1964).
7. Thomas, D. G., *J. Colloid Sci.*, **20**, 267 (1965).
8. Lee, D. I., *Trans. Soc. Rheol.*, **13**, 273 (1969).
9. Maron, S. H., Madow, P. and Krieger, I. M., *J. Colloid Sci.*, **6**, 584 (1957).
10. Saunders, F. L., *J. Colloid Sci.*, **16**, 13 (1961).
11. Barsted, S. J., Nowakowska, L. J., Wagstaff, I. and Walbridge, D. J., *Trans. Faraday Soc.*, **67**, 3598 (1971).
12. Woods, M. E. and Krieger, I. M., *J. Colloid Interface Sci.*, **34**, 91 (1970).
13. Papir, Y. S. and Krieger, I. M., *J. Colloid Interface Sci.*, **34**, 126 (1970).
14. Wagstaff, I. and Waters, J. A., ICI Paints Division, unpublished results, 1968.
15. Krieger, I. M. and Dougherty, T. J., *Trans. Soc. Rheol.*, **3**, 137 (1959).
16. Krieger, I. M., *Advan. Colloid Interface Sci.*, **3**, 111 (1972).
17. Overbeek, J. Th. G., *Colloid Science*, Vol. I (Ed. Kruyt, H. R.), Elsevier Publishing Company, Amsterdam, 1952.
18. Mysels, K. J., *Introduction to Colloid Chemistry*, Interscience Publishers, New York, 1959, pp. 63–64, 160–164.
19. Vold, M. J., *J. Colloid Sci.*, **18**, 684 (1963).
20. Vanderhoff, J. W., *Brit. Polym. J.*, **2**, 146 (1970).
21. Broutman, L. J. and Krock, R. H. (Eds.), *Modern Composite Materials*, Addison-Wesley Publishing Company, Reading, Massachusetts, 1967.
22. Ashton, J. E., Halpin, J. C. and Petit, P. H., *Primer on Composite Materials: Analysis*, Technomic Publishing Co. Inc., Stamford, Conn., 1969.
23. Sarvetnick, H. A. (Ed.), *Plastisols and Organosols*, Van Nostrand Reichold Co., New York, 1972.
24. Dillon, R. E., Matheson, L. A. and Bradford, E. B., *J. Colloid Sci.*, **6**, 108 (1951).
25. Henson, W. E., Taber, D. A. and Bradford, E. B., *Ind. Eng. Chem.*, **45**, 735 (1953).
26. Brown, G. L., *J. Polym. Sci.*, **22**, 423 (1956).
27. Talen, H. W., *J. Oil Colour Chemists Assoc.*, **45**, 387 (1962).
28. Voyutskii, S. S., *J. Polym. Sci.*, **32**, 528 (1958).
29. West, E. J. and Greig, J. R., ICI Paints Division, unpublished results, 1970.
30. Cousens, R. H., ICI Paints Division, unpublished results, 1962.
31. Eveson, G. T., *J. Oil Colour Chemists Assoc.*, **40**, 456 (1957).
32. Imperial Chemical Industries, *British Patent*, 1,317,249; 1,319,448; 1,319,781 (1973); Waters, J. A. and Baker, A. S., ICI Paints Division, unpublished results, 1970.
33. Tolman, R. J., *MSc. Thesis*, Bradford University, 1969.
34. Molau, G. E. in *Block Polymers* (Ed. Aggarwal, S. L.), Plenum Press, New York, 1970, pp. 79–106.
35. Sadron, C. and Gallot, B., *Makromol. Chem.*, **164**, 301 (1973).
36. Sadron, C., *Angew. Chem. Internat. Ed.*, **2**, 248 (1963).

# CHAPTER 7
# *Applications of Polymer Dispersions Prepared in Organic Media*

M. W. THOMPSON

| | |
|---|---|
| 7.1. GENERAL FEATURES OF APPLICATION OF POLYMER DISPERSIONS IN ORGANIC LIQUIDS.................................................. | 273 |
| 7.2. PRODUCTION OF POLYMER POWDERS................................ | 277 |
| 7.3. USE IN PAINT FORMULATIONS...................................... | 278 |
|     7.3.1. Comparison of dispersions with solution systems................ | 281 |
|     7.3.2. Void-filling formulations................................... | 283 |
|     7.3.3. Active solvent formulations................................. | 283 |
|     7.3.4. Combined dispersion and solution systems.................... | 284 |
| 7.4. GENERAL SURFACE COATING APPLICATIONS........................... | 285 |
|     7.4.1. Adhesives.................................................. | 285 |
|     7.4.2. Polishes.................................................... | 285 |
|     7.4.3. Textile impregnants......................................... | 286 |
|     7.4.4. Encapsulants............................................... | 286 |
|     7.4.5. Fibre coatings.............................................. | 286 |
| 7.5. MISCELLANEOUS APPLICATIONS..................................... | 287 |
|     7.5.1. Thermoplastic rubbers...................................... | 287 |
|     7.5.2. Coloured polymer particles.................................. | 287 |
|     7.5.3. Lubricants................................................. | 288 |
| 7.6. REFERENCES....................................................... | 289 |

## 7.1. GENERAL FEATURES OF APPLICATION OF POLYMER DISPERSIONS IN ORGANIC LIQUIDS

So far, the main application of the polymer dispersions prepared by dispersion polymerization in organic media has been for film-forming compositions in the surface coatings field and for paints, particularly stoving compositions for automotive applications (Table 7.2, Section 7.3, below). This is only to be expected since the dispersion polymerization process was evolved specifically for this type of end-product. In addition, a number of related surface coating applications such as adhesives, textile impregnants and fibre coatings have also been reported. However, before describing such applications in more detail, it is convenient first to discuss the general attributes of the polymer dispersions in order to define the potential areas of application more closely.

Dispersion polymerization produces organic polymer, usually in the form of spherical and smooth particles from 0·01 to 10 µm in size, dispersed at 20–70% polymer solids by weight in an organic liquid which is either a nonsolvent for the polymer or at the most is only a moderate swellant for it. Examples of various polymer dispersions which have been prepared in aliphatic hydrocarbon diluent are shown in Figure 7.1. On economic grounds, the organic liquid used as diluent should preferably be an aliphatic hydrocarbon (boiling range, 70–230 °C) which has the added advantage of a low latent heat of

Figure 7.1. Examples of polymer dispersions prepared in aliphatic hydrocarbons: 1, Nylon-11; 2, poly(acrylonitrile-co-acrylic acid) (75:25, w/w); 3, poly(propylene); 4, terylene; 5, poly(vinyl chloride); 6, poly(methyl methacrylate)

evaporation. However, suitable diluents can be selected from a wide range of organic liquids including those which are not inflammable. The stabilized polymer particles have amphipathic graft copolymers either permanently or temporarily attached to their surfaces. One component of the graft copolymer is associated with the polymer particle and the other is soluble in the organic diluent. The surrounding layer of graft copolymer is therefore, very different in its chemical composition from the polymer particle it stabilizes. If the graft copolymer is only adsorbed onto the surface of the polymer particle it can be displaced subsequently by heat or pressure, allowing the polymer in the particles to fuse together. The addition of special coalescing solvents or plasticizers assist this process. Naturally, if the graft copolymer is permanently attached to the surface of the polymer particles, no integration of the particles will take place and the particles will not fuse together although the polymer within may be extracted by the use of suitable solvents.

The nature and properties of the polymer within the particle will also influence its potential uses. Cross-linked polymer particles will normally remain as such even though heat or pressure subsequently applied can distort them from their original spherical shape to an extent depending on the level of cross-linking used. As might be expected, cross-linked polymer particles in which the main monomer normally produces a polymer of low glass transition temperature, will be elastic in nature and the particles can act as microgels imparting impact modification when incorporated into a hard polymer matrix.

The organic diluent is usually readily removed from the dispersion since it has no polymer dissolved in it except the low concentration of the soluble portion of the graft dispersant. Similarly, the viscosity of the polymer dispersion is little different from that of the original liquid diluent without the graft dispersant or polymer particles present except when a close approach to critical packing is reached. The mobility of a polymer dispersion in an organic diluent is illustrated in Figure 7.2. This contrasts the effect of dropping a weight into a container of poly(methyl methacrylate) dispersed in petrol at 40% solids (a) compared with the behaviour of a solution of the polymer at the same solids level (b). Such factors as high volatility and low viscosity are of great importance in the application of polymer dispersions as surface coatings or for impregnating fibrous materials, such as textiles. When an aliphatic hydrocarbon is used as the organic diluent, its low latent heat of evaporation is a considerable asset and is also of value when polymer powders are prepared from the dispersions in a continuous process. On substrates which are water-sensitive, such as glass fibre or wood, the application of dispersions in aliphatic hydrocarbons has advantages in that the diluent does not cause distortion, cracking or swelling and can be employed for adhesives as well as for coating glass fibres.

Polymer dispersions in aliphatic hydrocarbons, which employ graft dispersants with one component soluble in the hydrocarbon, can also be dispersed in solid as well as liquid aliphatic materials, such as paraffin wax, petroleum jelly, fats, vegetable oils and waxes. Such compositions are of use as cosmetics and polishes. Dispersions of polymers in liquids of a more polar nature can be prepared, for example, poly(acrylonitrile) dispersed in methyl ethyl ketone or in monomers such as methyl methacrylate. The latter can be used as a pourable composition for the preparation of polyphase composites.

The addition of particulate material insoluble in aliphatic liquids, such as pigments, insecticides, metal salts, etc., to a dispersion polymerization reaction mixture can lead to their encapsulation within the dispersed polymer particles. In this way, the contents of the polymer particles can be protected from the action of the outside environment or can be slowly released to it.

Dispersion polymerization in organic liquids, especially aliphatic hydrocarbons free of reactive groups, is ideal for the employment of ionic, especially anionic modes of polymerization. This method of polymerization is very suitable for the preparation of AB, ABA or ABC block copolymers of closely-defined composition. If the B or C components are insoluble in the organic liquid, fine polymer dispersions result. These can be used to produce thermoplastic rubbers when the appropriate monomers are used. If the insoluble portion is cross-linked, whilst the soluble portion is capable of vulcanizing

Figure 7.2. Comparison of fluidity of poly(methyl methacrylate) as a dispersion in an organic diluent (a) and as a solution (b) at 40% solids

with a continuous phase of monomer which also yields a rubber, a multiphase rubber can be produced.

The high phase volumes (60–75%) which are possible by deliberately producing crops of particles of different sizes, give dispersions which have low levels of organic diluent present. This makes it possible to prepare a solution lacquer of 40% solids from a polymer dispersion of 75–80% (w/w) solids in aliphatic hydrocarbon without the necessity of subsequently removing the organic liquid diluent. In the formation of certain addition polymers, such as poly(methyl methacrylate), the rate of dispersion polymerization is so rapid that the method is an economic way of producing polymer for use in solution lacquers or thermosetting systems. The subsequent dissolution process is greatly assisted by the fine state of division of the polymer produced.

## 7.2. PRODUCTION OF POLYMER POWDERS

For polymers with glass transition temperatures above room temperature, dispersion polymerization provides a rapid and simple method of low cost for preparing the polymers as finely-divided powders. The low latent heat of evaporation of the diluent allows its ready removal by distillation in cases where the particle size is too small for filtration. The continuous process (Section 5.6.1) developed for the preparation of polymer powders introduces the dispersion formed continuously into an evaporator and is designed to overcome the region of critical packing as the dispersion evaporates. The product is a fine, lightly aggregated powder and most of the organic diluent can be recovered for eventual re-use. The polymer powders obtained from addition dispersion polymerization produce, for example, poly(acrylates) which are suitable for injection moulding of plastics by standard procedures. Powders made by condensation polymerization in dispersion can also be prepared in a wide range of particle sizes. The larger particles can be used for 'flatting' or giving a 'matt' appearance to surface coatings and as fillers. Medium-sized, spherical particles with smooth oleophilic surfaces can be used as powders for cosmetic applications. Examples of polymer powders prepared by dispersion polymerization in organic media are shown in the Frontispiece.

The continuous production of acrylic polymer powder by dispersion polymerization in organic liquids has been described in detail elsewhere (Section 5.6.1). The polymers are obtained in aggregates of particles from 0·1 to 10 µm in size of constant $\overline{M}_n$ and $\overline{M}_w$, the most usual range being 50–100,000 $\overline{M}_w$ with an $\overline{M}_w/\overline{M}_n$ ratio of 2. The polymer powder produced tends to be electrostatically charged and the continuous apparatus must be extensively earthed.

The polymer powders can be dissolved in the usual solvents to produce acrylic surface coatings or compacted into granules for use in injection moulding of plastics by standard techniques. By a suitable choice of polymerization conditions (temperature, time, nature of diluents etc.) and evaporation conditions (diluent boiling point, evaporation temperature) a range of polymer powders can be produced. These have included poly(methyl methacrylate), poly(acrylates) and their copolymers, poly(acrylonitrile), poly(vinyl chloride) and poly(vinyl acetate). Two-phase polymer particles of poly(methyl methacrylate) polymerized within preformed poly(ethyl acrylate) particles have also been prepared continuously and subsequently converted into powders. The melt-flow properties of polymers made by dispersion polymerization are substantially below those obtained from polymer prepared by bulk polymerization[1] (Table 7.1).

Table 7.1. Comparison of melt-flow properties of polymers prepared by dispersion and bulk polymerization[1]. (From *British Patent* 956,454 (1964) with permission of the Controller, H.M. Stationery Office)

| Property | Dispersion | Bulk |
|---|---|---|
| Reduced specific viscosity | 0.047 | 0.045 |
| Molecular weight | 100,000 | 100,000 |
| Residual monomer | 1.52% | 2.1% |
| ASTM flow temperature | 135 °C | 158 °C |

Poly(styrene) powders, which are easy to isolate and are suitable for injection moulding, have been prepared by the dispersion polymerization of styrene using butyl lithium in heptane in the presence of a very small amount of crepe rubber[2,3]. The mouldings prepared from these powders were optically clear and had bending moduli of 450,000 lbf/in$^2$ with a Rockwell M hardness of c. 60.

## 7.3. USE IN PAINT FORMULATIONS

The development of dispersion polymerization in organic liquids has so far been primarily orientated towards the application of the polymer dispersions formed as surface coating materials (Table 7.2). An example of the use of a paint system, based on polymer dispersed in organic media, for application to motor cars is shown in Figure 7.3. The advantages of an organic diluent over an aqueous diluent for this type of application have already been described (Section 1.1). Ideally, both from the point of view of economy and on grounds of reducing pollution to the atmosphere, it would be better to eliminate entirely the need for a solvent or a diluent carrier for the polymer since neither are utilized in the final paint film. New types of film-forming systems have already been envisaged which involve the application of (solvent-less) powder coatings directly[4] or 'high solids' systems cured by radiation[5]. However, such formulations are not yet sufficiently advanced in development to provide a practical method for applying paints to motor cars and similar applications.

In some ways, particularly from the pollution aspect, the wide use of polymer dispersions in water for such applications would be preferred to those dispersed in organic diluents. However, as before, practical formulations are not available for their widespread use and in many cases, potentially toxic organic additives need to be used in the aqueous phase in order to approach an adequate balance of application and performance properties.

Table 7.2. Examples of published patents describing paint compositions based on non-aqueous polymer dispersions (for details, see Tables 8.2 and 8.1, Appendix)

| Company | Patent | | Year of Publication |
|---|---|---|---|
| Badische Anilin- & Soda-Fabrik | United States | 3,551,525 | 1970 |
| | British | 1,240,317 | 1971 |
| | British | 1,240,320 | 1971 |
| Celanese Coatings | British | 1,243,274 | 1971 |
| | United States | 3,654,201 | 1972 |
| Cook Paint & Varnish | British | 1,134,997 | 1968 |
| | Belgian | 782,158 | 1972 |
| Dai-Nippon | Japanese | 15435 | 1972 |
| | Japanese | 16526 | 1972 |
| | Japanese | 18995 | 1972 |
| | Japanese | 34684 | 1972 |
| | Japanese | 34685 | 1972 |
| du Pont | British | 1,181,316 | 1970 |
| | British | 1,210,161 | 1970 |
| | German | 2,103,849 | 1971 |
| | United States | 3,660,537 | 1972 |
| Ford-Werke | German | 2,151,782 | 1972 |
| | German | 2,151,824 | 1972 |
| | German | 2,209,875 | 1972 |
| | German | 2,209,889 | 1972 |
| Imperial Chemical Industries | British | 930,919 | 1963 |
| | British | 978,484 | 1964 |
| | British | 980,633 | 1965 |
| | British | 1,019,361 | 1966 |
| | British | 1,084,054 | 1967 |
| | British | 1,114,142 | 1968 |
| | British | 1,156,012 | 1969 |
| | British | 1,161,402 | 1969 |
| | British | 1,242,054 | 1971 |
| | German | 2,158,821 | 1972 |
| | German | 2,205,171 | 1972 |
| | British | 1,305,715 | 1973 |
| Morningstar–Paisley | French | 1,486,760 | 1967 |
| PPG Industries | Canadian | 875,617 | 1971 |
| | Belgian | 768,650 | 1971 |
| | United States | 3,666,710 | 1972 |
| Rohm and Haas | United States | 3,218,287 | 1965 |
| | United States | 3,711,449 | 1973 |
| Shell International | British | 1,209,409 | 1970 |
| Sommerfeld, E.G. | United States | 3,676,526 | 1972 |
| Toa Synthetic Chemicals | Japanese | 34396 | 1970 |
| Union Carbide | United States | 3,701,746 | 1972 |

Figure 7.3. Application of paint based on non-aqueous dispersion polymer ('DISPERSYMER'*) on the production line at Ford, Dagenham. (Reproduced by permission of Ford Motor Co., Dagenham)

The use of inert materials, like aliphatic hydrocarbons, as diluents in practical surface coating materials therefore does represent an advance over conventional solution systems. The latter utilize mainly aromatic solvents, which after their release into the atmosphere, undergo various photochemical reactions which produce toxic by-products and 'smogs'[6]. Consequently, the use of aromatic solvents as constituents of surface coating materials is now increasingly being restricted by legislation, particularly in North America and Japan. As a result of this trend, surface coatings based on polymers dispersed in aliphatic materials are now tending to displace the more conventional systems utilizing aromatic solvents.

The formulation of practical paint systems using polymer dispersions in aliphatic hydrocarbons can range from compositions with a high disperse polymer phase with little or no polymer in solution except for the soluble portion of the graft copolymer dispersant to compositions consisting entirely of polymer in solution. Such materials are prepared by dispersion polymerization in an aliphatic hydrocarbon and then blended with a sufficient amount of a strong solvent to dissolve all or a portion of the polymer. The solubility of the copolymer is, of course, dependent on its composition. It is possible to prepare a polymer 'seed' which is relatively insoluble on to which is grown various amounts of a different polymer which is more soluble. In this way, graduated

* 'DISPERSYMER' is the ICI Registered Trade Mark for its non-aqueous dispersion polymers.

particle compositions can be obtained which produce a range of materials from polymer in true solution, through solvent-swollen gel, to stable but unswollen polymer particles. The presence of these types of materials has a profound effect on the evaporation rate of the solvents used and on the rate of increase in viscosity during evaporation under varying temperature conditions. These effects can be used to give a wide control over the formation of thick coatings by spray application. Little sagging and good control of the 'lay-down' of flake pigments, such as metallics, is obtained.

Useful effects during application are also obtained with solutions of acrylic polymers prepared from dispersion polymers in organic liquids which are then dissolved up and formulated as a solution acrylic lacquer[7]. In this case, all of the polymer is apparently in solution but the application properties are different from polymer of the same molecular weight but made by solution polymerization. Part of this difference can be ascribed to the presence of graft dispersant in the dissolved dispersion polymer since it is known that the addition of graft copolymer to polymer made by solution polymerization has some effect on application properties[8]. However, this does not appear to account for the effect entirely. In fact, the swelling of particulate polymer by solvent, followed by dissolution, may never reach the same degree of solvation as polymer prepared directly in solution. Attempts to demonstrate this unequivocally by physical measurements such as viscometry have not given clear-cut results though the effects on spray application itself are real. The presence of very small amounts of high molecular weight polymer in solution are known to improve flow properties[9]. On the other hand, larger quantities of high molecular weight polymer can lead to the formation of fibrous material, known as 'webbing', when some acrylic lacquer compositions are applied by spraying.

### 7.3.1. Comparison of dispersions with solution systems

Polymers used for surface coatings are usually applied as solutions by a range of processes including spray, dip, roller-coating, etc. All of these methods of application are sensitive to the rheological characteristics of the polymer solution at all stages of the application. Low viscosities are required during the application so that high fluid delivery rates can be achieved. After coating the substrate, good flow is then required to give a smooth glossy finish. However, too much flow leads to sagging of thick coats. Thick coats are required in order to cover the substrate effectively and to fill in the indentations with as few coats of paint as possible. This is particularly important in industrial applications, such as in spray-line assemblies used in car manufacturing plants.

However, the molecular weight of the polymer required to give the necessary mechanical and physical properties, whether in a lacquer or in a thermosetting system, exerts effects on the rheology which are generally at variance with the requirements for easy application. This effect arises from the interactions between the extended, solvated polymer chains. An appropriate choice of solvent mixture to control the rheology of the composition during the various stages of application can minimize these effects and result in a higher solids level at application. This can be achieved by the use of a suitable blend of solvents, which is believed to lessen the interaction of polymer chains either by reducing their degree of solvation or by producing chain coiling. Since most of the polymers used for surface coating applications are generally insoluble in aliphatic hydrocarbons, it is common practice to add aliphatic hydrocarbons to achieve this end. However, there are limits to the improvements which can be obtained by these methods and the relatively low solids content of solution acrylic lacquers in particular leads to a large evolution of solvent on drying. For this reason, the formation of coatings by solvent evaporation is now being increasingly criticized both on grounds of pollution and of cost.

Coatings based on aqueous dispersions of polymers, produced by aqueous emulsion polymerization, have found a wide application in decorative markets as flat or semi-gloss finishes, although no really glossy coatings with good resistance properties have been developed so far. However, stoving compositions based on polymer dispersions in organic liquids give good gloss and resistance properties when cross-linked. The coatings produced have ranged from systems where nearly all the polymer is in dispersion to coatings in which most of the polymer is in solution. They also include systems where the polymer in dispersion is swollen to varying degrees by the solvent blend and systems where the disperse polymer particles are virtually unswollen by the solvent or diluent blend. Such systems, of course, have widely varying properties of application and performance. It is, therefore, an important additional formulating parameter that stable polymer particle dispersions are available for use in surface coatings in conjunction with solution polymers.

Surface coatings based solely on disperse polymers produce films which contain voids left between the adjacent spherical polymer particles. Some auxiliary process during application is therefore required to fill the voids or to soften the polymer in the particle so that the particles can distort and pack together more efficiently. The film-forming process also involves the displacement of the graft dispersant attached to the polymer particle to allow direct interaction between the polymer particles to take place (Section 6.3). It is important to realize however, that coatings formed from polymer dispersions do not undergo complete coalescence as do solution polymers but largely retain the identity of the original polymer particles.

Disperse systems can be prepared which have a very high polymer content at application by virtue of the fact that the dispersed polymer exerts little or no effect on the rheological properties of the paint composition during spray application. By using a blend of different particle sizes, including pigment, very high packing densities can be achieved. Very sharp increases in the apparent viscosity can also be attained after application to the substrate by the dispersion reaching critical packing conditions during evaporation of the diluent phase. The presence of a relatively small amount of polymer in solution implies also that the solvent/diluent can readily evaporate. It is possible therefore, in this way, to form coatings at a very high rate. This factor is particularly important in applications such as coating glass fibres or in other continuous plant operations.

### 7.3.2. Void-filling formulations

Discrete acrylic polymer particle dispersions of 0·1–0·2 µm in size in aliphatic hydrocarbons at 40–50% (w/w) solids when dried out in a thin film leave close-packed spheres touching one another. However, unless the polymer composition is such that the glass transition temperature ($T_g$) is low, no coalescence will occur unless the film is heated to well above its $T_g$. If no additional 'filling' material is present between the polymer spheres in the film before the particles are melt-flowed together by heat, discontinuities appear in the coating (commonly referred to as 'mud-cracking') which is caused by the differential strains set up in the film by the drying process.

In order to avoid this effect and to achieve coalescence of the polymer particles more readily and produce a more coherent film, formulations were developed which contained blends of various aliphatic hydrocarbons with both high and low boiling points. During evaporation, the high boiling hydrocarbon is retained to fill the voids between the particles until finally driven off by applied heat. Plasticizers were also selected so that there was no interaction with the polymer particles until the diluent had evaporated and heat was applied. A combination of these methods thus prevented the 'mud-cracking' effect during the rapid evaporation (or 'flash-off') of the low boiling diluent and kept the film intact until the heat allowed the plasticizer to diffuse into the polymer particles. This reduced the $T_g$ of the polymer and allowed the polymer particles to flow together sufficiently to form shapes which interlocked with each other and to displace the graft dispersant from the adjoining surfaces (see Section 6.3).

The type of plasticizer used depends on the nature of the disperse polymer. Poly(methyl methacrylate) particles generally require a rather weaker plasticizer than the butyl benzyl phthalate usually used. For this reason, butyl sextyl phthalate is used to avoid premature plasticization. Preferably the plasticizer should be soluble in the organic diluent. The subsequent incorporation of pigment into the polymer dispersion produced no major problems since the pigment particles are usually of the same order of size as those of the polymer. Thus, provided that the dispersing agent used for the pigment is compatible with the polymer particles and did not cause flocculation, stable and uniform dispersions were obtained. The final paint film has pigment uniformly embedded within it.

### 7.3.3. Active solvent formulations

It is clearly an advantage in the film formation of disperse polymer systems to have agents present in the dispersion which will temporarily reduce the $T_g$ of the polymer and promote flow after the diluent has evaporated. However, the incorporation of active solvents for the polymer into the formulation can modify the effectiveness of the stabilizing chain of the dispersant due to its lower solubility in the more polar diluent mixture. In extreme cases, the graft dispersant can actually be displaced from the surface of the particles by the use of strong solvents. In either case, the stability of the dispersion is reduced and flocculation usually ensues.

A suitable balance can be achieved by the use of graft dispersants which are covalently bound to the surface of the polymer particle (Section 3.5.3). This effectively prevents the removal of the dispersant whilst the more polar diluent is present but thereafter allows coalescence to take place when heat is applied and the material in the particle is mobile enough to allow the polymer particles to be deformed and the graft dispersant to be displaced. However, if too many covalent links are formed between the graft dispersant and the surface of the polymer particles, the stabilizing barriers are less easily displaced during the film formation process. Formulations can be developed which, providing a correct balance between these factors has been achieved, can give good storage stability and effective film formation on heating. These formulations include both thermoplastic and thermosetting polymer systems into which metallic flake pigments can be successfully incorporated to give metallic finishes of the required properties.

### 7.3.4. Combined dispersion and solution systems

Thermosetting, surface coating compositions have been prepared containing a major proportion of dispersed polymer with reactive groups either attached to the surface of the dispersed polymer particles or as a component of the soluble chain of the graft dispersant. The composition was cured by reaction of complementary reactive groups incorporated into a soluble polymer or oligomer present in the continuous phase.

For example, acrylic copolymer dispersions having a hydroxyl group attached to the soluble chain of the dispersant which react with melamine/formaldehyde resin in solution, have been used as thermosetting, film-forming compositions. A dispersion of a cross-linked polyester with hydroxyl groups present on the surface of the particles has also been used as the reactive dispersed component. In this type of system during film formation, the soluble polymer present fills the voids between the polymer particles and co-reaction between them takes place on curing. A suitable choice of the particle size of the disperse polymer in relation to the size of pigment used, can lead to the formation of a close-packed structure which requires minimal amounts of soluble polymer to fill the voids.

Very satisfactory thermosetting systems have also been developed in which a major portion of the polymer used is in solution (or apparently in solution) and only a minor portion of the polymer (20–25%) is in the disperse state. Such compositions have been used as the basis of two-coat paint systems for motor cars which have a considerable latitude for variable conditions of application. The use of a carefully selected blend of solvents, diluents and polymers can give a controlled rate of evaporation and a corresponding viscosity increase during application. This prevents sagging of the coating and orientates the metallic flake within it to give the required property of sharp reflection changes with respect to its angle of inspection ('flip-tone').

Undercoats with special impact resistance for automotive coatings have been formulated from dispersions of cross-linked polyester particles (at 50% solids) with a melamine/formaldehyde resin in solution in the continuous phase. In this type of system, the polyester particles containing reactive groups behave as macromolecules of very high functionality which readily react with a melamine/formaldehyde resin present in the continuous phase. The resultant film consists of a matrix of hard, cross-linked melamine/formaldehyde with elastic cross-linked polyester particles dispersed within it. Such systems have a high impact resistance.

## 7.4. GENERAL SURFACE COATING APPLICATIONS

### 7.4.1. Adhesives

Both pressure-sensitive and thermosetting types of adhesive materials can be based on polymer dispersions in organic liquids. Dispersions of poly(vinyl acetate) prepared in aliphatic hydrocarbons at 50% solids can be converted to corresponding dispersions of poly(vinyl alcohol) by hydrolysis with sodium in methanol. This material can be used as an adhesive[10].

Adhesive compositions for flock coatings, based on polymer dispersions containing a dissolved alkylated aminoplast have been described[11]. Adhesive compositions have also been claimed, based on dispersions of copolymers of methyl methacrylate with an amino-hydroxyalkyl methacrylate[12].

Dispersions of poly(chloroprene) can be prepared by ionic polymerization in aliphatic hydrocarbons[13]. The dispersions are claimed as good adhesives for textiles, wood, paper etc. and have the advantage of a low viscosity for spraying applications compared with the same polymer applied from solution which shows 'webbing' or 'stringing' effects.

Polyamic acid adhesives have been prepared by the dispersion polymerization of benzophenone tetracarboxylic acid anhydride and 4:4-methylene dianiline in tetrahydrofuran at 40% solids. Fibre-glass tape coated with the dispersion gave a dry polyamic acid content of 71% after heating for 1 h at 28 °C and 5 min at 135 °C. The tapes were placed between two steel strips and heated at 260 °C under pressure (50 lbf/in$^2$) during which a good flow of the polyamic acid was noted. Subsequently a force of 3000 lbf/in$^2$ was required to separate the steel strips[14,15].

### 7.4.2. Polishes

Polymers or copolymers, with a $T_g$ just above ambient temperature and dispersed in organic liquids, have been used for the formulation of polishes. Copolymers of ethyl acrylate and methyl methacrylate as dispersions in an aliphatic hydrocarbon will form a powdery coating when first applied to a substrate as the diluent evaporates but by rubbing with a cloth, sufficient heat can be generated for coalescence to take place. This process can be assisted by the presence of some waxes. A typical composition[16] was based on a 1:1 methyl methacrylate/ethyl acrylate copolymer dispersion, heated with Carnauba wax until it had dissolved. On cooling, the paste-like material contained fine polymer particles dispersed within the wax/diluent mixture and gave a glossy protective coating to furniture after a light buffing.

### 7.4.3. Textile impregnants

A cross-linkable polymer dispersion in aliphatic hydrocarbon (boiling range 80–140 °C) was prepared by dispersion polymerization from a mixture of butanediol monomethacrylate, ethyl acrylate, methyl acrylate, and acrylonitrile in the presence of a graft dispersing agent based on 2-ethylhexyl acrylate, butanediol monomethacrylate and toluene di-isocyanate[17]. A nylon-6 cloth was treated with the dispersion and subsequently heated at 90 °C for 5 min. It is reported that after the treatment, the cloth was soft to handle and was outstandingly resistant to solvents. The modification of fibrous materials with a stabilized dispersion of an acid polymer in a chlorinated hydrocarbon has also been described recently[18].

### 7.4.4. Encapsulants

Dispersions of either inorganic or organic particles in organic diluents can be coated with a layer of polymer precipitated from the continuous phase. For example, a copolymer of methyl methacrylate and methacrylic acid can be precipitated from solution and onto the particle surfaces by the addition of a non-solvent in the presence of a graft dispersing agent. This polymer 'strike layer' is then grown on by a dispersion polymerization process in which the coated particles take the place of the normal polymer 'seed' (see Section 5.6.4).

In this way, a variety of organic and inorganic pigments, including red iron oxide, thio indigo red, vegetable carbon black and rutile titanium dioxide have been encapsulated by polymer as stable particulate dispersions in organic liquids[19]. A uniform common surface is thus imparted to materials of widely different surface characteristics and which helps to prevent many of the pigmentation difficulties (e.g. 'flooding' or 'floating') which originate from these differences. The use of pigments encapsulated in this way, results in surface coating materials which have better gloss and weather-resistance properties than those prepared with untreated pigments.

Dispersions of condensation polymers, e.g. poly(ethylene terephthalate), nylon-6, nylon-6:6 and nylon-11, can be pigmented during the dispersion polymerization process. A dispersion of the required pigment is prepared in the reactants, preparatory to the emulsification and polymerization stages. Using this method, titanium dioxide, iron oxide and copper phthalocyanine pigments have been incorporated within the dispersed condensation polymers[20].

### 7.4.5. Fibre coatings

Fibrous materials, both organic and inorganic, in either the form of a monofilament or as strand or roving, can be coated readily by dispersions of suitable polymers in organic liquids. The coating can be applied from a bath or by a suitable spraying process. The mobility of the polymer dispersions allows uniform and efficient penetration between the filaments of the fibre and voids are readily filled. The subsequent drying operation can be achieved by the continuous passage of the fibre through heated zones. The diluent is first removed by vaporization at an appropriate temperature and residence time; this process is assisted by the use of a diluent with a low latent heat of evaporation. Passage through a second heated zone at an appropriate temperature completes particle integration or promotes cross-linking processes. A range of polyamides, polyesters, polyacrylates etc. dispersed in aliphatic hydrocarbons has been used for this purpose[21].

The fibres coated in this manner, are particularly useful in composites where the fibrous reinforcement needs to be protected from abrasion and attack by chemicals. The coating also forms a useful interlayer between the matrix of the composite and the fibre. The polymeric interlayer improves the adhesion between the two phases and also provides a buffer-layer to reduce crack propagation to the fibre.

Most of the work reported has been concerned, so far, with inorganic fibres which have previously been 'sized' or coated with conventional aqueous emulsion polymers and similar adhesion promoters. Virgin glass fibre in both the mono- and multi-filamentary form can be coated with dispersion polymer immediately after its manufacture to prevent the attack of water or inorganic ions on the glass as occurs with conventional agents[21]. Further, the application of a ductile polymer layer to the surface of the glass fibre before any surface cracks have had time to develop allows the fibre to retain a large proportion of its initial strength.

## 7.5. MISCELLANEOUS APPLICATIONS

### 7.5.1. Thermoplastic rubbers

The preparation of AB and ABC block copolymers by the anionic dispersion polymerization of such monomers as styrene, p-t-butyl styrene, isoprene, butadiene, acrylonitrile etc. described previously (Section 5.4.1) can give particles of polymer with thermoplastic outer-layers and elastic or cross-linked inner-layers. These can be moulded in powdered form to produce hard particles embedded in a matrix of an elastic polymer or elastic polymer particles contained in a hard, thermoformable matrix[22].

The preparation of a typical example, an ABC block copolymer of poly(t-butyl styrene), poly(isoprene) and poly(styrene), has been described in detail elsewhere (Section 5.7.6). The resulting powder was moulded for 5 min at 150 °C under a pressure of 2500 lbf/in$^2$ to form a $\frac{1}{8}$-in thick translucent sheet having a notched impact strength of 23 kg/cm$^2$.

### 7.5.2. Coloured polymer particles

Dispersions of polymer particles can be prepared by the free radical initiated dispersion polymerization of monomers in aliphatic hydrocarbon in the presence of a copolymer which is largely soluble in the diluent but also contains polar groups which will absorb onto the surface of the polymer particles as they are formed. By including a coloured monomer in the feed, that is an unsaturated group (e.g. an acrylate) attached to a suitable organic chromophore, coloured polymer particles can be produced[23]. Cross-linkable monomers, such as butoxymethyl acrylamide, have also been used to produce coloured cross-linked polymer particles in a similar manner. The dispersions of coloured polymer particles in organic liquids have been used to replace aqueous dispersions used in surface finishes, where the substrates are particularly water-sensitive or water-repellant.

Dispersions of coloured addition polymer particles can also be prepared by the incorporation of organic dyestuffs which are soluble in the monomer used but insoluble in the organic liquid diluent. The dyestuff is dissolved in the monomer feed and is incorporated into the polymer particle as it grows during dispersion polymerization[23]. It is also possible to dissolve certain stable dyestuffs into a monomer melt[20] used in the condensation polymerization in dispersion described previously (Section 5.5.2). Molten bishydroxyethyl terephthalate can dissolve certain dyestuffs or alternatively pigments can be dispersed within it[24].

Emulsification of the appropriate reactants in aliphatic hydrocarbon, followed by condensation polymerization, can yield fine particle dispersions of coloured polyester particles[20]. With oligomers or polymers of low molecular weight, coloured particles can be produced by melting the polymer with the pigment or dyestuff, emulsifying the mixture and then cooling to obtain the coloured polymer dispersion[24]. The coloured particles produced can be used directly as dispersions or the polymer recovered by centrifugation or spray-drying to obtain coloured particulate polymer powders. If the particle size is coarse (2 µm or more) with no fines present, it is then possible to recover the polymer particles directly by filtration.

Examples of coloured polymer powders produced by these methods are shown in the Frontispiece. Such materials can be used as cosmetics, powder coatings and for powder impregnation processes or as pigments in their own right if the particle size is less than 0·5 µm[20,24].

### 7.5.3. Lubricants

Dispersions of poly(tetrafluoro ethylene-co-hexafluoro propylene) in water have been transferred into an organic diluent by azeotropic distillation. The organic diluents used have been fluorocarbons, carbon tetrachloride, aromatic hydrocarbons, and ketones[25]. The final dispersions (at 30–35% solids) had particle sizes in the range 0·01–3 µm and were applied as coatings for metal panels. After this treatment and drying, the panels exhibited a low coefficient of friction. Other resins, such as oil-modified alkyds, polyurethanes, poly-acrylates, epoxy resins, nitrocellulose, aminoplasts and polyamides may also be blended with the dispersions to give highly adherent coatings again with low coefficients of friction.

The preparation of dispersions of vinylidene chloride copolymers by dispersion polymerization, using alkyl/allyl acrylate copolymers as dispersants, to produce spherical particles for improving the slip properties of film bases has also been described[26].

## 7.6. REFERENCES

1. Rohm and Hass, *British Patent*, 956,454 (1964).
2. Firestone Tire and Rubber, *British Patent*, 1,007,476 (1965).
3. Firestone Tire and Rubber, *British Patent*, 1,008,188 (1965).
4. Levinson, S. B., *J. Paint Technol.*, **44**, No. 570, 37 (1972).
5. Levinson, S. B., *J. Paint Technol.*, **44**, No. 571, 27 (1972).
6. Levy, A. and Miller, S. E., *Final Technical Report on the Role of Solvents in Photochemical-smog Formation*, National Paint, Varnish and Lacquer Association, Inc., Washington, 1970.
7. Imperial Chemical Industries, *British Patent*, 1,104,403 (1968).
8. du Pont, *United States Patent*, 3,505,256 (1970).
9. du Pont, *United States Patent*, 3,060,148 (1962).
10. Union Carbide, *British Patent*, 1.199,651 (1970).
11. Rohm and Haas, *United States Patent*, 3,533,826 (1970).
12. du Pont, *United States Patent*, 3,547,766 (1970).
13. du Pont, *British Patent*, 1,272,890 (1972).
14. du Pont, *British Patent*, 1,147,515 (1969).
15. du Pont, *United States Patent*, 3,441,532 (1969).
16. Rohm and Haas, *British Patent*, 993,794 (1965).
17. Badische Anilin- & Soda-Fabrik, *British Patent*, 1,203,427 (1970).
18. Rohm and Haas, *British Patent*, 1,268,132 (1972).
19. Imperial Chemical Industries, *British Patent*, 1.156,653 (1969).
20. Imperial Chemical Industries, *British Patent Application*, 28838 (1973).
21. Imperial Chemical Industries, *German Patent*, 2,139,315 (1972).
22. Imperial Chemical Industries, *South African Patent*, 72/7635 (1973).
23. Badische Anilin- & Soda-Fabrik, *British Patent*, 1,200,216 (1970).
24. Imperial Chemical Industries, *German Patent*, 2,152,515 (1971).
25. du Pont, *British Patent*, 1,064,840 (1967).
26. du Pont, *United States Patent*, T904,001 (1972).

# APPENDIX

# *Patents on Polymers Dispersed in Organic Media*

K. E. J. BARRETT

A.1. ABSTRACTS OF PATENTS ON POLYMERS DISPERSED IN ORGANIC LIQUIDS IN ORDER OF APPLICATION DATE.............................................. 291
A.2. PATENTS ON POLYMERS DISPERSED IN ORGANIC MEDIA LISTED ACCORDING TO NATIONALITY........................................................ 306
A.3. GROWTH IN THE NUMBER OF PATENT APPLICATIONS ON DISPERSION POLYMERS ISSUED TO PRINCIPAL PATENTEES....................................... 308

Abstracts of patents on polymers dispersed in organic media are listed in order of application date under Section A.1. The country of origin is included where the nationality of the patent quoted is different from that of the patentee. The publication date of the patent is also given.

The patents are also listed according to nationality in order of patent number in Section A.2.

The growth in the number of patent applications issued is summarized in Section A.3 and illustrated in Figure A.1.

## A.1. ABSTRACTS OF PATENTS ON POLYMERS DISPERSED IN ORGANIC LIQUIDS IN ORDER OF APPLICATION DATE

1. *British* 385,970: British Thomson-Houston, 23.4.1930, *USA*; 22.12.1932. Dispersions of resin condensates (alkyds, phenol- or urea-formaldehydes) stabilized in organic media with dissolved rubber.
2. *British* 434,783: I.G.Farbenindustrie, 1.12.1933 (*Germany* 11.12.1932); 11.6.1934. Dispersion polymerization in the presence of high molecular weight emulsifying agents such as rubber.
3. *British* 797,346: British Rubber Producers' Research Association, 9.8.1954; 2.7.1958. Preparation of polymer dispersions by dissolution of graft copolymers (e.g. methyl methacrylate–rubber) in a common solvent and subsequently adding a non-solvent for one component of the graft.
4. *British* 893,429: Imperial Chemical Industries, 27.5.1957; 11.4.1962. Rubber-stabilized acrylic polymer dispersions.
5. *British* 934,038: Rohm & Haas, 4.9.1958, *USA*; 14.8.1963. Polymer dispersions stabilized with amphipathic graft copolymers.
6. *British* 941,305: Imperial Chemical Industries, 5.11.1958 *et seq.*; 6.11.1963. Polymer dispersions stabilized with block and graft copolymers.

7. *British* 956,453: Rohm & Haas, 15.5.1959, *USA*; 29.4.1964. Polymer dispersions stabilized with a copolymer of an alkyl methacrylate and a nitrogen-containing monomer.
8. *British* 956,454: Rohm & Haas, 15.5.1959, *USA*; 29.4.1964. Dispersion polymerization using a copolymer of an alkyl methacrylate and a nitrogen-containing monomer as stabilizer and with a mercaptan as chain transfer agent.
9. *British* 958,023: Imperial Chemical Industries, 31.7.1959; 13.5.1964. Copolymer dispersions with polar components in outer layer of particle.
10. *British* 971,885: Imperial Chemical Industries, 18.12.1959; 7.10.1964. Addition of cross-linking agents or plasticizers during preparation of polymer dispersions.
11. *British* 978,484: Imperial Chemical Industries, 7.4.1960; 23.12.1964. Addition of polysiloxanes to coating compositions based on polymer dispersions.
12. *British* 980,633: Imperial Chemical Industries, 4.5.1960; 13.1.1965. Incorporation of reactive plasticizers into surface coating compositions based on polymer dispersions.
13. *British* 990,154: Imperial Chemical Industries, 4.5.1960; 28.4.1965. Formation of rubber-graft copolymer stabilizer in dispersion polymerization controlled by free radical scavenger.
14. *British* 994,867: Imperial Chemical Industries, 4.5.1960; 10.6.1965. Termination of seed stage by free radical scavenger in preparation of rubber-stabilized acrylic polymer dispersions.
15. *United States* 3,218,302: Rohm & Haas, 8.6.1960; 16.11.1965. Suspension polymerization with a copolymer of an alkyl methacrylate and a nitrogen-containing monomer as granulating agent.
16. *British* 987,751: Imperial Chemical Industries, 25.7.1960; 31.3.1965. Random dispersion copolymerization of acrylates with maleic anhydride.
17. *British* 992,635: Rohm & Haas, 1.9.1960, *USA*; 19.5.1965. Dispersion polymerization using as stabilizer a graft copolymer with an oxidatively degraded rubber as the soluble component.
18. *British* 1,011,482: Imperial Chemical Industries, 7.12.1960; 1.12.1965. Control of formation of graft copolymer as stabilizer by successive use of peroxide and azo initiators.
19. *British* 992,636: Rohm & Haas, 13.2.1961, *USA*; 19.5.1965. Dispersion polymerization with a fluoro- or chloro-substituted aliphatic hydrocarbon as the continuous phase.
20. *British* 992,637: Rohm & Haas, 13.2.1961, *USA*; 19.5.1965. Dispersion polymerization in a medium reactive towards the dispersed polymer (e.g. acrylate and air-drying alkyd).
21. *British* 993,794: Rohm & Haas, 13.2.1961, *USA*; 2.6.1965. Compositions based on mixtures of polymer dispersions with wax as polishes or sizes.

22. *British* 992,638: Rohm & Haas, 23.2.1961, *USA*; 19.5.1965. Surface coating compositions based on polymer dispersions mixed with coal tar or coumarone-indene resins.
23. *British* 1,017,931: Imperial Chemical Industries, 12.4.1961; 26.1.1966. Dispersion polymerization using low molecular weight degraded rubber as a stabilizer precursor.
24. *British* 930,919: Imperial Chemical Industries, 10.5.1961; 10.7.1963. Addition of solvent to coating compositions based on polymer dispersions to aid application.
25. *British* 1,019,361: Imperial Chemical Industries, 10.5.1961; 2.2.1966. Balance of thinners required for spray-application of paints based on polymer dispersions.
26. *British* 1,007,476: Firestone Tire & Rubber, 30.6.1961 *et seq.*, *USA*; 13.10.1965. Ionically initiated dispersion polymerization with a rubber as dispersant.
27. *British* 1,008,188: Firestone Tire & Rubber, 30.6.1961 *et seq.*, *USA*; 27.10.1965. Ionically initiated dispersion polymerization of vinyl aromatics with a rubber as dispersant.
28. *British* 1,009,004: Rohm & Haas, 7.9.1961, *USA*; 3.11.1965. Polyvinyl ethers as stabilizer precursors in dispersion polymerization.
29. *British* 1,015,393: Rohm & Haas, 28.9.1961, *USA*; 31.12.1965. Preparation of dispersions of condensation polymers with a swellant for the disperse polymer.
30. *British* 1,049,772: Imperial Chemical Industries, 4.10.1961; 30.11.1966. Coating particles by precipitation of block or graft copolymers.
31. *United States* 3,232,903: Rohm & Haas, 19.10.1961; 1.2.1966. Soluble polymers based on long-chain alkyl methacrylates as stabilizer precursors for use in dispersion polymerization.
32. *British* 1,053,791: Imperial Chemical Industries, 12.12.1961; 4.1.1967. Dispersion polymerization using adhesion-promoting comonomers.
33. *British* 1,002,493: Rohm & Haas, 30.1.1962, *USA*; 25.8.1965. Drying oil-modified polyesters as stabilizer precursors in dispersion polymerization.
34. *British* 1,007,723: Rohm & Haas, 1.5.1962, *USA*; 22.10.1965. Drying oil-modified polyesters or their copolymers with acrylic esters as stabilizers in dispersion polymerization.
35. *British* 1,008,001: Firestone Tire & Rubber, 7.5.1962, *USA*; 22.10.1965. Dispersion polymerization of a lactam using an alkaline catalyst with rubber as a dispersant.
36. *British* 1,052,241: Imperial Chemical Industries, 22.6.1962; 21.12.1966. Dispersion polymerization using a synthetic reactive polymer as a stabilizer precursor or as a preformed graft.
37. *United States* 3,304,279: Mono-Sol, 11.3.1963; 14.2.1967. Monomer, initiator and organic dispersant, which is a swelling agent for the polymer produced, polymerized and ball-milled to produce polymer dispersions.

38. *British* 1,084,054: Imperial Chemical Industries, 15.3.1963; 20.9.1967. Coating compositions based on polymer dispersions suitable for brush application.
39. *British* 1,095,931: Imperial Chemical Industries, 16.5.1963; 20.12.1967. Preparation of dispersions of polyesters, polyamides and polyethers etc. using a stabilizer precursor in the dispersion polymerization.
40. *British* 1,095,932: Imperial Chemical Industries, 16.5.1963; 20.12.1967. Preparation of dispersions of polyesters, polyamides and polyethers etc. using preformed block or graft copolymers as stabilizer in the dispersion polymerization.
41. *British* 1,095,288: Imperial Chemical Industries, 1.8.1963; 13.12.1967. Thermosetting coating compositions based on polymer dispersions containing cross-linking groups in different phases.
42. *British* 1,096,912: Imperial Chemical Industries, 6.8.1963; 29.12.1967. Stabilizer precursors or preformed grafts based on monofunctional polymers with a terminal reactive group prepared with matched initiator and transfer agent.
43. *British* 1,114,142: Imperial Chemical Industries, 25.11.1963; 15.5.1968. Coating compositions based on polymer dispersions using a mixture of high and low-boiling liquids.
44. *United States* 3,218,287: Rohm & Haas, 24.2.1964; 16.11.1965. Surface coating compositions based on mixtures of polyesters with acrylic polymer dispersions.
45. *British* 1,104,403: Imperial Chemical Industries, 13.3.1964; 28.2.1968. Preparation of polymer dispersions and their conversion to solutions.
46. *British* 1,093,081: Kao Soap, 3.4.1964, *Japan*; 29.11.1967. Dispersing agent for resins and pigments based on an octyl half-amide of a di-isobutylene-maleic acid copolymer and related structures.
47. *French* 1,438,003: Metallgesellschaft, 26.6.1964, *Germany*; 6.5.1966. Dispersion of finely-divided polymers into a butadiene–styrene copolymer or its solution using epoxy resins, polyisobutylene etc. as dispersing agents.
48. *British* 1,101,983: du Pont, 16.7.1964 *et seq.*, *USA*; 7.2.1968. Preparation of graft copolymers using hydrogen abstracting initiators with a polymeric backbone for use as stabilizer in dispersion polymerization.
49. *British* 1,122,397: Imperial Chemical Industries, 4.8.1964; 7.8.1968. Polymer dispersions stabilized with amphipathic copolymers of a 'comb'-like structure.
50. *British* 1,123,611: Imperial Chemical Industries, 4.8.1964; 14.8.1968. Polymer dispersions stabilized with amphipathic copolymers containing low molecular weight soluble chains.
51. *Belgian* 669,261: Imperial Chemical Industries, 7.9.1964, *UK*; 7.3.1966. Dispersion polymerization using ionic initiation.
52. *British* 1,095,728: du Pont, 16.10.1964 *et seq.*, *USA*; 21.12.1967. Organosol compositions based on polyfluorocarbons prepared with polymers which undergo hydrogen abstraction in the presence of free radicals.

53. *British* 1,049,088: Pennsalt Chemicals, 23.10.1964, *USA*; 23.11.1966. Polyvinylidene fluoride dispersions prepared by mechanical agitation in the presence of an acrylic polymer.
54. *British* 1,064,840: du Pont, 23.11.1964, *USA*; 12.4.1967. Tetrafluoroethylene-hexafluoropropylene copolymer dispersions.
55. *British* 1,101,984: du Pont, 29.1.1965, *USA*; 7.2.1968. Dispersion polymerization using an amphipathic graft copolymer, one component of which is swollen by the main monomer.
56. *United States* 3,255,135: Rohm & Haas, 5.2.1965; 7.6.1966. Dispersion polymerization of acrylic monomers in the presence of a solution of oxidized vegetable or animal oils.
57. *British* 1,143,404: Imperial Chemical Industries, 12.2.1965; 19.2.1969. Polymer dispersions with stabilizer anchored to surface of particle by strong specific interactions.
58. *United States* 3,419,515: Rohm & Haas, 17.3.1965; 31.12.1968. Use of a stabilizer precursor in dispersion polymerization.
59. *British* 1,156,012: Imperial Chemical Industries, 27.5.1965; 25.6.1969. Thermosetting compositions based on polymer dispersions containing cross-linking groups.
60. *British* 1,156,652: Imperial Chemical Industries, 10.6.1965; 2.7.1969. Stabilization of particles in organic liquids by coating with polymer in the presence of a stabilizing graft copolymer.
61. *British* 1,156,653: Imperial Chemical Industries, 10.6.1965; 2.7.1969. Encapsulation of particles by dispersion polymerization in the presence of a stabilizer for the polymer dispersion using a 'strike-layer' technique.
62. *French* 1,486,670: Morningstar–Paisley, 23.7.1965, *USA*; 30.6.1967. Thermosetting coating compositions based on a polyvinyl chloride plastisol mixed with epoxy resins.
63. *British* 1,157,630: Imperial Chemical Industries, 29.7.1965; 9.7.1969. Dispersion polymerization to yield very high solids polymer dispersions containing particles of at least two different sizes.
64. *British* 1,161,402: Imperial Chemical Industries, 29.7.1965; 13.8.1969. Partially coalesced film formed from a polymer dispersion and coalescence completed in a subsequent stage.
65. *British* 1,157,615: Imperial Chemical Industries, 4.8.1965; 9.7.1969. Dispersion polymerization to form a polymer of a swelling factor 20–25% from a polymer seed of a swelling factor 15%.
66. *British* 1,151,518: Rohm & Haas, 22.10.1965, *USA*; 7.5.1969. Dispersions of polyesters prepared by dispersion of reactants in a non-solvent in the presence of a swellant for the disperse polymer and by-products of condensation removed as reaction proceeds.
67. *British* 1,147,515: du Pont, 6.12.1965 *et. seq.*, *USA*; 2.4.1969. Preparation of organosols of polyamide acids.
68. *United States* 3,441,532: du Pont, 14.1.1966; 29.4.1969. Dispersions of polyamic acids as heat-sealing adhesives and curable coatings.

69. *British* 1,141,225: Pennsalt Chemicals, 17.1.1966, *USA*; 29.1.1969. Polyvinylidene fluoride dispersions prepared with a quarternary ammonium salt as dispersing agent.
70. *British* 1,174,391: Imperial Chemical Industries, 31.1.1966; 17.12.1969. Dispersion polymerization in the presence of a preformed graft copolymer free of ungrafted components.
71. *British* 1,178,233: Imperial Chemical Industries, 18.2.1966; 21.1.1970. Dispersion polymerization to produce dispersions of polymer in its own monomer (vinyl chloride, vinylidene chloride, acrylonitrile).
72. *French* 1,580,115: Lubrizol, 17.3.1966 *et. seq.*, *USA*; 5.9.1969. Mixture of polymeric resin with a metal dispersion stabilized with a soluble organic compound with a hydrophobic component and a polar substituent.
73. *British* 1,156,235: PPG Industries, 31.3.1966, *USA*; 25.6.1969. Dispersion polymerization with ethylene–dicyclopentadiene copolymers as graft stabilizer precursors.
74. *British* 1,199,651: Union Carbide, 11.7.1966, *USA*; 22.7.1970. Dispersion polymerization of vinyl acetate with ethylene–vinyl acetate copolymer as stabilizer and subsequent hydrolysis to produce dispersion of polyvinyl alcohol.
75. *French* 1,531,022: Union Carbide, 11.7.1966, *USA*; 28.6.1968. Polyvinyl ethers as stabilizers in dispersion polymerization.
76. *British* 1,198,052: Imperial Chemical Industries, 20.7.1966; 8.7.1970. Stabilizer anchored to polymer particles by strong specific interactions used for redispersion or flushing of polymers into different media.
77. *British* 1,202,207: BALM Paints, 19.9.1966; 12.8.1970. Dispersion polymers containing condensable groups solubilized by reacting with a component containing a complementary group.
78. *British* 1,211,532: Imperial Chemical Industries, 9.11.1966; 11.11.1970. Emulsions stabilized with amphipathic graft copolymers with one component soluble in the disperse liquid and the other soluble in the continuous phase.
79. *British* 1,206,398: Imperial Chemical Industries, 14.11.1966; 23.9.1970. Redispersion of polymer using amphipathic graft copolymer as stabilizer in which the anchor component is flexible under the conditions used.
80. *British* 1,134,997: Cook Paint & Varnish, 15.11.1966, *USA*; 27.11.1968. Thermosetting compositions based on acrylic polymer dispersions containing cross-linkable groups (hydroxyl, carboxyl) with a solution of a melamine–formaldehyde condensate.
81. *British* 1,211,344: BALM Paints, 25.11.1966; 4.11.1970. Amphipathic copolymers as dispersants for the preparation of polymer or pigment dispersions in both water and in non-polar organic liquids.
82. *United States* 3,547,766: du Pont, 25.11.1966; 15.12.1970. Adhesive composition based on dispersions of copolymers of methyl methacrylate with an amino-hydroxyalkyl methacrylate.

83. *British* 1,200,216: Badische Anilin- & Soda-Fabrik, 26.11.1966, *Germany*; 29.7.1970. Dispersions of coloured, cross-linkable copolymers based on unsaturated pigments reacted with monomers containing $N$-methylol or $N$-alkoxymethyl groups.
84. *British* 1,203,427: Badische Anilin- & Soda-Fabrik, 3.12.1966, *Germany*; 26.8.1970. Acrylic polymer dispersions cross-linked with isocyanates for coating and impregnating textiles.
85. *British* 1,165,840: Hercules, 30.12.1966 *et. seq.*, *USA*; 1.10.1969. Dispersion polymerization of propylene or ethylene-propylene with Ziegler–Natta catalyst in the presence of polyoctene-1.
86. *Belgian* 712,716: Badische Anilin- & Soda-Fabrik, 25.3.1967, *Germany*; 25.9.1968. Dispersion polymerization of acrylic monomers in the presence of polyvinyl ethers as stabilizers.
87. *British* 1,181,316: du Pont, 20.4.1967, *USA*; 11.2.1970. Coating compositions based on polymer dispersions, metallic pigments and a coalescing agent.
88. *British* 1,210,161: du Pont, 23.5.1967, *USA*; 28.10.1970. Coating compositions based on dispersions of graft copolymers with two different types of side chains and prepared from polymer backbone containing active grafting sites.
89. *British* 1,220,695: du Pont, 2.6.1967, *USA*; 27.1.1971. Blending dispersions of two different types of graft copolymers capable of cross-linking reactions.
90. *British* 1,242,054: Imperial Chemical Industries, 8.6.1967; 11.8.1971. Coating compositions containing dispersed rubbery particles (microgels) prepared by dispersion polymerization.
91. *British* 1,234,395: Imperial Chemical Industries, 30.6.1967; 3.6.1971. Continuous dispersion polymerization in an unstirred reactor.
92. *British* 1,231,614: BALM Paints, 6.7.1967, *Australia*; 12.5.1971. Dispersion polymerization using reactive stabilizer which forms covalent links with the disperse polymer.
93. *United States* T904,001: du Pont, 12.7.1967; 7.11.1972. Preparation of dispersions of vinylidene chloride copolymers by dispersion polymerization using alkyl/allyl acrylate copolymer as dispersant to produce spherical particles for improving the slip properties of film bases.
94. *United States* 3,533,826: Rohm & Haas, 10.8.1967; 13.10.1970. Adhesive composition for flock coatings based on polymer dispersions containing dissolved alkylated aminoplast with a trialkylamine salt of a strong acid as catalyst.
95. *British* 1,206,442: Blundell–Permoglaze, 11.8.1967; 23.9.1970. Acrylic ester–modified alkyds as stabilizer precursors and alkyd grafts as stabilizers for dispersion polymerization.
96. *British* 1,242,734: du Pont, 22.9.1967, *USA*; 11.8.1971. Graft copolymers with long chain polyvinyl esters (e.g. vinyl $\alpha,\alpha$-dimethyl octanoate) as backbone for use as stabilizers in dispersion polymerization.

97. *British* 1,248,972 : BALM Paints, 27.10.1967, *Australia*; 6.10.1971. Polymer dispersions containing unsaturated groups capable of cross-linking by autoxidation.
98. *British* 1,253,387: BALM Paints, 27.10.1967, *Australia*; 10.11.1971. Preparation of cross-linked rubbery particles (microgels) by dispersion polymerization.
99. *British* 1,186,819: du Pont, 16.11.1967, *USA*; 8.4.1970. Dispersions of preformed polymer particles with soluble polymers attached through functional groups.
100. *United States* 3,551,525: Badische Anilin- & Soda-Fabrik, 24.11.1967, *Germany*; 29.12.1970. Thermosetting surface coating compositions based on dispersed acrylic copolymers containing etherified $N$-methylolamides.
101. *Japanese* 70/34396: Toa Synthetic Chemicals, 27.11.1967; 5.11.1970. Dispersion copolymerization of acrylonitrile–acrylic esters in the presence of oil-soluble polymers to produce dispersions used for preparation of frosted varnishes.
102. *British* 1,240,317: Badische Anilin- & Soda-Fabrik, 6.12.1967, *Germany*; 21.7.1971. Polymer dispersions containing hydroxyl groups mixed with polyisocyanates to form polyurethane coatings.
103. *British* 1,240,320: Badische Anilin- & Soda-Fabrik, 13.12.1967, *Germany*; 21.7.1971. Thermosetting coating compositions based on dispersions of copolymers containing etherified $N$-methylolamide groups
104. *British* 1,220,777: Gulf Oil of Canada, 5.1.1968, *Canada*; 27.1.1971. Continuous dispersion polymerization process, especially suitable for aqueous emulsion polymerization but also applicable to non-aqueous dispersion polymerization (bulk vinyl chloride or vinyl chloride in benzene–methanol).
105. *British* 1,212,165; Union Carbide, 7.2.1968, *USA*; 11.11.1970. Dispersion polymerization of vinyl acetate–acrylic esters to form stable dispersions of copolymers.
106. *British* 1,259,516: BALM Paints, 9.2.1968, *Australia*; 5.1.1972. Polymer dispersions incorporating swelling agents for the polymer.
107. *British* 1,223,343: du Pont, 7.5.1968, *USA*; 24.2.1971. Dispersion polymerization in the presence of graft copolymers prepared from an insoluble polymer backbone containing comonomer units providing sites for grafts which are soluble in the medium.
108. *British* 1,209,409: Shell International, 29.5.1968, *USA*; 21.10.1970. One-pack thermosetting compositions based on dispersions of epoxide resin and an anhydride or phenol–formaldehyde resin together with a curing catalyst and surfactant.
109. *Canadian* 887,122: du Pont, 21.6.1968, *USA*; 30.11.1971. Polytetrafluoroethylene organosols containing heat-stable auxiliary materials.
110. *British* 1,228,438: du Pont, 19.7.1968, *USA*; 15.4.1971. Preparation of fluorocarbon resin organosols in the presence of auxiliary film-

forming materials (epoxy resins, polyorganosiloxanes, acrylic polymers etc.).

111. *British* 1,275,434: BALM Paints, 7.8.1968, *Australia*; 24.5.1972. Separate incorporation of plasticizers in the continuous and the disperse phase of polymer dispersions.
112. *British* 1,268,692: BALM Paints, 7.8.1968, *Australia*; 29.3.1972. Polymer dispersions in which reactive groups in the stabilizer or disperse polymer are used to modify its properties.
113. *British* 1,243,274: Celanese Coatings, 8.8.1968, *USA*; 18.8.1971. Thermosetting coatings based on dispersed copolymers of acrylonitrile and methyl methacrylate prepared with polybutadiene or polystyrene–butadiene as stabilizer precursor.
114. *British* 1,269,964: BALM Paints, 16.9.1968, *Australia*; 12.4.1972. Dispersion polymerization using a copolymerizable polymer stabilizer in the seed stage.
115. *German* 1,946,688: FMC, 17.9.1968, *USA*; 26.3.1970. Dispersions of polyesters prepared by mechanical comminution of polyesters with amorphous zones hydrolysed.
116. *Japanese* 72/19404: Nippon Carbide Industries, 18.9.1968; 3.6.1972. Stable dispersions of vinyl polymers prepared by solution polymerization of vinyl monomer in the presence of ethylene–vinyl ester copolymers.
117. *British* 1,272,565: Dunlop Holdings, 1.10.1968; 3.5.1972. Dispersions of elastomeric polymers in non-aqueous media which contain a soluble polymer in the presence of a surface active agent.
118. *Japanese* 71/17123: Nippon Carbide Industries, 8.11.1968; 12.5.1971. Stable dispersions of vinyl polymers prepared in absence of stabilizer. After precipitation of polymer, polymerization continued in the presence of a suitable comonomer.
119. *British* 1,276,528: Tokuyama Soda, 25.11.1968, *Japan*; 1.6.1972. Copolymerization of monomers dissolved in an organic medium, in which the resulting polymer is insoluble, in the presence of a dissolved linear polymer (e.g. polymers based on maleic anhydride, styrene, divinyl benzene and atactic polypropylene in benzene) to produce finely divided polymers which are redispersible in organic liquids.
120. *British* 1,307,355: Plastic Coating Research, 10.2.1969; 21.2.1973. Coating compositions based on thermoplastic polymers dispersed in a solution of the same or different polymer, e.g. poly(ethylene) dispersed in a solution of chlorinated poly(ethylene).
121. *Japanese* 73/29551: Toray Industries, 11.3.1969; 11.9.1973. Preparation of dispersions of acrylonitrile copolymers in organic liquids by polymerization in the presence of ethylene–vinyl acetate copolymers.
122. *Canadian* 875,617: PPG Industries, 14.3.1969, *USA*; 13.7.1971. Free hydroxyl groups in hydroxystearic acid–glycidyl methacrylate stabilizer precursor reacted with acetic anhydride to overcome film embrittlement on ageing.

123. *United States* 3,632,669: Union Carbide, 1.4.1969; 4.1.1972. Dispersion polymerization of cyclic esters (ε-caprolactone) in the presence of amphipathic block or graft copolymers.
124. *United States* 3,646,170: Union Carbide, 1.4.1969; 29.2.1972. Dispersion copolymerization of cyclic esters with alkylene oxides in the presence of amphipathic block or graft copolymers.
125. *British* 1,268,132: Rohm & Haas, 3.4.1969, *USA*; 22.3.1972. Modification of fibrous materials with a stabilized dispersion of an acid polymer in a chlorinated hydrocarbon.
126. *United States* 3,586,654: National Distillers & Chemicals, 15.4.1969; 22.6.1971. Control of shape and size-distribution of polymer particles by agitation of their suspension in a non-solvent at a temperature above the softening point of the polymer and in the presence of a non-ionic surfactant.
127. *United States* 3,634,303: Hercules, 2.6.1969; 11.1.1972. Dispersion polymerization of epoxides in the presence of a swollen polyepoxide microgel.
128. *British* 1,294,645: Celanese Coatings, 4.6.1969, *USA*; 1.11.1972. Dispersion polymerization of acrylic monomers in the presence of polylauryl methacrylate as stabilizer and t-butyl peroctoate as initiator.
129. *British* 1,305,715: Imperial Chemical Industries, 4.7.1969; 14.11.1972. Coating compositions based on polymer dispersions using a stabilizer containing reactive groups which cross-link on application and improve adhesion.
130. *Netherlands* 70/10007: Dow Chemical, 9.7.1969, *USA*; 12.1.1971. Dispersions of immiscible polymer solutions stabilized with AB or ABA block copolymers.
131. *British* 1,317,249: Imperial Chemical Industries, 14.8.1969; 16.5.1973. Preparation of dispersions of amphipathic copolymers by dissolution in a common solvent mixture which is subsequently converted to a hemi-solvent by distillation.
132. *British* 1,319,448: Imperial Chemical Industries, 14.8.1969; 6.6.1973. Preparation of dispersions of amphipathic copolymers by heating in a hemi-solvent at a temperature above the glass transition of the insoluble component of the copolymer.
133. *British* 1,319,781: Imperial Chemical Industries, 14.8.1969; 6.6.1973. Coating compositions containing cross-linkable microparticles which are aggregates of amphipathic copolymers.
134. *United States* 3,666,710: PPG Industries, 8.9.1969; 30.5.1972. Acrylic dispersion polymers reacted with imines, especially 2-hydroxyethylethyleneimine, to improve adhesion of surface coatings.
135. *British* 1,272,890: du Pont, 7.10.1969, *USA*; 3.5.1972. Adhesives based on polychloroprene dispersions prepared by dispersion polymerization with chloroprene–2-ethylhexyl methacrylate graft copolymers as stabilizers.

136. *British* 1,331,087: Ceskoslovenska Akademie, 24.10.1969, *Czechoslovakia*; 19.9.1973. Suspension polymerization of glycol mono(methyl)acrylates in higher hydrocarbon media, such as liquid paraffin, in the presence of poly(isobutylene), poly(lauryl methacrylate), etc. as stabilizers.
137. *United States* 3,644,255: du Pont, 9.12.1969; 22.2.1972. Silicone-based block copolymer as an anti-flocculating agent for polymer and pigment dispersions.
138. *United States* 3,645,959: Union Carbide, 5.1.1970; 29.2.1972. Non-aqueous dispersion polymerization of vinyl esters in the presence of up to 1% of a diene to yield high molecular weight polymer.
139. *Japanese* 73/27746: Dai-Nippon, 12.1.1970; 25.8.1973. Non-aqueous dispersion polymerization of monomers in presence of polyesters prepared from poly(butadiene) and aliphatic alcohols.
140. *United States* 3,654,201: Celanese Coatings, 19.1.1970; 4.4.1972. Dispersion polymerization of acrylic monomers using a fatty acid-modified vinyl oxazoline polymer as stabilizer.
141. *German* 2,103,849: du Pont, 28.1.1970, *USA*; 12.8.1971. Self-stabilized organosols based on graft copolymer dispersions as surface coating compositions.
142. *United States* 3,660,537: du Pont, 28.1.1970, *USA*; 2.5.1972. Dispersions of graft copolymers containing allyl methacrylate and diethylaminoethyl methacrylate as surface coating compositions.
143. *United States* 3,676,526: E. G. Sommerfeld, 2.2.1970; 11.7.1972. Dispersions of two different types of graft copolymer in an organic liquid as coating composition for cans. One graft copolymer has a polymeric diene as backbone and a nitrile graft, the other has an unsaturated polymer as backbone with a vinyl graft.
144. *German* 2,008,991: Rohm & Haas, 26.2.1970; 23.9.1971. Graft copolymers prepared from polymers with pendant azo or peroxide groups used as stabilizers in dispersion polymerization.
145. *British* 1,329,062: Rohm GmbH, 27.2.1970, *Germany*; 5.9.1973. Suspension polymerization of aqueous monomer (acrylamide etc.) solutions dispersed in a continuous organic phase in the presence of an amphipathic stabilizer.
146. *United States* 3,701,746: Union Carbide, 10.4.1970; 31.10.1972. Dispersions of polymers in organic liquids, in presence of high-boiling solvents and plasticizers of critical volatility and solvency, as film-forming compositions.
147. *British* 1,311,835: Celanese Coatings, 20.4.1970, *USA*; 28.3.1973. Dispersion polymerization in the presence of cellulose esters and ethers as stabilizer precursors.
148. *British* 1,301,068: du Pont, 20.4 and 25.11.1970, *USA*; 29.12.1972. Preparation of polymer dispersions in organic liquids by treating an aqueous dispersion of polymer with an organic solution of an ion-exchange agent to convert the emulsifier. The aqueous phase is then removed.

149. *United States* 3,686,111 : PPG Industries, 22.6.1970; 22.8.1972. Dispersion polymerization in the presence of an imine and an active solvent for the polymer to produce non-aqueous 'pseudo'-dispersions of polymer particles stable to active or coalescing solvents. 'Pseudo'-dispersions useful as protective coatings and dispersing agents for pigments.
150. *United States* 3,637,569: du Pont, 1.7.1970; 25.1.1972. Conversion of aqueous dispersions of polyfluorocarbons to non-aqueous dispersions by azeotropic removal of water.
151. *United States*, 3,661,831: du Pont, 1.7.1970; 9.5.1972. Preparation of fluorocarbon copolymer dispersions in organic liquids by direct dispersion of powdered polymer or by azeotropic distillation of aqueous dispersions mixed with organic liquids.
152. *South African* 71/4337: du Pont, 8.7.1970, *USA*; 11.6.1971. Three stage process for preparation of reversible, self-stabilized organosols consists of: (1) preparation of graft copolymer in solution, (2) inversion by addition of non-solvent, (3) post-reaction of residual monomers.
153. *German* 2,139,315 : Imperial Chemical Industries, 5.8.1970, *UK*; 23.3.1972. Coating glass fibres with polymer dispersions (polyamides, etc.).
154. *German* 2,140,135 : Imperial Chemical Industries, 10.8.1970, *UK*; 17.2.1972. Heterogeneous polymer particles prepared by dispersion polymerization of a monomer in the presence of another disperse polymer.
155. *Japanese* 72/9036: Dulux Australia, 17.8.1970, *Australia*; 11.5.1972. Coating compositions based on non-aqueous dispersions of self-stabilized, gelled urethane polymers.
156. *United States* 3,742,091 : Phillips Petroleum, 17.8.1970; 26.6.1973. The agglomeration of vinyl polymers is substantially inhibited when the liquid phase polymerization is carried out in the presence of polymeric additives (e.g. hydrogenated butadiene–styrene copolymer) which are soluble in the liquid monomer.
157. *German* 2,142,598: Dow Corning, 25.8.1970, *USA*; 2.3.1972. Dispersions of organo-siloxane copolymers.
158. *United States* 3,786,010: Dai-Nippon, 8.9.1970, *Japan*; 15.1.1974. Coating compositions based on dispersions of block or graft copolymers of oil-modified alkyd resins in organic liquids.
159. *South African* 71/6030: du Pont, 16.9.1970 *et. seq., USA*; 9.8.1971. Composition of graft copolymer dispersions defined by solubility parameter differences between backbone and graft or continuous phase.
160. *United States* 3,711,449 : Rohm & Haas, 18.9.1970; 16.1.1973. Copolymers containing sulpho-alkyl acrylate/methacrylate units prepared in non-aqueous media by bulk, solution or dispersion polymerization. Used as basis for coating compositions with improved adhesion properties.
161. *German* 2,151,782 : Ford Werke, 19.10.1970, *USA*; 20.4.1972. Thermosetting film-forming composition based on a dispersion of an acrylic copolymer containing an epoxide group and an alkylated melamine–formaldehyde resin.

162. *German* 2,151,824: Ford Werke, 19.10.1970, *USA*; 20.4.1972. Dispersion polymerization of monomers, including one containing an epoxide group, in the presence of a melamine–formaldehyde resin and the addition of a solution copolymer containing reactive groups to produce thermosetting film-forming compositions.
163. *German* 2,152,515: Imperial Chemical Industries, 21.10.1970, *UK*; 21.10.1971. Powder coatings prepared by emulsification of coating materials, followed by pigmentation and solidification of the emulsion and finally removal of the liquid (usually non-polar) continuous phase.
164. *British* 1,340,179: du Pont, 25.11.1970, *USA*; 12.12.1973. Polymeric materials and dispersions containing them.
165. *German* 2,158,821: Imperial Chemical Industries, 27.11.1970, *UK*; 29.6.1972. Thermosetting polymer dispersions containing melamine formaldehyde as a stabilizer precursor.
166. *United States* 3,778,403: du Pont, 22.12.1970; 11.12.1973. Organosols of aziridinyl alkyl acrylate or methacrylate graft copolymers.
167. *Japanese* 72/15435: Dai-Nippon, 25.1.1971; 22.8.1972. Thermosetting coating compositions based on a mixture of a solution of an amino resin precondensate with a dispersion of a block or graft copolymer (based on alkyds, vinyl monomers, e.g. acrylonitrile etc.) in an organic liquid.
168. *Japanese* 72/16526: Dai-Nippon, 1.2.1971; 2.9.1972. Thermosetting coating compositions based on a mixture of a solution of a formaldehyde resin condensate with a vinyl copolymer dispersion containing a hydroxyl group in the solvatable component.
169. *Belgian* 768,650: PPG Industries, 1.2.1971, *USA*; 3.11.1971. Thermoplastic acrylic composition for automobiles using a mixture of a solvent and non-solvent diluent (hexane or heptane).
170. *German* 2,205,171: Imperial Chemical Industries, 8.2.1971, *UK*; 24.8.1972. Thermoplastic coating compositions based on polymer dispersions using a dispersion stabilizer anchored by covalent links and a continuous phase containing a high proportion of a solvent for the polymer.
171. *Japanese* 72/18995: Dai-Nippon, 15.2.1971; 19.9.1972. Polymer dispersions based on block or graft copolymers of an unsaturated oil-modified alkyd resin with acrylonitriles and other vinyl monomers.
172. *German* 2,209,875: Ford Werke, 1.3.1971, *USA*; 14.9.1972. Non-aqueous dispersions of thermosetting copolymers based on ethylenic monomers as film-formers.
173. *German* 2,209,889: Ford Werke, 1.3.1971, *USA*; 14.9.1972. Non-aqueous dispersions of thermosetting copolymers based on amides and unsaturated acids as film-formers.
174. *Japanese* 72/34480: Japan Atomic Energy Research Institute, 23.3.1971; 21.11.1972. Preparation of stable dispersions of poly(vinyl esters) etc. in organic liquids by $\gamma$-irradiation polymerization in the presence of ethylene–vinyl acetate copolymers.

175. *German* 2,215,732: Imperial Chemical Industries, 30.3.1971, *UK*; 7.12.1972. Formation of condensation polymer dispersions in non-aqueous media in the presence of a graft copolymer stabilizer by condensation polymerization in dispersion with at least one insoluble reactant in a liquid form.
176. *United States* 3,702,288: du Pont, 31.3.1971; 7.11.1972. Process for finishing metal substrates by electrodepositing a primer composition and applying an acrylic 'self-stable' organosol coating composition.
177. *Japanese* 72/34684: Dai-Nippon, 6.4.1971; 22.11.1972. Polymer dispersions based on block or graft copolymers of a urethane oil (reaction products of di-isocyanates with mono- or di-glycerides of fatty acids) with vinyl monomers.
178. *Japanese* 72/34685: Dai-Nippon, 6.4.1971; 22.11.1972. Polymer dispersions based on block or graft copolymers of epoxy-ester resins or epoxy-modified alkyd resins with vinyl monomers.
179. *United States* 3,775,327: du Pont, 12.4.1971; 27.11.1973. Polyesters with terminal polyhydroxy groups as dispersing agents for solid particles in organic liquids.
180. *Netherlands* 72/6366: Mobil Oil, 12.5.1971, *USA*; 14.11.1972. Preparation of dispersions of poly(styrene) in aliphatic hydrocarbons by anionic polymerization in the presence of AB or ABA block copolymers (based on styrene and t-butyl styrene) as dispersing agents.
181. *German* 2,223,324: Imperial Chemical Industries, 13.5.1971, *UK*; 21.12.1972. The tendency of a liquid hydrocarbon fuel to spatter on impact is reduced by the dissolution of a non-aqueous polymer dispersion which may be obtained by dispersion polymerization in the required organic liquid.
182. *British* 1,336,188: du Pont, 4.6.1971, *USA*; 7.11.1973. Stabilization of polybutadiene graft copolymer dispersions in aliphatic hydrocarbons by addition of dialkyl hydroxylamines.
183. *United States* 3,765,924: du Pont, 29.6.1971; 16.10.1973. Finishing compositions based on stabilized non-aqueous polymer dispersions which undergo only partial coalescence on drying under ambient conditions for abrasive treatment prior to full coalescence.
184. *German* 2,136,585: Rohm GmbH, 22.7.1971; 2.11.1972. Solid polymer particles are produced by (a) dispersing a solution of a polymer in a volatile polar solvent in a non-polar organic medium which is immiscible with the polar solvent and not a solvent for the polymer, and then (b) removing the solvent by distillation in the presence of an amphipathic graft copolymer as dispersing agent.
185. *Belgian* 782,158: Cook Paint and Varnish, 16.8.1971, *USA*; 31.7.1972. Coating compositions based on copolymer dispersions prepared by copolymerizing unsaturated monomers in solution of alkylated amine–aldehydes.
186. *Japanese* 73/42027: Dai-Nippon, 28.9.1971; 19.6.1973. Stable low-temperature curing coating compositions prepared from amino resins

and copolymer dispersions in organic liquids in the presence of alkyd, epoxy or urethane resins modified with fatty acids and containing hydroxyl groups.
187. *United States* 3,745,137: Celanese Coatings, 26.10.1971; 10.7.1973. Preparation of non-aqueous dispersions of hydroxy-containing copolymers by a two-stage process (initially solution followed by dispersion polymerization). Products can be used with aminoplast resins as thermosetting coating compositions.
188. *South African* 72/7635: Imperial Chemical Industries, 1.11.1971, *UK*; 25.5.1973. Block copolymers prepared by anionic dispersion polymerization and containing one component insoluble in the diluent and forming the particle core and one component soluble in the diluent which stabilizes the particles.
189. *United States* 3,689,593: du Pont, 19.11.1971; 5.9.1972. Preparation of chain transfer-linked urethane graft copolymers as self-stabilized organosols.
190. *United States* 3,723,571: du Pont, 19.11.1971; 27.3.1973. Preparation of spherical particulate dispersions (particle diameter 1–5 μm) of vinylidene chloride/acrylic copolymers in aliphatic diluents by dispersion polymerization in the presence of a soluble copolymer of 2-ethylhexyl acrylate and allyl acrylate as stabilizer precursor.
191. *Japanese* 73/60180: Kansai Paint, 30.11.1971; 23.8.1973. Non-aqueous polymer dispersions prepared by polymerizing a mixture of monomers, e.g. methyl methacrylate and methyl acrylate, in the presence of an oil-modified alkyd in an aliphatic hydrocarbon liquid. The graft copolymer formed in the process acts as the dispersion stabilizer.
192. *German* 2,262,463: Imperial Chemical Industries, 20.12.1971, *UK*; 26.7.1973. Thermosetting coating compositions based on polymer dispersions using a stabilizer anchored by covalent links and a continuous phase containing a high proportion of a solvent for the polymer.
193. *German* 2,301,202: PPG Industries, 14.1.1972, *USA*; 2.8.1973. Surface coatings for cataphoretic electrodeposition processes based on non-aqueous dispersions of polymers containing basic nitrogen atoms (e.g. imine-modified polymers). Amphipathic copolymers capable of cross-linking with the dispersed polymer are used as dispersion stabilizers.
194. *French* 2,173,258: Esso Research, 25.2.1972, *USA*; 23.2.1973. Thermoplastic polymer suspensions (ionomer, graft or block copolymer) in involatile liquid for ease of coating or moulding.
195. *German* 2,314,982: Ford-Werke, 3.4.1972, *USA*; 18.10.1973. Non-aqueous dispersions of thermosetting film-forming polymers.
196. *German* 2,319,089: Imperial Chemical Industries, 14.4.1972, *UK*; 31.10.1973. Non-aqueous dispersions of polyamides.
197. *Netherlands* 73/5444: Dow Chemical, 24.4.1972, *USA*; 26.10.1973. Production of non-aqueous dispersions of poly(vinylidene chloride) and

copolymers using as dispersion stabilizer the reaction product of poly-12-hydroxystearic acid with unsaturated cyclic imines.
198. *Netherlands* 73/7182: Imperial Chemical Industries, 24.5.1972, *UK*; 27.11.1973. Non-aqueous polymer dispersions for use in paints and printing inks.
199. *German* 2,333,216: Canadian Industries, 30.6.1972, *UK*; 31.1.1974. Cellulose acetate butyrate in thermosetting NAD metallic finishes.

## A.2. PATENTS ON POLYMERS DISPERSED IN ORGANIC MEDIA LISTED ACCORDING TO NATIONALITY

(Corresponding number in Section A.1 given in parentheses)

1. *Belgian*

| | | | |
|---|---|---|---|
| 669,261 (51); | 712,716 (86); | 768,650 (169); | 782,158 (185) |

2. *British*

| | | | |
|---|---|---|---|
| 385,970 (1); | 434,783 (2); | 797,346 (3); | 893,429 (4); |
| 930,919 (24); | 934,038 (5); | 941,305 (6); | 956,453 (7); |
| 956,454 (8); | 958,023 (9); | 971,885 (10); | 978,484 (11); |
| 980,633 (12); | 987,751 (16); | 990,154 (13); | 992,635 (17); |
| 992,636 (19); | 992,637 (20); | 992,638 (22); | 993,794 (21); |
| 994,867 (14); | 1,002,493 (33); | 1,007,476 (26); | 1,007,723 (34); |
| 1,008,001 (35); | 1,008,188 (27); | 1,009,004 (28); | 1,011,482 (18); |
| 1,015,393 (29); | 1,017,931 (23); | 1,019,361 (25); | 1,049,088 (53); |
| 1,049,772 (30); | 1,052,241 (36); | 1,053,791 (32); | 1,064,840 (54); |
| 1,084,054 (38); | 1,093,081 (46); | 1,095,288 (41); | 1,095,728 (52); |
| 1,095,931 (39); | 1,095,932 (40); | 1,096,912 (42); | 1,101,983 (48); |
| 1,101,984 (55); | 1,104,403 (45); | 1,114,142 (43); | 1,122,397 (49); |
| 1,123,611 (50); | 1,134,997 (80); | 1,141,225 (69); | 1,143,404 (57); |
| 1,147,515 (67); | 1,151,518 (66); | 1,156,012 (59); | 1,156,235 (73); |
| 1,156,652 (60); | 1,156,653 (61); | 1,157,615 (65); | 1,157,630 (63); |
| 1,161,402 (64); | 1,165,840 (85); | 1,174,391 (70); | 1,178,233 (71); |
| 1,181,316 (87); | 1,186,819 (99); | 1,198,052 (76); | 1,199,651 (74); |
| 1,200,216 (83); | 1,202,207 (77); | 1,203,427 (84); | 1,206,398 (79); |
| 1,206,442 (95); | 1,209,409 (108); | 1,210,161 (88); | 1,211,344 (81); |
| 1,211,532 (78); | 1,212,165 (105); | 1,220,695 (89); | 1,220,777 (104); |
| 1,223,343 (107); | 1,228,438 (110); | 1,231,614 (92); | 1,234,395 (91); |
| 1,240,317 (102); | 1,240,320 (103); | 1,242,054 (90); | 1,242,734 (96); |
| 1,243,274 (113); | 1,248,972 (97); | 1,253,387 (98); | 1,259,516 (106); |
| 1,268,132 (125); | 1,268,692 (112); | 1,269,964 (114); | 1,272,565 (117); |
| 1,272,890 (135); | 1,275,434 (111); | 1,276,528 (119); | 1,294,645 (128); |
| 1,301,068 (148); | 1,305,715 (129); | 1,307,355 (120); | 1,311,835 (147); |
| 1,317,249 (131); | 1,319,448 (132); | 1,319,781 (133); | 1,340,179 (164); |
| 1,329,062 (145); | 1,331,087 (136); | 1,336,188 (182) | |

3. *Canadian*
   875,617 (122);   887,122 (109)

4. *French*
   1,438,003 (47);   1,486,670 (62);   1,531,022 (75);   1,580,115 (72);
   2,173,258 (194)

5. *German*
   1,946,688 (115);   2,008,991 (144);   2,103,849 (141);   2,136,585 (184);
   2,139,315 (153);   2,140,135 (154);   2,142,598 (157);   2,151,782 (161);
   2,151,824 (162);   2,152,515 (163);   2,158,821 (165);   2,205,171 (170);
   2,209,875 (172);   2,209,889 (173);   2,215,732 (175);   2,223,324 (181);
   2,262,463 (192);   2,301,202 (193);   2,314,982 (195);   2,319,089 (196);
   2,333,216 (199)

6. *Japanese*
   70/34396 (101);   71/17123 (118);   72/9036 (155);   72/15435 (167);
   72/16526 (168);   72/18995 (171);   72/19404 (116);   72/34480 (174);
   72/34685 (178);   73/27746 (139);   73/29551 (121);   72/34684 (177);
   73/42027 (186);   73/60180 (191)

7. *Netherlands*
   70/10007 (130);   72/6366 (180);   73/5444 (197);   73/7182 (198)

8. *South African*
   71/4337 (152);   71/6030 (159);   72/7635 (188)

9. *United States*
   3,218,287 (44);   3,218,302 (15);   3,232,903 (31);   3,255,135 (56);
   3,304,279 (37);   3,419,515 (58);   3,441,532 (68);   3,533,826 (94);
   3,547,766 (82);   3,551,525 (100);   3,586,654 (126);   3,632,669 (123);
   3,634,303 (127);   3,637,569 (150);   3,644,255 (137);   3,645,959 (138);
   3,646,170 (124);   3,654,201 (140);   3,660,537 (142);   3,661,831 (151);
   3,666,710 (134);   3,676,526 (143);   3,686,111 (149);   3,689,593 (189);
   3,701,746 (146);   3,702,288 (176);   3,711,449 (160);   3,723,571 (190);
   3,742,091 (156);   3,745,137 (187);   3,765,924 (183);   3,775,327 (179);
   3,778,403 (166);   3,786,010 (158)
   T904,001 (93)

## A.3. GROWTH IN THE NUMBER OF PATENT APPLICATIONS ON DISPERSION POLYMERS ISSUED TO PRINCIPAL PATENTEES

| Patentee (>2 Applications) | pre-1955 | 1955–9 | 1960–4 | 1965–9 | 1970 onwards | Total |
|---|---|---|---|---|---|---|
| Badische Anilin | | | | 6 | | 6 |
| BALM, Dulux Australia (see also ICI) | | | | 9 | 1 | 10 |
| Celanese | | | | 2 | 3 | 5 |
| Dai-Nippon | | | | | 7 | 7 |
| du Pont | | | 3 | 15 | 16 | 34 |
| Firestone | | | 3 | | | 3 |
| Ford-Werke | | | | | 5 | 5 |
| ICI (see also BALM, Dulux Australia) | | 4 | 22 | 18 | 11 | 55 |
| PPG | | | | 3 | 3 | 6 |
| Rohm & Haas/ Rohm GmbH | | 3 | 12 | 5 | 4 | 24 |
| Union Carbide | | | | 5 | 2 | 7 |
| *Others* (<3 Applications) | 3 | | 4 | 17 | 13 | 37 |
| Totals | 3 | 7 | 44 | 80 | 65 | 199 |

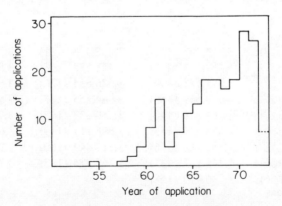

Figure A.1. Number of annual patent applications for dispersion polymers (1950–1972)

# Author and Patentee Index

Page numbers in bold type indicate that the author's name is cited in the text. Page numbers in ordinary type refer to the author's work only. Numbers in parentheses denote the page on which the complete reference is listed.

Adank, G., 53(112)
Ahn, Y., 215(241)
Albers, W., **21**(43)
Alexander, A. E., 48(110)
Alfrey, T., 192(200)
Allport, D. C., 103, 104, 105(113)
Altier, M. W., **129**(198)
Ashton, J. E., 253(271)

Backhouse, M. P., **94**, 95(113)
Badische Anilin- & Soda-Fabrik, 84(113); 209(240); 231, 232(241); **(279)**, 286, 288(289); **(297)**, **(298)**
Baker, A. S., 264, 265, 270(271)
Ball, R. T., **(22)**
Ballegooijen, H. van, 125(198)
BALM Paints, 69, 102(112); 102(113); 231(241); **(296)**, **(297)**, **(298)**, **(299)**; see also Dulux Australia
Bamford, C. H., 3, 4(6); 97(113); 182, 190, 191(200); 214(241)
Banderet, A., 67(112)
Barb, W. G., 3, 4(6); 182, 191(200); 214(241)
Barrett, K. E. J., 4(6); 116, **(117)**, **(178)**, 184, 185, 187, **(188)**, **(189)**, **(190)**, 194, **(197)**; 177, 186, 187, 189, 190, 191, 194(200)
Barsted, S. J., 60, 61, 62, 103(112); **(245)**, (271)
Battaerd, H. A. J., 5(7)
Baylis, R. L., 5(7); 205(240)
Becher, P., 75(112)
Becker, R., **146**, **151**(199)
Berryman, D., 5(6)
Beynon, K. I., 90, 91(113); 193(200)
Black, W., 203(240)
Bluestone, S., **35**(44)
Blundell, D. J., 59, 60(112)
Blundell–Permoglaze **(297)**
Boer, J. H. de, **12**(43)
Bohrer, J. J., 192(200)
Bort, D. N., 137, 152, 178(199)

Bovey, F. A., 2(6); 202(240)
Boyer, R. F., 125(198)
Bradford, E. B., 255(271)
Bradley, R. S., **12**(43)
Bristow, G. M., 125(198)
British Rubber Producers' Research Association, **81**(113); 202(240); **(291)**
British Thomson–Houston, 220(241); **(291)**
Brodnyan, J. G., 143, 171(199)
Broutman, L. J., 253(271)
Brown, G. L., 255(271)
Bueche, F., 5(6); **86**(113); 140(199)
Burrell, H., 120, 122, 123(198)

Cala, J. A., 143, 171(199)
Cameron, G. G., 182(200)
Cameron, J., 182(200)
Canadian Industries, **(306)**
Cantow, H. J., 131(199)
Cantow, M. I. R., 128(198)
Carpenter, D. K., 126, **128**(198)
Casimir, H. B. G., **14**, **17**(43)
Celanese Coatings, 84(113); 231, 232(241); **(279)**; **(299)**, **(300)**, **(301)**, **(305)**
Celanese Corporation, 89(113)
Ceresa, R. J., 5(7); 81, 103(113); 103(114)
Ceskoslovenska Akademie, **(301)**
Clarke, J. T., 190(200)
Clayfield, E. J., **26**, **27**, **35**(43)
Cobbold, A. J., 132(199)
Colange, J., 197(200)
Collie, B., 203(240)
Collins, E. A., 131(199)
Cook Paint & Varnish, 84(113); 231, 232(241); **(279)**; **(296)**, **(304)**
Cooke, D. D., 131(199)
Copolymer Rubber and Chemical, 84(113)
Cornell, S. W., 227(241)
Cornet, C. F., 125(198)
Cornish, G. R., **99**, 100(113)
Cotten, G. R., 125(198)
Cousens, R. H., **258**(271)
Crowley, J. D., 123(198)

309

Dai-Nippon, (**279**); (**301**), (**302**), (**303**), (**304**)
Dainton, F. S., 3(6)
Davidson, J. A., 131(199)
Davies, J. T., 71(112)
Dawkins, J. V., 144(199)
Debye, P., **10**(42)
Derbin, G. M., 5(7)
Deryaguin, B. V., 19, 39(43)
Dietschy, H., 53(112)
Dillon, R. E., 255(271)
Dolan, A. K., (**36**)
Döring, W., **146**, **151**(199)
Doroszkowski, A., **29**, (**30**), 31, **33**, 37(43); 56, 60, (**61**), 62, 71(112)
Doty, P., 125(198)
Dougherty, T. J., **247**(271)
Dowbenko, R., 5(7); 102(113)
Dow Chemical, 86(113); (**300**), (**305**)
Dow Corning, (**302**)
Downing, S. B., 106(114)
Duell, E. G., 93(113)
Dulux Australia, (**302**); *see also* BALM Paints
Dunlp Holdings, (**299**)
Dunn, A. S., 146(199)
du Pont (E. I. du Pont de Nemours), 83, 87, 91, 98(113), 140(199); 202, 203(240); 210, 214, 220(241); (**279**), 281, 285, 288, 289(289); (**294**); (**295**), (**296**), (**297**), (**298**), (**300**), (**301**), (**302**), (**303**), (**304**), (**305**)
Dzyaloshinskii, I. E., 17(43)

Eaton, R. S., 3(6)
Edwards, S. F., (**36**)
Einstein, A., **243**(270)
Elias, H.-G., 33(44); (**53**), 74(112)
Esso Research (**305**)
Etter, O., 53(112)
Evans, R., (**18**), **39**, (**41**), (44); 52(112)
Eveson, G. T., **260**(271)
Ewart, R. H., 118, **119**, **143**, **151**, **171**, 185(197)
Eyring, H., 148(199)

Firestone Tire & Rubber, 84(113); 217. 220(241); 278(289); (**293**)
Fink, W., 125, 157(198)
Fischer, E. W., **28**, **32**(43)
Fisher, J. C., 154(199)
Fitch, R. M., 5(6); 57, (**58**), 60, 101(112); 131, 142, **144**, **146**, **147**, **148**, 151, (**159**), 160, **161**, **162**, **163**, **164**, 166.

Fitch, R. M.—*continued*
168, **170**, **171**, (**172**), **173**, **174**(199); 204(240)
Fløgstad, H., 175(199)
Flory, P. J., **23**, **27**, **30**, 33(43); 33(44); **120**, 121, 124, 125, 126, **166**(198)
FMC Corporation, 202(240); (**299**)
Ford-Werke, 84(113); (**279**); (**302**), (**303**), (**305**)
Fox, T. G., 125(198)

Gaines, G. L., 154, 176(199)
Gallot, B., 74(112); 269(271)
Gardon, J. L., 123, 125, **129**, 178, 184, 185(198); 143, 165, 171(199)
Gerrens, H., 125, 157(198)
Gershberg, D., 143, 171(199)
Gibbs, J. W., 129(198)
Gilmour, R. E., 132(199)
Glasstone, S., 148(199)
Goldsmith, H. L., 244, 247(270)
Graham, N. B., 144(199)
Gregory, J., 12, 15, 17(43)
Greig, J. R., 257, 258(271)
Gruber, U., 53(112)
Gulf Oil of Canada, 227(241); (**298**)
Guyot, P., 197(200)

Halicioğlu, T., **17**(43)
Haller, H. S., 131(199)
Halpin, J. C., 253(271)
Hamaker, H. C., **12**, **14**, **15**(43)
Hansen, C. M., 123(198)
Harborth, G., 182, (**183**), (200)
Harkins, W. D., **118**, **119**, **146**(197)
Harriott, P., 185(200)
Hart, D. P., 5(7); 102(113)
Hashimoto, T., 74(112)
Haskell, V. C., (**91**)
Haward, R. N., 118(198)
Hawkins, W. L., 5(7)
Haydon, D. A., 17, **31**, **32**(43)
Henson, W. E., 255(271)
Hercules, 90(113); 105(114); 216, (217), 218(241); (**297**), (**300**)
Hertzberg, T., 175(199)
Hess, P. H., 206(240)
Hesselink, F. Th., **26**, **27**, **34**(43); **32**(44)
Hibbin, B. C., 131(198)
Hildebrand, J. H., **120**, 123(198)
Holden, H. W., 144(199)
Hölscher, F., 3(6); 131(199)
Howard, R. O., 190(200)
Howland, L. H., 131(199)

Hoy, K. L., 122, 123(198)
Huggins, M. L., **120, 121, 124**(198)
Hunter, R. J., 5(6); 10, 20(42); 46(110)

Ibrahim, F. W., 53(112)
I. G. Farbenindustrie, 81(112); **(291)**
Immergut, B., 90, 91(113); 122(198); 193 (200)
Immergut, E. H., 90, 91(113); 193(200)
Imperial Chemical Industries, 4(6); 49, 65, 68, 70, 74, 79, 102(112); 81, 88, 89, 91, 92, 93, 96, 98, 99, 101, 102, 104(113); 103, 104, 105, 110(114); 196, 197(200); 202, 203, 206, 209, 210(240); 211, 212, 213, 214, 215, 216, 218, 220, 221, 222, 225, 229, 231, 232, 233, 234(241); 264, 265, 270(271); **(279)**, 281, 286, 287, 288(289); **(291)**, **(292)**, **(293)**, **(294)**, **(295)**, **(296)**, **(297)**, **(300)**, **(302)**, **(303)**, **(304)**, **(305)**, **(306)**
Inoue, T., 74(112)
Isemura, T., 144(199)

J ckel, K., **34, 39**(44)
James, D. G. L., 3(6)
Janes, W. H., 103, 104, 105(113)
Japan Atomic Energy Research Institute, **(303)**
Jenkins, A. D., 3, 4(6); 97(113); 182, 190, 191(200); 214(241)
Johnston, R., 190(200)

Kaizerman, S., **129**(198)
Kamath, Y. K., 57, **(58)**, 60, 101(112)
Kambara, S., 106(114)
Kansai Paint, **(305)**
Kao Soap **(294)**
Kawai, H., 74(112)
Kay, D., 131(199)
Keesom, W. H., **10**(42)
Kelley, E. L., 143, 171(199)
Kerker, M., 131(199)
King, J. M., **(55)**; 56(112); 138(199)
Kitchener, J. A., 96(113)
Klein, J., 152(199)
Koehnlein, E., 125, 157(198)
Koelmans, H., 18, 21(43); 47(110)
Kokes, R. J., 125(198)
Kolthoff, I. M., 2(6); 202(240)
Konen, T., 143, 171(199)
Koningsveld, R., **128**(198)
Krause, S., 144(199)

Krieger, I. M., **20**(43); 244, **245**, **247**(271)
Krigbaum, W. R., **128**(198)
Krock, R. H., 253(271)
Krupp, H., **11**(43)
Kruyt, H. R., 6(7)

La Mer, V. K., 152(199)
Laidler, K. J., 148(199)
Lambourne, R., **29**, **(30)**, 31, **33**, 37(43); 56, 60, **(61)**, 62(112)
Landau, L. D., 19, 39(43)
Langbein, D., **17**(43)
Lee, D. I., **244**(271)
Levinson, S. B., 278(289)
Levy, A., 280(289)
Lifshitz, E. M., **17**(43)
Lilley, H. S., **99**, 100(113)
London, F., **11**(43)
Long, F. A., 125(198)
Long, J., 42(44)
Lowe, J. W., 123(198)
Lubrizol, **(296)**
Lumb, E. C., **26**, **27**, **35**(43)
Lyklema, J., 10, 20(42)

McClellan, A. L., 71(112); 211(241)
McGeary, R. K., **227**(241)
McLachlan, A. D., **14, 17**(43)
Mackor, E. L., **(25)**, **26, 38**(43)
Madow, P., 244(271)
Mandelkern, L., 152, 154(199)
Manley, R. St. J., 244, 247(271)
Mark, H., 90, **91**(113); 192, 193(200)
Maron, S. H., 125(198); 244(271)
Mason, S. G., **244**, **247**(270); **244**, 247(271)
Matheson, L. A., 255(271)
Medalia, A. I., 2(6); 202(240)
Meehan, E. J., 2(6); 202(240)
Meier, D. J., **26, 27, 34**(43)
Meredith, C. L., 84(113)
Messer, W. E., 131(199)
Metallgesellschaft, **(294)**
Miller, S. E., 280(289)
Mitchell, D. J., 17, **(18)**, (43)
Mobil Oil, 105(114); 217(241); **(304)**
Moelwyn-Hughes, E. A., 12(43)
Moilliet, J. L., 203(240)
Molau, G. E., 67(112); 269(271)
Monk, C. J. H., 71(112)
Mono-Sol, **(293)**
Moody, A. G., 92(113)
Morningstar–Paisley, **(279)**; **(295)**
Morton, J., 135, 139(199)

Morton, M., **129**(198)
Mysels, K. J., 249(271)


Nakagawa, T., 144(199)
Nakajima, N., 125(198)
Napper, D. H., 5(6); 10, (**18**), 20(42); **29**, **33**(43); 33, **39**, (41), (44); **46**, 48, 50, (**51**), (**52**), 56, (**57**), 65(110); 50, 52, (**53**), (112)
National Distillers and Chemicals, 202 (240); (**300**)
Neyer-Gomez, M. J., 67(112)
Nicks, P. F., **67**, 68(112); **221**(241)
Ninham, B. W., 13, **17**, (**18**), **19**(43)
Nippon Carbide Industries, (**299**)
Nisonoff, A., 131(199)
Norrish, R. G. W., **118**(198)
North, A. M., 182(200)
Nowakowska, L. J., 60, 61, 62, 103(112); (**245**), (271)


Onyon, P. F., 3, 4(6); 182, 191(200); 214(241)
Orofino, T. A., 125(198)
Osborn, P. G., **221**(241)
Osmond, D. W. J., 16(43); 37, 42(44); 47, 49, 54, (**55**), 57, (**58**), 60, 92(110); **67**, 68(112); 82, 83(113); 143(199)
Ottewill, R. H., 10, 35(42); 20, (**28**), 37(43); 71(112)
Overbeek, J. Th. G., 10, **14**, 19, 21, 39(42); 18, **21**, 26, 34(43); 47(110); 248(271)


Papir, Y. S., 245(271)
Parfitt, G. D., (**22**); 47(110)
Parker, H. P., 206(240)
Parsegian, V. A., 13, **17**, **19**(43)
Patat, F., 152(199)
Pennsalt Chemicals, 203(240); (**295**), (**296**)
Periard, J., 67(112)
Petit, P. H., 253(271)
Phillips Petroleum, (**302**)
Pietro, A. R. Di, 125(198)
Pimentel, G. C., 71(112); 211(241)
Pitaevskii, L. P., 17(43)
Plastic Coating Research, (**299**)
Polder, D., **14**, **17**(43)
Pont, J. D., 143, 173(199)
PPG Industries, 70(112); (**279**); (**296**), (**299**), (**300**), (**302**), (**303**), (**305**)
Puddington, I. E., 125(198)


Rae, W. N., 72(112)
Raymond, F. L., 144(199)
Reginato, G., 131(198)
Rehner, J., 125(198)
Reilly, J., 72(112)
Rhind-Tutt, A. J., 82, 83(113)
Rideal, E. K., 71(112)
Riess, G., 67(112)
Robb, I. D., 146(199)
Roe, C. P., **118**, **146**, **151**(198)
Rohm GmbH, (**301**), (**304**)
Rohm & Haas, 4(6); 81, 83, 88, 104(113); 202, 205, 209, 210, 221(240); 220(241); (278), (**279**), 285, 286(289); (**291**), (**292**), (**293**), (**294**), (**295**), (**297**), (**300**), (**301**), (**302**)
Rutgers, I. R., 244(270)
Ryabov, A. V., 137, 178(199)


Sadron, C., 74(112); (**268**), 269(271)
Sarvetnick, H. A., 5(6); 203(240); 254(271)
Saunders, F. L., 244(271)
Scatchard, G., **120**(198)
Schildknecht, C. E., 3, (**4**), (6)
Schulz, A. R., 125(198)
Schulz, G. V., 182, (**183**), **187**(200)
Scott, R. L., 120, 123, **128**(198)
Seaman, P. H., 3(6)
Semchikov, Yu. D., 137, 178(199)
Shaw, J. N., 20(43)
Shell International, (**279**); (**298**)
Sherman, P., 243(270)
Shibata, M., 5(7)
Shinoda, K., 144(199)
Sinanoğlu, O., **17**(43)
Sirianni, A. F., 125(198)
Slavnitskaya, N. N., **137**, 178(199)
Small, P. A., **122**(198)
Smith, E. R., (**18**)
Smith, H. M., 66(112)
Smith, R. R., **118**(198)
Smith, W. V., 118, **119**, **143**, **151**, **171**, 185(197)
Smoluchowski, M. V., 148, 165(199)
Soen, T., 74(112)
Solomon, D. H., 1(6)
Sommerfeld, E. G., (**279**); (**301**)
Spencer, R. S., 125(198)
Stampa, G. B., 84(113); (**196**), (200); 217(241)
Stockmayer, W. H., **99**(113); 190(200)
Suminoe, T., 106(114)
Sund, E., 175(199)

Taber, D. A., 255(271)
Tabor, D., 17(43)
Talen, H. W., 255(271)
Tamamushi, B., 144(199)
Tanford, C., **23**(43)
Tankey, H. W., 76(112); 135, 137, 138(199)
Taylor, J. L., 17, 31(43)
Teague, G. S., 123(198)
Thomas, D. G., **244**(271)
Thomas, H. R., 5(6); 116, (**117**), (**178**), 184, 185, 187, (**188**), (**189**), (**190**), 194(197); 186, 187, 189, 190, 191, 194(200)
Thompson, L. J., 125(198)
Thompson, M. W., 49, 68, 102(112); **96**(113); 135, 139(199); 197(200)
Tiffany, J. M., 71(112)
Toa Synthetic Chemicals, (**279**), (**298**)
Tokuyama Soda, (**299**)
Tolman, R. J., 179, 189, 191(200); 264(271)
Toray Industries, (**299**)
Tregar, G. W., 5(7)
Tsai, C. H., 131, 142, **144**, **147**, **148**, 151, (**159**), **160**, **161**, **162**, **163**, **164**, **166**, **168**, **170**, **171**, (**172**), **173**, **174**(199)..
Turnbull, D., 154(199)

Ugelstad, J., 175(199)
Union Carbide, 209(240); 213, 219, 220(241); (**279**), 285(289); (**296**), (**298**), (**300**), (**301**)

Vand, V., **244**, 247(270)
Vanderhoff, J. W., 250, 252(271)
Vanzo, E., 3(6)
Vervey, E. J. W., 10, 14, 19, 39(42)
Vincent, B., 10(42); 13, 16(43); 42(44)
Vishnevskaya, I. N., 152(199)
Visser, J., 12, (**13**), (43)
Vold, M. J., **16**(43); **35**(44); **250**(271)

Volmer, M., **146**, **151**(199)
Von Weimarn, P. P., **151**(199)
Voyutskii, S. S., **256**(271)
Vrij, A., 26, 34(43)

Waals, J. H., van der, 26, **38**(43)
Waghorn, M. J., 5(7)
Wagstaff, I., 60, 61, 62, 78, 103(112); **245**, 246, 247(271)
Wainman, J. E., 178, 179(200)
Waite, F. A., 16, 30, 33(43); (**55**); 64, 67, 71, 72, 73, 74, 82, 91, 92, **96**, 98, 99(112); 101, 102(113); 208(240)
Walbridge, D. J., 49, 54, (**55**), 57, (**58**), 60, 92(110); 50, 57, 60, 61, 62, 101, 103(112); 93, 101(113); 245(271)
Wales, M., 131(199)
Walker, J. F., 231(241)
Walker, T., (**28**), 37(43)
Warson, H., 2(6)
Waters, J. A., 50, 59, 60, 67, 71, 72, 73, 74, 101(112); **245**, 246, 247, 264, 265, 270(271)
Watson, W. F., 125(198)
West, E. J., 257, 258(271)
White, T., 118(198)
Wigglesworth, D. J., 5(7)
Winter, B., 227(241)
Winterton, R. H. S., 17(43)
Wolf, B. A., 126(198)
Wood, J. A., (**22**)
Woods, M. E., 245, 247(271)

Yamazaki, N., 106(114)
Yearsley, F., 131(198)
Yerazunis, S., **227**(241)
Young, L. J., 90, 91(113); 193(200)

Zable, H. S., 125(198)

# Subject Index

ABS, 84, 258
Acrylic copolymer dispersions, 211–212
Active solvent paints, 283–284
Addition polymer dispersions, 207–216, 216–218
Adhesion, 70, 84, 102, 258, 259, 263
Adhesives, 212, 285
Adsorbed polymers, desorption of, 36–38
  hydrodynamic thickness of, 60–63
  mobility of, 36–38
  non-equilibrium situations for, 38–39
  surface concentration of, 54–60
Agglomerate, 6
Agglomeration, 142–143
  dispersant-limited, 151, 171–173
Aggregation, 6, 249–250, 269
Aggregative nucleation, 144–146
  and diffusion capture, 162
Alkyd resins, 1, 47, 83, 288
Allyl glycidyl ether, 109
Aluminium trialkyls, 218
Amino-hydroxyalkyl-methacrylate, 285
Aminoplasts, 285, 288
Amino-undecanoic acid, 223–224
Ammonium persulphate, 203
Amphipathic copolymer dispersants, 5, 149, 205, 267, 269
  solution properties, 264, 269–270
Anchor, soluble group balance (ASB), 75–76, 263–264, 266, 267, 269
Anchoring, by acid: base interactions, 70–74
  by adsorption, 64–67
  by covalent links, 68–70
  in emulsions, 67–68
  multi-point, 70–71
  single point, 71–74
Anchoring component, selection of, 63–74
Anchor polymer flexibility, 264
Applications of dispersion polymers, 273–289
  adhesives, 212, 285
  coloured polymer particles, 287–288
  cosmetics, 275, 277, 288
  fibre coatings, 286–287
  general features of, 273–277
  glass fibre coatings, 287
  lubrication, 288–289

Applications of dispersion polymers—*continued*
  pigment encapsulation, 275, 286
  paints, 278–285
  polishes, 285–286
  powders, 277–278
  processes, 281
  rubbers, 287
  textile treatments, 286
Aqueous dispersions, 20, 39, 250–251, 252, 256, 260, 278, 282
  surfactants, 63, 75, 252, 256
Athermal solvent, 23, 35
Attractive forces between particles, 10–19
  colloidal stability and, 18–19
  direct measurement of, 17
Auto-accelerated polymerization, 3, 116, 178, 182, 183
Auto-adhesion, 256, 262
Auto-protolytic ions, 20
Autoxidation, 1
Azeotropic distillation, 203, 222, 223
Azobis-cyanopentanoic acid, 96–97
Azobis-isobutyronitrile, 208

Benzoyl peroxide, 84, 208
Bimodal particle size distribution, 227–229
Bishydroxyethyl terephthalate, 288
Block copolymers, as dispersants, 48–49, 79, 103–166, 265–266
  solution properties, 268–269
  synthesis by condensation, 104
    by free radical methods, 103–104
    by ionic methods, 104–106, 110–111
  terminology, 6
Boron trifluoride, 89, 218–219, 220
Brownian collisions, 38
  diffusion rate, 249
  motion, 21, 247
Bulk modulus, 39
Butoxymethyl acrylamide, 288
Butylbenzyl phthalate, 283
Butyl rubber, 86–87
Butylsextyl phthalate, 283

Capillary stress, 255–256, 258–259, 261
Chain termination, 3, 182
Chain transfer, 194, 210, 213, 214

315

Chain transfer agent, 97–98, 182, 194, 210
Charge stabilization, 10, 19–22, 39
  failure in non-polar media, 19–22
Chromophores, 288
Coagulation, 6
Collisions, two-body, 244
  three-body, 244
Colloidal dispersions, 4–5
  stability, 9–10, 18–19, 39, 249–250
  attractive forces in, 18–19
  reviews on, 6, 10
Coloured polymer particles, 287–288
'Comb' graft copolymer, 76, 77, 78, 99–103, 108–109, 137, 140, 211, 214, 215, 222, 265–266, 267
Comminution, 202
Condensation dispersion polymerization, 220–221
Condensation polymer dispersions, 220–225
Condensation polymerization of dispersed reactants, 221–225
Continuous dispersion polymerization, 225–227, 277
  apparatus, 226–227
  kinetics of, 195–196
Continuum electrodynamic model, 16–18
Copolymer terminology, 5–6
Copolymerization parameters, 192–194, 212
Cosmetics, 275, 277, 288
Coulombic repulsion, 20, 21, 248
Counter-ions, 20
Counter-surface, 34
Critical flocculation temperature, 52–53
  and molecular weight, 52
  and surface coverage, 56–57
Critical flocculation volume, 51
  micelle concentration (CMC), 77, 105, 269, 270
  packing, 243, 245, 255, 258, 260, 275, 277, 282
Cross-linking agents, 233
  reactions, 230–232, 233
Cyclic acetals, 218–219
  amides, 218–219
  esters, 218–219
  ethers, 218–219

Debye attraction, 10–11
Desorption of polymer, 36–38
Dialysis, 20
Diazobicyclo-octane, 232
Dielectric constant, 14, 20
Differential scanning calorimeter, 187, 233

Diffusion capture, 147–148, 159–166
  Fick's laws, 165, 170
Di-isopropyl peroxydicarbonate, 213, 214, 215
Dilatancy, 230, 250
Diluents, 205–206, 208, 209, 211, 222, 274–275, 277
Dimensions of steric barrier, 60–63
Diphenylol propane, 65, 104, 264
Dispersant-limited agglomeration, 151, 171–173
Dispersant-limited nucleation, 150–151, 170–171
Dispersant precursors, 210–211
  reactivity of, 140
Dispersants, anchoring component, 63–74
  block copolymers as, 48–49, 103–106
  design of, 79–103
  for pigments, 47–48
  graft copolymers as, 48–49, 79–103, 209
  recipes for preparation of, 106–110
  role of, 45–49
  soluble component, 49–54
  synthesis of, 79–106
Dispersion polymerization, 3–4, 115–119
  apparatus for, 209, 236
  characteristics of, 116–119
  condensation, 220–225
  continuous processes, 195–196, 225–227, 277
  copolymerization parameters, 192–194, 212
  copolymer preparation, 211–212
  diluents for, 205–206, 208, 231, 274–275, 277
  dispersants for, 45–113
  free radical addition, 207–216
    acrylic acids, 209–212
    acrylic esters, 187–189, 209–212
    acrylonitriles, 190–191, 214–215
    basic procedures, 207–208
    $N$-vinyl pyrrolidone, 215–216
    vinyl aromatics, 213–214
    vinyl chloride, 191–192, 215
    vinyl esters, 189–190, 212–213
    vinylidene chloride, 191–192, 215
  general features of, 3–4, 115–119
  initiation of, 208
  ionic initiation, 216–220, 275
    ionic addition, 216–218
    ring-opening, 218–220
  kinetic model for, 184–187
  kinetics of, 115–197
  mechanism of, 115–197

Dispersion polymerization—*continued*
  molecular weight control in, 194–195
  non-free radical mechanisms, 196–197
  particle formation, 130–177
  particle growth, 177–197
  recipes for, 234–240
  role of dispersants in, 45–49
  scope of, 205–207
  sequential, 233
  special techniques of, 225–234
  terminology of, 4–6
Dispersion polymers (*see also* Polymer dispersions)
  melt-flow properties, 278
  need for in organic media, 1–2
  preparation, 201–240
  properties, 243–271
  technology of, 1–6
  terminology of, 4–6
Dispersion stability, 249–250
DLVO theory, 19–20, 39
Double 'comb' graft copolymer, 102, 206, 222
Doublet formation, 247
Dyestuffs, 288

Effective Hamaker constant, 15, 17, 19
Einstein coefficient, 243–246
Elasticity of polymer layers, 34
Electron spin resonance, 179
Electrostatic stabilization, 20, 22
Electro-viscous effect, 248–249
Emulsification, 202, 221–223, 239–240, 288
Emulsion paints, 2, 212, 250, 282
  polymerization, 2–3, 118–119, 129, 166, 170, 174, 178, 184–185, 203, 205, 212
Emulsions, 67–68, 216
Encapsulation, 234, 275, 286
Enthalpic stabilization, 47
Enthalpy of mixing, 23, 120
Entropic stabilization, 47
Entropy of mixing, 23, 120–121, 124
EPDM rubbers, 84
Epichlorhydrin, 65, 104, 264
Epoxy resins, 288
Equilibrium capture, 148–149
Excluded volume, 27, 50

Fibre coatings, 286–287
Fick's laws of diffusion, 165, 170
Film formation, 208, 250–263, 282, 283, 284
  aqueous dispersions, 252, 255, 256
  capillary stress, 255–256, 258–259, 261
  cavitation, 259
  deflocculated hard particles

Film formation—*continued*
  in volatile diluent, 258–260, 261–262
  deflocculated particles, 253–255, 256–258
  deflocculated particles in volatile diluent, 255–256
  deformation in, 252, 259
  diluent evaporation, 258–259
  flocculating hard particles in volatile diluent, 260–261
  panel coalescing systems, 262–263
  plasticization in, 254–255
  problem of, 251–253
  sterically-stabilized particles, 256–258, 261–262
Flocculants, 94–96
Flocculation, 6, 33, 49–50, 94–96, 142, 173, 284
  orthokinetic, 21
  weak, 42
Flocs, 6
  structure of, 250
Flory–Huggins polymer solution theory, 124–126
  interaction parameters, 125
Fluidity, 275–276
Forces of attraction between particles, 10–19
  and colloidal stability, 18–19
  direct measurement of, 17
Forces of repulsion, determination of, 31–32
  generated by charge, 19–22
  generated by soluble polymer, 22–39
Free energy changes, 23
  of mixing, 24, 25, 27–29, 33, 37
Free radical addition, dispersion polymerization, 207–216
  characteristics of, 116–119
  polymerization, kinetics of, 179–183

Gel effect, 182–183, 187, 190, 213, 227
Geometrical (Hamaker) function, 12, 13–14
Glass fibre coatings, 287
Glass transition temperature, 69, 78, 79, 103, 179, 190, 202, 203, 213, 233, 254, 260, 263, 270, 277, 283, 285
Graft copolymers, as dispersants, 48–49, 79–103
  as flocculants, 94–96
  behaviour in dispersion polymerization, 74–79
  degree of adsorption on polymer particles, 54–56
  selected recipes for, 106–110

Graft copolymers, as dispersants—
*continued*
   synthesis of, 79–103
      grafting by random copolymerization, 88–96
      grafting by terminal copolymerization, 96–103
      random grafting, 80–84
      transfer grafting, 84–88
   terminology of, 6, 80
Grafting reactions, 80–103

Hamaker constant, 12–13, 15, 16
   and polarizability, 13
   values of, 12–13
Hamaker integration, 12–14, 16
   applied to liquid media, 14–16
   applied to layered particles, 16
Heat of mixing, 23, 120
   neutralization, 71–72
   polymerization, 187, 227
   vaporization, 120, 277
Heisenberg uncertainty principle, 11
Hemi-solvent, 74, 264, 267, 268, 269, 270
Heterogeneous polymerization, 3–4, 119
   polymer particles, 232–234
High solids polymer dispersions, 227–230, 277
   paints, 278
Homogeneous nucleation, theory of, 151–158, 174
Hydrodynamic drag, 244, 248
Hydrogen abstraction, 80, 82, 85, 87, 208, 212
   bonding, 78, 101, 211
Hydrophile, lipophile balance (HLB), 75
Hydrosols, 5

Impact resistance, 84, 285
   strength, 287
Injection moulding, 277, 278
Inorganic dispersions, 206–207
Interfacial tension, 129, 152, 157, 169, 176, 260
Intermolecular association, 264
   forces, 10, 13, 16
Interpenetration of polymer chains, 34
Intramolecular association, 264
Ionic addition dispersion polymerization, 216–218, 275
Isobutyryl peroxide, 91
Isocyanate polymerization, 218

Keesom attraction, 10–11

Kinetic model for dispersion polymerization, 184–187
Kinetics of dispersion polymerization, 115–197
   copolymerization parameters, 192–194, 212
   continuous processes, 195–196
   free radical addition, 177–196
   molecular weight control, 194–195
   non-free radical mechanisms, 196–197
   particle formation, 130–177
   particle growth, 177–196
   results of studies on, 187–192
      acrylonitrile, 190–191, 214
      $\alpha$-methyl styrene, 196
      methyl methacrylate, 187–189
      vinyl acetate, 189–190
      vinyl chloride, 191–192
      vinylidene chloride, 191–192
Kinetics of free radical addition polymerization, 179–183

Lewis acids, 218
Lithium butyl, 105, 217
'Living' polymer, 104–105, 217–218
London attraction, 11–12, 13, 16, 21
   interaction constant, 12
Lubrication, 288–289
Lyophilic colloids, 5
Lyophobic colloids, 5

Macro-monomer, 88, 98, 100, 101
Matched transfer agent, 97
Maxwell's equations, 17, 22
Mechanism of dispersion polymerization, 115–197
Medium effects, 14–15, 17, 19
Melamine formaldehyde resins, 83, 231–232, 284, 285
Melt-flow properties, 278
Metallic coatings, 284
   dispersions, 206–207
Micellar dispersions, 103, 264–270
Micellization, 38, 48, 74–75, 101, 146, 202, 257–258, 262, 264–270
   number, 265–269
Micro-bulk polymerization, 4, 179, 227
Microcalorimetry, 187
Microgels, 204, 211, 218, 232–234, 275
Minimum film-forming temperature, 260
Mixing term, 28, 34
Mobility of adsorbed polymer, 36–38
Molecular weight control, 194–195
'Mud-cracking', 261–262, 283

Nitrocellulose, 1, 288
Non-aqueous dispersions (NAD), 5, 205
Non-equilibrium situations, 38–39
Nucleation, aggregative, 144–146
   control by diffusion capture, 162
   control by oligomer capture, 159–170
   dispersant, effect on, 137–139
   dispersant-limited, 170–171
   from micelles, 146
   self, 144–145
   suppression of, 147, 158
   theory of homogeneous, 151–158
Nylons, 223, 225, 226, 286

Octyl mercaptan, 210
Optical path length, 14
Organosols, 5, 46, 87, 98, 103, 202, 203–204, 261
Orthokinetic flocculation, 21
Osmotic pressure, 24, 27, 46, 50–51
Oleic acid, 71
Oligomer capture, 147, 159–170
   comparison of models of, 170

Packing fraction, 227, 248
Paints
   active solvent formulations, 283–284
   combined dispersion and solution systems, 284–285
   comparison with solution systems, 281–282
   from polymer dispersions, 87–88, 102, 278–285
   industrial stoving, 254
   lacquers, 87
   requirements for, 1–2
   thermoplastic, 1, 284
   thermosetting, 1, 83, 230–231, 277, 284
   undercoats, 285
   void-filling formulations, 283
   water-based, 2
Pair-wise additivity, 11, 12, 16–17
Particle formation, 130–177
   control by dispersants, 149–151
   control of particle number, 134–143
   control of particle size, 134–143
   methods of investigation, 131–134
      electron microscopy, 131
      light scattering, 131–132
   period of, 134
   theories of, 143–151, 173–177
      equilibrium systems, 143–144
      dispersant-limited agglomeration, 171–173

Particle formation—*continued*
   dispersant-limited nucleation, 170–171
   homogeneous nucleation, 151–158, 174
   irreversible processes, 144
   nucleation control by oligomer capture, 159–170
   present status of, 173–177
Particle growth, 177–196
Particle number, control of, 134–143
   constancy of, 134
Particle size, control of, 134–143
   critical, 21–22
   distribution, bimodal, 227–229
      polymodal, 229–230
   effect on rheology, 247–249
Particles, attraction between, 10–19
Partition coefficient, 178, 184, 191
Partition of monomer in polymer dispersions, 128
   effect of particle size on, 129–130
Patent abstracts, 291–306
Patents, 291–308
   growth in, 308
   nationality of, 306–307
Phase separation, in polymer solutions, 126–128
   in presence of solvent or monomer, 128–129
Phosphorus pentafluoride, 220
Pigmentation, 203, 283, 284, 286, 288
Pigment dispersants, 47–48
   dispersion, 49, 79, 221
   encapsulation, 234, 286
   flushing, 203
Plasticization, 79, 183, 231, 254–255, 262–263, 274, 283
Plastisols, 5, 202, 203–204, 254, 256
Polarizability, 13
Polishes, 275, 285–286
Pollution, atmospheric, 278–280
Poloids, 5, 86
Poly(acrylates), 69, 277, 287, 288
Poly(acrylic acid), 211
Poly(acrylonitrile), 49, 65, 66, 123, 214–215, 233, 237–238, 275, 277
Poly(acrylonitrile-co-acrylic acid), 215
Poly(acrylonitrile-co-methyl methacrylate), 215
Poly(aluminium phosphate), 207
Polyamic acids, 220, 285
Polyamides, 222–223, 232, 287, 288
Poly[bis(chloromethyl)oxacyclobutane], 219
Poly(butadiene), 84, 217, 220, 232

Poly(butadiene-co-styrene), 217
Poly(butadienyl) lithium, 217
Poly(butyl acrylate), 125
Poly(ε-caprolactam), 220
Poly(ε-caprolactone), 219–220
Poly(chloroprene), 285
Poly(dimethyl siloxane), 86
Poly(epichlorhydrin), 219
Polyepoxides, 90, 218
Polyesters, 102, 211, 221, 223, 232, 234, 237, 287
Poly(ethyl acrylate), 76–77, 142, 232, 233, 278
Poly(ethylhexyl acrylate-co-allyl acrylate), 90, 140
Poly(ethylhexyl methacrylate-b-styrene), 214
Poly(ethylene), 106, 216
Poly(ethylene-b-propylene), 106, 216
Poly(ethylene-co-propylene), 217
Poly(ethylene-co-vinyl acetate), 213
Poly(ethylene oxide), 40, 42, 104, 219
Poly(ethylene oxide-b-propylene oxide, 214
Poly(ethylene terephthalate), 49, 222–223, 239–240, 286
Poly(formaldehyde), 89, 220
Poly(glycidyl acetate), 219
Poly(glycidyl stearate-b-phenyl glycidyl ether), 219
Poly(hexamethylene adipamide), 223
Poly(hexamethylene diamine adipate), 223
Poly(hexamethylene diamine azelate), 223
Poly(hexamethylene terephthalamide), 223
Poly(hydroxystearic acid), 31, 49, 55, 56, 57, 59, 60, 61, 62, 76, 78, 98, 101, 104, 108, 244, 264
Poly(hydroxystearic acid-g-glycidyl methacrylate), 102, 108–109
Poly(hydroxystearic acid-g-methyl methacrylate), 55, 68
Poly(isobutene), 47, 140
Poly(isobutylene), 86–87, 125
Poly(isoprene), 217
Poly(lauryl methacrylate), 47, 57, 58, 59, 60, 82–83, 89, 90, 98, 107, 123, 209, 211, 220
Poly(lauryl methacrylate-co-diethyl-aminoethyl methacrylate), 220
Poly(lauryl methacrylate-co-glycidyl methacrylate), 74, 89, 92, 95, 210, 213, 218, 220
Poly(lauryl methacrylate-co-methacrylic acid), 109

Poly(lauryl methacrylate-g-methyl methacrylate), 92
Poly(lauryl methacrylate-g-vinyl pyrrolidone), 219
Poly(methacrylic acid), 211, 237
Poly(methacrylonitrile), 214
Poly(methyl methacrylate), 47, 49, 51, 56, 58, 61, 62, 76, 78, 79, 82–83, 86, 87, 123, 125, 230, 233, 234, 235, 244, 245, 247, 249, 264, 270, 275, 276, 277, 278, 283
Poly(methyl methacrylate-co-ethyl acrylate), 76, 285
Poly(methyl methacrylate-co-ethyl acrylate-co-dimethylaminoethyl methacrylate), 70, 203
Poly(methyl methacrylate-co-glycidyl methacrylate), 67, 76, 211
Poly(methyl methacrylate-co-methacrylic acid), 213, 231, 234–237, 286
Poly(methyl methacrylate-g-2 ethyl hexyl acrylate), 203
Poly(methyl methacrylate-g-isoprene), 202–203
Poly(α-methyl styrene), 84, 217
Poly(N-vinyl pyrrolidone), 215–216
Poly(octene-1), 105, 217
Poly(p-oxybenzoate), 222
Poly(phenyl glycidyl ether), 219
Poly(propylene), 105, 106, 216–217
Poly(styrene), 42, 59, 105, 123, 125, 214, 217, 233, 245, 278
Poly(styrene oxide), 219
Poly(t-butyl styrene-b-styrene), 105, 110, 217
Poly(t-butyl styrene-b-isoprene-b-styrene), 104, 218, 238–239, 287
Poly(tetrafluoroethylene-co-hexafluoropropylene), 288
Poly(trioxane), 220
Poly(undecanoamide), 223–224
Poly(ureidoethyl vinyl ether), 221
Polyurethanes, 288
Poly(vinyl acetate), 3, 40, 48, 65, 69, 78, 142, 212–213, 234, 278, 285
Poly(vinyl alcohol), 3, 125, 213, 285
Poly(vinyl alkyl ether), 84, 217
Poly(vinyl chloride), 47, 79, 82–83, 125, 203, 215, 227, 234, 238, 277
Poly(vinyl chloride-co-lauryl methacrylate), 220
Poly(vinylidene chloride), 203, 204, 215, 289
Poly(vinylidene fluoride), 203

Poly(vinyl toluene-b-methyl methacrylate), 104
Polymer colloids, 4–5
Polymer composites, 253
Polymer desorption, 36–38
Polymer dispersions, applications of, 273–289
  heterogeneous particles, 232–234
  high solids, 227–230
  microgels, 232–233, 275
  need for in organic media, 1–2
  paint formulations of, 278–285
  patents on, 291–308
  preparation of, 201–240
  properties of, 243–271
  reactive polymers, 230–232
  recipes for, 234–240
  solutions from, 281
  steric stabilization of, 39–42
  surface coatings from, 285–287
  technology of, 1–6
  terminology of, 4–6
Polymer emulsification, 202
Polymer interaction volumes, 31
Polymer matrix, polymerization in, 179–183
Polymer mobility, 36–38
Polymer particles, coloured, 287–288
Polymer powders, 225, 227, 277–278
Polymer precipitation, 119–130, 202
Polymer seed, 87, 138, 208, 229, 280, 286
Polymer segment density distribution, 34
Polymer solubility, 50–51, 119–130
  effect of added monomer, 128–129
    molecular weight, 127
    particle size, 129–130
    solvency of medium, 134–137
    parameters, 121–124
Polymer solution, phase separation in, 126–129
  theory, 23–24, 50–51, 124–126
Polymer 'strike' layer, 286
Polymer swelling, by monomer, 177–178
  in emulsion polymerization, 129, 178
Polymodal particle size distribution, 229–230
Potential energy of attraction, 11–12, 14
  of repulsion, 19–22
  particle separation relations, 39
Powder coatings, 278, 288
Powders, coloured polymer, 287–288
  polymer, 277–278
Precipitation of polymers, 119–130, 202
Precipitation polymerization, 3–4, 118–119, 137

Preparation of polymer dispersions, 201–240
  displacement methods, 203
  experimental details, 234–240
  indirect methods, 201–205
Primary maximum, 39–40
Primary medium effect, 14
Primary minimum, 40
Primary surface, 24
Properties of polymer dispersions, 243–271
  aggregative, 249–250
  evaporation rate, 277, 281, 283
  film-formation, 250–263
  fluidity, 275–276
  rheological, 229, 243–249, 275–276, 281–282
  sedimentation, 249–250
  settlement, 249–250
  ultra-colloidal systems, 263–270
  viscosity, 275–276, 281, 282

Radiation curing, 278
Radiation field, 11
Random copolymer terminology, 5
  grafting, 80–84, 267
Reactive polymer dispersions, 230–232
Reactivity ratios, 192–193, 212
Recipes for dispersant preparation, 106–110
  dispersion polymerization, 234–240
Redispersion of polymer powders, 203–204
Renucleation, 140–142
Repulsive forces, determination of, 31–32
  generated by charge, 19–22
  generated by soluble polymer, 22–39
Retardation effect, 14, 17, 19
Reverse emulsification, 222, 223, 240
Rheological properties, 229, 243–249
  effect of particle size on, 247, 249
  effect of steric barrier on, 244–249
  formation of doublets, 247
  Newtonian behaviour, 243–244, 245
  shear stress, 247, 248
    thickening, 245–249
    thinning, 245–249
  thixotropy, 245
Ring-opening polymerization, 218–220
Rubber, butyl, 86
  crepe, 217, 278
  multiphase, 277
  natural, 81, 82, 83, 84, 85, 125, 209, 212, 215, 220
  thermoplastic, 275, 287

Scope of dispersion polymerization, 205–207

Secondary medium effect, 15, 19
Secondary minimum, 39–40, 42
Secondary surface, 24
Sedimentation, 249–250
  rate, 249
Self nucleation, 144–145
Sequential dispersion polymerization, 233
Settlement of dispersions, 249–250
Shear stress, 247, 248
  thickening, 245–249
  thinning, 245–249
Solubility of polymers, 119–130
Solubility parameters, 54, 120–124
  as criteria of miscibility, 120–121
  of polymers, 121–124
Soluble component of graft dispersant, 49–54
Solution lacquers, 281
Solution paints, combined with dispersions, 284
  comparison of dispersion paints with, 281–282
Solvated layer thickness, 47
Solvency coefficient, 40
  of medium, 32, 33
Stabilization, 5
  by charge, 19–22
  by steric forces, 39–42
Stabilizers, 5
Stearic acid, 71
Steric barrier, dimensions, 60–63
  effect on rheology, 244–249
  strength of, 60–63
Steric stabilization, 9, 22–42, 46–47
  current view of, 34–36
  enthalpic component, 46–47
  entropic component, 46–47
  in theta solvents, 33–34, 50
  models for, 23, 24–33, 34–36
  models with interacting polymer chains, 27–33
  nature of, 46–47
  non-equilibrium situations in, 38–39
  polymer desorption in, 36–38
  polymer dispersions in organic media, 39–42
  polymer mobility in, 36–38
  redistribution models, 29–31, 37
  volume restriction models, 25–27

Steric stabilization—*continued*
  terminology of, 24–25
Stokes' law, 249
Stoving paints, 254, 282
Strength of steric barrier, 60–63
Supersaturation, 152–155
Surface balance, 31
  coatings, 1–2, 277, 278–287
  coverage, 54–60
Suspension polymerization, 4, 118–119, 205
'Sweat layer', 263

t-Butyl peroxypivalate, 87
Tetramethylene sulphone, 220
Textile treatments, 286
Thermodynamic stability, 249
Thermoplastic paints, 1, 284
  rubbers, 275, 287
Thermosetting paints, 1, 83, 230–231, 277, 284
Theta solvent, 33–34, 50–51
  temperature, 24, 52, 125
Thixotropy, 245
Titanium dioxide, 49, 60–61, 62
Toluene sulphonic acid, 232
Transfer grafting, 84–88
Trapped free radicals, 179, 182, 192
Tyndall effect, 116

Ultra-colloidal polymer dispersions, 263–270
Undercoats, 285
Urea-formaldehyde resins, 232

Van der Waals–London attraction, 10–12
Viscosity, 229, 275–276, 281, 282
Void-filling paints, 282, 283
Vold effect, 16
Volumeless polymer chain, 34
Volume restriction, 25–27, 33, 34

Water, as paint medium, 2
Weak flocculation, 42, 94

Zero-point energy, 11
Ziegler–Natta catalysts, 105, 106, 216–217
Zinc dibutyl, 219